T0207444

Communications
in Computer and Information Science   1950

Rationale

The CCIS series is devoted to the publication of proceedings of computer science conferences. Its aim is to efficiently disseminate original research results in informatics in printed and electronic form. While the focus is on publication of peer-reviewed full papers presenting mature work, inclusion of reviewed short papers reporting on work in progress is welcome, too. Besides globally relevant meetings with internationally representative program committees guaranteeing a strict peer-reviewing and paper selection process, conferences run by societies or of high regional or national relevance are also considered for publication.

Topics

The topical scope of CCIS spans the entire spectrum of informatics ranging from foundational topics in the theory of computing to information and communications science and technology and a broad variety of interdisciplinary application fields.

Information for Volume Editors and Authors

Publication in CCIS is free of charge. No royalties are paid, however, we offer registered conference participants temporary free access to the online version of the conference proceedings on SpringerLink (http://link.springer.com) by means of an http referrer from the conference website and/or a number of complimentary printed copies, as specified in the official acceptance email of the event.

CCIS proceedings can be published in time for distribution at conferences or as post-proceedings, and delivered in the form of printed books and/or electronically as USBs and/or e-content licenses for accessing proceedings at SpringerLink. Furthermore, CCIS proceedings are included in the CCIS electronic book series hosted in the SpringerLink digital library at http://link.springer.com/bookseries/7899. Conferences publishing in CCIS are allowed to use Online Conference Service (OCS) for managing the whole proceedings lifecycle (from submission and reviewing to preparing for publication) free of charge.

Publication process

The language of publication is exclusively English. Authors publishing in CCIS have to sign the Springer CCIS copyright transfer form, however, they are free to use their material published in CCIS for substantially changed, more elaborate subsequent publications elsewhere. For the preparation of the camera-ready papers/files, authors have to strictly adhere to the Springer CCIS Authors' Instructions and are strongly encouraged to use the CCIS LaTeX style files or templates.

Abstracting/Indexing

CCIS is abstracted/indexed in DBLP, Google Scholar, EI-Compendex, Mathematical Reviews, SCImago, Scopus. CCIS volumes are also submitted for the inclusion in ISI Proceedings.

How to start

To start the evaluation of your proposal for inclusion in the CCIS series, please send an e-mail to ccis@springer.com.

Nguyen Thai-Nghe · Thanh-Nghi Do ·
Peter Haddawy

Editors

Intelligent Systems and Data Science

First International Conference, ISDS 2023
Can Tho, Vietnam, November 11–12, 2023
Proceedings, Part II

 Springer

Editors
Nguyen Thai-Nghe ⓘ
Can Tho University
Can Tho, Vietnam

Thanh-Nghi Do ⓘ
Can Tho University
Can Tho, Vietnam

Peter Haddawy ⓘ
Mahidol University
Salaya, Thailand

ISSN 1865-0929   ISSN 1865-0937 (electronic)
Communications in Computer and Information Science
ISBN 978-981-99-7665-2   ISBN 978-981-99-7666-9 (eBook)
https://doi.org/10.1007/978-981-99-7666-9

This Springer imprint is published by the registered company Springer Nature Singapore Pte Ltd.
The registered company address is: 152 Beach Road, #21-01/04 Gateway East, Singapore 189721, Singapore

Paper in this product is recyclable.

Preface

This proceedings contains the papers from the first International Conference on Intelligent Systems and Data Science (ISDS 2023), held at Can Tho University, Vietnam from November 11–12, 2023. ISDS 2023 provided a dynamic forum in which researchers discussed problems, exchanged results, identified emerging issues, and established collaborations in related areas of Intelligent Systems and Data Science.

We received 123 submissions from 15 countries to the main conference and a special session. To handle the review process, we invited 100 expert reviewers. Each paper was reviewed by at least three reviewers. We followed the single-blind process in which the identities of the reviewers were not known to the authors. This year, we used the EasyChair conference management service to manage the submission and selection of papers. After a rigorous review process, followed by discussions among the program chairs, 35 papers were accepted as long papers and 13 as short papers, resulting in acceptance rates of 28.46% and 10.57% for long and short papers, respectively.

We are honored to have had keynote talks by Kazuhiro Ogata (Japan Advanced Institute of Science and Technology, Japan), Masayuki Fukuzawa (Kyoto Institute of Technology, Japan), Dewan Md. Farid (United International University, Bangladesh), Mizuhito Ogawa (JAIST, Japan), and Nguyen Le Minh (JAIST, Japan).

The conference program included four sessions: Applied Intelligent Systems and Data Science for Agriculture, Aquaculture, and Biomedicine; Big Data, IoT, and Cloud Computing; Deep Learning and Natural Language Processing; and Intelligent Systems.

We wish to thank the other members of the organizing committee, the reviewers, and the authors for the immense amount of hard work that has gone into making ISDS 2023 a success. The achievement of the conference was also contributed to by the kind devotion of many sponsors and volunteers.

We hope you enjoyed the conference!

Nguyen Thai-Nghe
Thanh-Nghi Do
Peter Haddawy
Tran Ngoc Hai
Nguyen Huu Hoa

Organization

Honorary Chairs

Ha Thanh Toan	Can Tho University, Vietnam
Nguyen Thanh Thuy	University of Engineering and Technology, VNU Hanoi, Vietnam

General Chairs

Tran Ngoc Hai	Can Tho University, Vietnam
Nguyen Huu Hoa	Can Tho University, Vietnam

Program Chairs

Nguyen Thai-Nghe	Can Tho University, Vietnam
Do Thanh Nghi	Can Tho University, Vietnam
Peter Haddawy	Mahidol University, Thailand

Publication Chairs

Tran Thanh Dien	Can Tho University, Vietnam
Nguyen Minh Khiem	Can Tho University, Vietnam

Technical Program Committee

Atsushi Nunome	Kyoto Institute of Technology, Japan
An Cong Tran	Can Tho University, Vietnam
Bay Vo	HCMC University of Technology, Vietnam
Bob Dao	Monash University, Australia
Binh Tran	La Trobe University, Australia
Bui Vo Quoc Bao	Can Tho University, Vietnam
Cu Vinh Loc	Can Tho University, Vietnam
Dewan Farid	United International University, Bangladesh

Daniel Hagimont	Institut de Recherche en Informatique de Toulouse, France
Diep Anh Nguyet	University of Liege, Belgium
Duc-Luu Ngo	Bac Lieu University, Vietnam
Duong Van Hieu	Tien Giang University, Vietnam
Hiroki Nomiya	Kyoto Institute of Technology, Japan
Hien Nguyen	University of Information Technology - VNU HCMC, Vietnam
Huynh Quang Nghi	Can Tho University, Vietnam
Ivan Varzinczak	Université Paris 8, France
Lam Hoai Bao	Can Tho University, Vietnam
Le Hoang Son	Vietnam National University Hanoi, Vietnam
Luong Vinh Quoc Danh	Can Tho University, Vietnam
Maciej Huk	Wroclaw University of Science and Technology, Poland
Mai Xuan Trang	Phenikaa University, Vietnam
Masayuki Fukuzawa	Kyoto Institute of Technology, Japan
Nicolas Schwind	AIST - Artificial Intelligence Research Center, Japan
Nghia Duong-Trung	German Research Center for Artificial Intelligence (DFKI), Germany
Ngo Ba Hung	Can Tho University, Vietnam
Nguyen Chanh-Nghiem	Can Tho University, Vietnam
Nguyen Chi-Ngon	Can Tho University, Vietnam
Nguyen Dinh Thuan	University of Information Technology, VNU HCMC, Vietnam
Nguyen Huu Hoa	Can Tho University, Vietnam
Nguyen Minh Khiem	Can Tho University, Vietnam
Nguyen Minh Tien	Hung Yen University of Technology and Education, Vietnam
Nguyen Thai-Nghe	Can Tho University, Vietnam
Nguyen Thanh Thuy	University of Engineering and Technology, VNU Hanoi, Vietnam
Nguyen-Khang Pham	Can Tho University, Vietnam
Nhut-Khang Lam	Can Tho University, Vietnam
Peter Haddawy	Mahidol University, Thailand
Pham Truong Hong Ngan	Can Tho University, Vietnam
Phan Anh Cang	Vinh Long University of Technology Education, Vietnam
Phan Phuong Lan	Can Tho University, Vietnam
Quoc-Dinh Truong	Can Tho University, Vietnam
Salem Benferhat	CRIL, CNRS & Artois University, France
Si Choon Noh	Namseoul University, South Korea

Thai Tran	Lincoln University, New Zealand
Thanh-Hai Nguyen	Can Tho University, Vietnam
Thanh-Nghi Do	Can Tho University, Vietnam
Thanh-Nghi Doan	An Giang University, Vietnam
Thanh-Tho Quan	University of Technology - VNU HCMC, Vietnam
The-Phi Pham	Can Tho University, Vietnam
Thi-Lan Le	Hanoi University of Science and Technology, Vietnam
Thuong-Cang Phan	Can Tho University, Vietnam
Thai Minh Tuan	Can Tho University, Vietnam
Thi-Phuong Le	EBI School of Industrial Biology, France
Tomas Horvath	Eötvös Loránd University, Hungary
Tran Cao Son	New Mexico State University, USA
Tran Hoang Viet	Can Tho University, Vietnam
Tran Khai Thien	HCMC University of Foreign Languages and Information Technology, Vietnam
Tran Nguyen Minh Thu	Can Tho University, Vietnam
Trung-Hieu Huynh	Industrial University of Ho Chi Minh City, Vietnam
Truong Minh Thai	Can Tho University, Vietnam
Truong Xuan Viet	Can Tho University, Vietnam
Van-Hoa Nguyen	An Giang University, Vietnam
Vatcharaporn Esichaikul	Asian Institute of Technology, Thailand
Vinh Nguyen Nhi Gia	Can Tho University, Vietnam
Van-Sinh Nguyen	International University, VNU HCMC, Vietnam
Wu-Yuin Hwang	National Central University, Taiwan
Yuya Yokoyama	Kyoto Prefectural University, Japan
Yoshihiro Mori	Kyoto Institute of Technology, Japan

Finance and Secretary

Phan Phuong Lan	Can Tho University, Vietnam
Lam Nhut Khang	Can Tho University, Vietnam
Dinh Lam Mai Chi	Can Tho University, Vietnam

Local Committee

Le Nguyen Doan Khoi	Can Tho University, Vietnam
Le Van Lam	Can Tho University, Vietnam

Huynh Xuan Hiep	Can Tho University, Vietnam
Ngo Ba Hung	Can Tho University, Vietnam
Nguyen Nhi Gia Vinh	Can Tho University, Vietnam
Pham Nguyen Khang	Can Tho University, Vietnam
Pham The Phi	Can Tho University, Vietnam
Thuong-Cang Phan	Can Tho University, Vietnam
Tran Nguyen Minh Thu	Can Tho University, Vietnam
Truong Minh Thai	Can Tho University, Vietnam
Quoc-Dinh Truong	Can Tho University, Vietnam

Contents – Part II

Deep Learning and Natural Language Processing

Contents – Part I

Deep Learning and Natural Language Processing

Intelligent Systems

Applied Intelligent Systems and Data Science for Agriculture, Aquaculture, and Biomedicine

Towards Automatic Internal Quality Grading of Mud Crabs: A Preliminary Study on Spectrometric Analysis

Nhut-Thanh Tran[1]([✉]), Hai-Dang Vo[2,3]([✉]), Chi-Thinh Ngo[1], Quoc-Huy Nguyen[1], and Masayuki Fukuzawa[3]

[1] Faculty of Automation Engineering, Can Tho University, Can Tho 94000, Vietnam
nhutthanh@ctu.edu.vn
[2] Faculty of Multimedia Communications, Can Tho University, Can Tho 94000, Vietnam
vhdang@ctu.edu.vn
[3] Graduate School of Science and Technology, Kyoto Institute of Technology, Kyoto 606-8585, Japan

Abstract. Towards automatic internal quality grading of mud crabs, a spectrometric analysis has been conducted in the visible and near-infrared range for essential components (shell, ovary, and meat) of a mud crab (*Scylla paramamosain*), which is an important product in the Mekong Delta region. Since the internal qualities of mud crabs, such as meat yield and ovarian fullness, are manually evaluated and graded at present, it is strongly desired to automate them by a nondestructive and quantitative technique. Spectrometry is one of the promising techniques, but it is difficult to develop a practical system because of the complexity of in-vitro study and consequent difficulties in dataset preparation for adopting machine-learning techniques. In this study, we proposed a research strategy of cumulative spectrometric analysis by developing various spectrometric systems optimized for three stages of in-vitro, semi-in-vivo and in-vitro conditions to enable an effective dataset preparation for developing internal quality models with machine-learning techniques. A preliminary experiment under in-vitro condition showed that the transmittance of the crab essential components revealed different spectrometric characteristics in the near-infrared wavelength region (at 940, 760, and 680 nm for shell, ovary, and meat, respectively). This result suggests that spectrometric analysis has a great potential for internal quality grading of mud crabs.

Keywords: Spectrometric Analysis · Dataset Preparation · Machine-learning · Crab Quality · Ovarian Fullness · Meat Yield

1 Introduction

The Mekong Delta, situated in the southern part of Vietnam, is renowned for its natural diversity and its pivotal role in rice cultivation. However, beyond the lush expanses of rice paddies, it is also a thriving hub for the aquaculture industry. Characterized by a complex network of rivers, estuaries, and mangrove forests, the delta provides an ideal environment for aquaculture, attracting various aquatic species, from shrimp to different types of fish. Aquaculture has become a vital contributor to both the regional and national economies, propelling Vietnam into a prominent position within the global seafood market. Within this diverse aquatic landscape, the farming of mud crabs, especially the Scylla paramamosain species, has gained significant attention in recent years.

Mud crabs are known as premium seafood with high economic value due to their distinctive flavor as well as its high protein levels [1, 2]. They are primarily distributed in tropical regions of the Indian Ocean and the Pacific Ocean, and belong to the genus Scylla, comprising four species: Scylla serrata, S. tranquebarica, S. paramamosain, and S. olivacea [3]. In the Mekong Delta region, Scylla paramamosain is a common species, typically inhabiting estuaries, mangroves, and coastal areas [1]. With the recent surge in demand for mud crab consumption, both domestically and in export markets, it has become a significant seafood product within the aquaculture industry. They have played a crucial role in the rapid growth of the aquaculture industry in recent years within the Mekong Delta region, particularly in coastal provinces, and are expected to contribute to sustainable development in that area.

It is extremely important to assess the crab quality because it directly affects product value, consumer satisfaction, and overall market competitiveness. With the increase in the production market of mud crabs, there is a strong demand for non-destructive, objective, and automated techniques to assess the quality of mud crabs without altering or damaging them.

The quality indicators of mud crabs include not only external ones such as weight, gender, number of claws and legs, and visible defects, but also internal ones such as ovarian fullness for females and meat yield for males. The external qualities are easily evaluated by physical measurements such as weighing and by visual inspection. There were also several attempts to automate the visual inspection such as grading [4], gender classification [5, 6], and molting identification [7] from the images acquired with the conventional camera using various machine-learning (ML) techniques. However, the internal qualities are difficult to be visually assessed because they are not always reflected on the appearance.

The internal quality of mud crabs is now still examined indirectly and empirically using visual and tactile senses of evaluators by touching the crab portion. A conventional technique was developed by C-AID Consultants, Australia [8]. This technique involves applying manual pressure on the mud crab to evaluate its meat yield, with the grading (A to C) determined based on the observed flexion of the crab's abdomen (for male crabs) or carapace (for female crabs) at certain points of pressure, as shown in Fig. 1. The "X" mark in Fig. 1 are commonly testing points for checking the meat yield of male crabs. However, this technique carries the risk of damaging the products and may yield subjective and unreliable evaluation results.

One of the promising approaches for objectifying and automating internal quality assessment is spectrometry, which provides various information about the internals of products such as chemical composition, freshness, and structural properties by analyzing their absorption, reflection, and scattering profiles. It has been employed in various applications to assess parameters such as moisture content, protein content, TVB-N content, and the presence of contaminants [9–14]. However, it is still difficult to assess

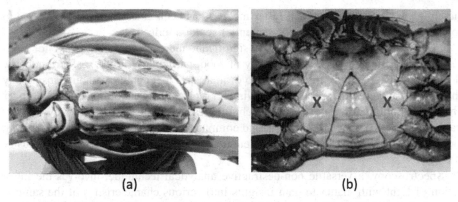

(a)   (b)

Fig. 1. Major quality assessment method for (a) female crab and (b) male crab.

the internal quality of crabs under in-vivo condition based on the spectrometry because conventional spectrometers were applicable only for in-vitro experiments and there is no established technique for analyzing the acquired spectra, especially for internal qualities.

Towards automatic internal quality grading of mud crabs, we propose a research strategy to develop a practical technique of internal quality assessment based on spectrometry. It includes repetitive prototyping of various spectrometric systems optimized for three stages of in-vitro, semi-in-vivo, and in-vivo conditions to enable an effective dataset preparation for developing internal quality models with machine-learning techniques. This paper introduces the details of this strategy. A preliminary result of an in-vitro experiment will be shown to discuss the potential and practical usefulness for internal quality grading of mud crabs.

2 Related Works

Traditional techniques have been employed for many years to evaluate the quality of seafood products. However, in an industrial setting, these methods have drawbacks such as being time-consuming, risking specimen damage, incurring high labor costs, and potentially leading to inconsistent or biased assessments.

Conventional methods for assessment of seafood products mainly rely on sensory evaluation, chemical analysis, and physical measurements [17–19]. Sensory evaluation relates to human perception of product attributes such as odor, texture, and taste, and it is often subjective and influenced by personal preferences and biases. Sensory analysis is carried out by experienced inspectors who assess external factors such as smell, taste, touch, and appearance using grading schemes.

Chemical and physical analysis techniques, such as measuring pH, total volatile basic nitrogen (TVB-N), K-value, and total viable count (TVC) provide objective measurements but can be time-consuming and require specialized equipment [9, 20, 21]. These traditional laboratory-based methods have been utilized for decades to assess food quality, but are less feasible in the seafood industry context due to the time-consuming, complex sample preparation, sample destruction, high labor costs, may result in inconsistent or biased assessments, and incompatibility in real-time operations. Therefore, the simpler, non-destructive, and faster method has been emerged as an alternative to conventional method of quality assessment of seafood industry.

In recent years, there has been a significant increase in the demand for non-destructive methods to assess the quality of food, particularly in the seafood industry. Non-destructive methods are designed to evaluate product quality without causing any changes or damage to the specimen, allowing for rapid and repeatable measurements over time while minimizing waste. Various technologies, including spectroscopy, hyperspectral imaging (HSI), machine learning, and computer vision, have emerged as promising tools for non-destructive quality assessment in the food production and seafood industry [10–16].

Spectroscopy, a versatile non-destructive analytical technique, involves the interaction of light with matter to gain insights into various characteristics of the sample. It enables the identification and quantification of specific compounds, nutrients, and contaminants in seafood products. By analyzing the absorption, emission, or scattering of light, spectroscopic techniques provide valuable information about attributes such as chemical composition, freshness, and structural properties. In the context of seafood quality assessment, spectroscopy has been employed in various forms, including Visible-near infrared (VIS-NIR) spectroscopy, near-infrared (NIR) spectroscopy, and Raman spectroscopy, to assess parameters such as moisture content, protein content, TVB-N content, and the presence of contaminants [9–13, 15, 16].

The numerous HSI systems developed in recent years by combining spectroscopy, image processing, and machine learning to evaluate the quality of seafood products without causing excessive damage specimens. Cheng et al. [9] proposed a system that utilizes HSI techniques, in conjunction with partial least squares regression (PLSR) and least squares support vector machines (LS-SVM) model, to evaluate the freshness of fish based on the TVB-N index. Similarly, Yu et al. [14] employed HSI techniques combined with deep learning, using the Stacked Auto-Encoders (SAEs) algorithm along with LS-SVM and multiple linear regression (MLR) models, achieving improved results compared to the Successive Projections Algorithm (SPA). In another study, Zhang et al. [22] investigated the feasibility of using near-infrared hyperspectral imaging (NIR-HSI) to predict the fat and moisture contents in salmon fillets. Qu et al. [23] explored the potential of a visible-near-infrared (Vis-NIR) hyperspectral imaging system to differentiate between fresh, cold-stored, and frozen-thawed shelled shrimp. Dixit et al. [10] evaluated the performance of a portable HSI system based on snapshot sensors for online assessment of salmon, comparing it to a conventional hyperspectral system. Kong et al. [13] demonstrated that NIR-HSI combined with convolutional neural network (CNN) calibration provides a promising technique for detecting marine fishmeal adulterated with low-cost processed animal proteins, achieving an accuracy of up to 99.37%. The findings by Shao et al. [24] confirmed the feasibility of HSI for non-destructive determination of freshness in small yellow croaker, offering valuable technical guidance for the storage and marketing of aquatic products.

Despite the comprehensive research conducted on quality assessment methods in the seafood industry, there is a conspicuous lack of studies dedicated to non-destructive assessing the internal quality of mud crabs. Our study introduces an innovative method for evaluating the internal quality of mud crabs, employing HSI and machine learning techniques. In the following section, we delineate our proposed three-stage strategy for non-destructive quality assessment of mud crabs, which combines spectrometry and machine learning techniques.

3 Research Strategy

Figure 2 shows a research strategy that we proposed in this study for internal quality evaluation of mud crabs using spectrometric technique. It consists of major three stages.

The first stage corresponds to so-called in-vitro studies. This stage aims to comprehensively understand the spectrometric characteristics of representative mud crab components, including the shell, meat, and ovary. The key point is to examine each component individually rather than whole crab, and it is worthwhile to obtain the unique characteristics of each component. The experiments are conducted using conventional spectrometers on the separated crab components or their aqueous solutions filled in a cuvette. Spectral data are acquired for each component in transmission, reflection, and interactance modes and then analyzed to establish basic correlations with component types. The feature of this stage is to focus on acquiring various know-hows in crab handling and component-specific spectrometry as well as the essential knowledge of crab components. The crab handling know-how must be helpful to develop crab portioning techniques to be required in the next stage. Furthermore, the component-specific spectrometry gives useful insights for designing a custom spectrometer in the next stage.

The second stage can be categorized as semi-in-vivo studies. This stage aims to establish a special spectrometric technique to obtain semi-in-vivo spectrum of a crab portion and to analyze its correlation with the contained components in the portion based on ML-based techniques. The feature of this stage is to examine each crab 'portion' which includes different components. It includes development of a custom spectrometer specialized for crab portions based on the findings of the previous stage. Since crab portion is not a commercial product, this spectrometer is of little practical use. However, it has a clear advantage in the development of a special spectrometer to cover the whole crab in the next stage because it provides with the spectrometry know-how specific to crab portions containing different components. It should also be noted that the crab portions are very easy to identify the contained components after individual spectrometric analysis under semi-in-vivo condition. On the other hand, whole crabs require retrospective portioning and destructive evaluation for component identification and still not suitable for semi-in-vivo spectrometry due to its complicated structure. Therefore, a lot of spectrum data can be efficiently prepared with annotation of contained components by using the crab portions, which realizes preliminary ML-based analysis even at this stage. It is also expected from the results of that preliminary analysis to obtain special knowledge about appropriate variety and granularity of annotation for each crab portion and its containing components. This knowledge must be useful in the next stage for further data preparation with more complicated annotation.

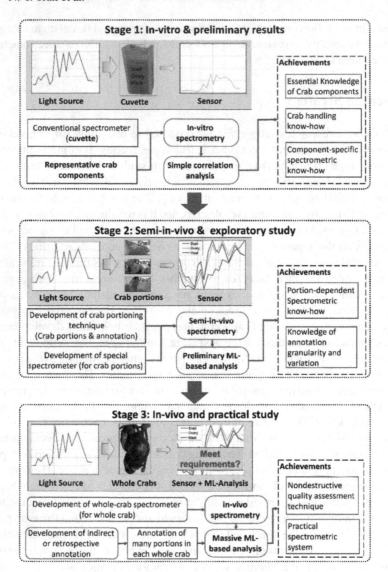

Fig. 2. A research strategy for internal quality evaluation of mud crabs using spectrometric technique.

In the third stage of the proposed strategy, an in-vivo and practical assessment of internal quality will be attempted for the whole crabs. The main focus of this stage is the development of a sophisticate spectrometer achieving in-vivo spectrometry for living mud crabs. It is also highly practical because it can cover the whole crabs. A new annotation technique is also developed for various portions contained in a whole crab by retrospective portioning and identification of contained components after the spectrometric evaluation, which is based on the knowledge of annotation variety and granularity for crab portion in the previous stage. By combining the in-vivo spectrometry for a whole crab, retrospective portioning, and annotation of contained components for each portion, a massive ML-based analysis is realized in this stage to establish a new technique for automatic and nondestructive quality assessment and grading of internal crab quality.

Table 1 shows technical specifications for each stage in the proposed strategy. It is clearly described that the spectroscopy, annotation, and analysis techniques are updated stage by stage, and an achievement in a stage becomes the basis for the next. Through these efforts, we finally aim to develop a practical spectrometric system for non-destructive quality assessment of mud crabs, optimizing the evaluation process and contributing to the aquaculture industry. The practical spectrometric systems and their developing methodology are the key to achieve high performance of quality evaluation, and also serve as a framework of this and similar studies.

Table 1. Technical specifications of each step in the proposed research strategy

Stage #	1	2	3
Type	In-vitro	Semi-in-vivo	In-vivo
Target	Crab Component (shell, meat, ovary)	Crab Portion	Whole Crab
Spectrometer	Conventional Optics, Generic Holder	Custom Optics, Optimized Holder	Sophisticated Optics, Automechanism
Annotation	Prospective, Direct	Prospective, Direct	Retrospective, Indirect
Analysis Technique	Simple Correlation	Preliminary ML-based	Massive ML-based
Expected Achievement	Component-specific spectrometry, Crab Handling Know-How	Portion-specific Spectrometry, Variation and Granularity of Annotation	Quality Assessment Technique, Practical Spectrometric system

4   In-vitro Experiments and Preliminary Results

4.1   Sample Preparation and Experimental Setup

In this study, a mature female mud crab (*Scylla paramamosain*) weighing 400 g was harvested from a crab pond in Bac Lieu province, located in the Mekong Delta region. The shell of this crab was then carefully removed to collect several essential components including the shell, ovary, and meat. Subsequently, the collected ovary and meat were mixed with distilled water at a ratio of 1:1 to get ovary and meat solutions, as presented in Fig. 3.

Figure 4 shows a diagram of typical spectrometric experiment and an in-vitro experimental setup employed in this study. Each of three crab component samples (ovary

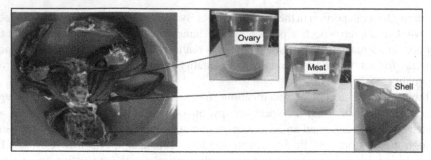

Fig. 3. Preparation of Crab components

solution, meat solution and shell) was fully filled into a square glass cuvette with a light path of 10 mm. A halogen lamp (12 VDC, 35 W) served as a light source and was controlled using a regulated power adapter.

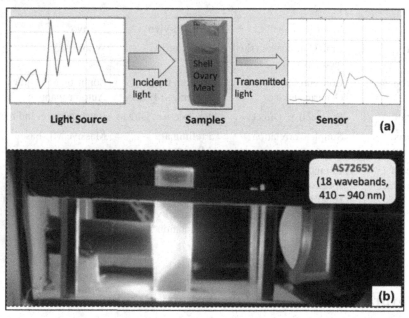

Fig. 4. Schematic diagram of typical spectrometric experiment (a) and in-vitro experimental setup (b).

After interacting with the samples, the intensity of transmitted light was collected by a multi-spectral sensor, named AS7265X. The sensor encompasses 18 wavebands, covering from visible to near-infrared regions (410–940 nm). This sensor was utilized in previous studies [25–29] and proved its effectiveness for various spectrometric analysis applications. In this study, the integration time of each scan was 400 ms, and the measurement was performed with the minimal duration to prevent heating of the solutions.

The experiment was conducted at room temperature, approximately 28 °C. The transmittance of the three crab component samples was normalized by dividing the intensity of the transmitted light by that of the incident light.

4.2 Results and Discussion

Figure 5 shows the transmittance of the three crab component samples at several wavebands collected by the AS7265X sensor. There was a similar trend in the transmittance of these component samples. It revealed relatively low transmittance in the visible region and did relatively high in the near-infrared. The lowest transmittance was recorded at 535 nm waveband for all three component samples. On the other hand, the highest transmittance wavebands of each crab component were different. They were at 940, 760, and 680 nm for shell, ovary, and meat, respectively. Furthermore, the wavelength band in which the transmittance was relatively high compared to other components was also different for each component, such as 410, 485, and 535 nm for shell, 610, 760, 810, 860 nm for ovary, and 585, 680 nm for meat.

The low transmission in the visible range is easily speculated to be due to the opacity and color of the sample. The opacity and color are one of primitive features to distinguish the components, but from the spectrometric viewpoint, a critical waveband of a component is overlapped with that of other components. Therefore, it would be difficult to evaluate crab components by focusing on a waveband of transmission bottom.

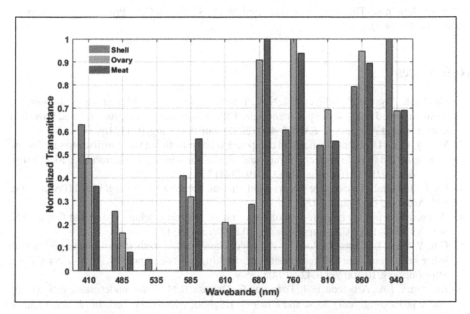

Fig. 5. Transmittance of three main components of mud crab.

It is well known that living cell and tissues reveals high transmittance in the near-infrared region, which is consistent with the results of the highest transmittance of this experiment. The experimental results showed that the waveband of maximum transmittance and the wavebands with relatively high transmittance differed significantly for each crab component, which suggests its usefulness in evaluating crab components. It should be noted that a combination of multiple wavebands is more robust to fluctuations in spectrometry than a single waveband. Although further study is still required, such waveband combination clearly suggested great potential for evaluating internal quality of mud crabs.

5 Conclusion

Towards automatic internal quality grading of mud crabs, we proposed a research strategy of cumulative spectrometric analysis by developing various spectrometric systems optimized for three stages of in-vitro, semi-in-vivo, and in-vitro conditions to enable an effective dataset preparation for developing internal quality models with machine-learning techniques. A spectrometric analysis has been conducted in the visible and near-infrared range for essential components (shell, ovary, and meat) of a mud crab (*Scylla paramamosain*). A preliminary experiment under in-vitro condition showed that the transmittance of the crab essential components revealed different spectrometric characteristics in the near-infrared wavelength region (at 940, 760, and 680 nm for shell, ovary, and meat, respectively). This result suggests that spectrometric analysis has a great potential for internal quality grading of mud crabs.

Acknowledgments. This work was supported by JSPS Core-to-Core Program (grant number: JPJSCCB20230005).

References

1. Van Ly, K., Arsa, C.P., Nguyen Thi, N.A., Tran Ngoc, H.: Use of different seaweeds as shelter in nursing mud crab, Scylla paramamosain: Effects on water quality, survival, and growth of crab. J. World Aquac. Soc. **53**, 485–499 (2022). https://doi.org/10.1111/jwas.12830
2. Wu, Q., et al.: Different biochemical composition and nutritional value attribute to salinity and rearing period in male and female mud crab Scylla paramamosain. Aquaculture **513** (2019). https://doi.org/10.1016/j.aquaculture.2019.734417
3. FAO. Mud Crab Aquaculture - A practical manual. https://www.fao.org/3/ba0110e/ba0110e.pdf. Accessed on 13 July 2023
4. Wang, H., et al.: Quality grading of river crabs based on machine vision and GA-BPNN. Sensors **23**, 1–19 (2023). https://doi.org/10.3390/s23115317
5. Cui, Y., Pan, T., Chen, S., Zou, X.: A gender classification method for Chinese mitten crab using deep convolutional neural network. Multimed. Tools Appl. **79**, 7669–7684 (2020). https://doi.org/10.1007/s11042-019-08355-w
6. Baluran, C.I.A., Arboleda, E.R., Dizon, M.G., Dellosa, R.M.: Crab gender classification using image processing, fuzzy logic and k nearest neighbor (KNN) classifier. Int. J. Sci. Technol. Res. **8**, 1458–1462 (2019)
7. Baharuddin, R.R., Niswar, M., Ilham, A.A., Kashihara, S.: Crab molting identification using machine learning classifiers. In: 2021 International Seminar on Machine Learning, Optimization, and Data Science ISMODE 2021, pp. 295–300 (2022). https://doi.org/10.1109/ISMODE 53584.2022.9743136

8. C-AID Consultants. Australian Industry Live Mud Crab Grading Scheme (2016). https://www.c-aid.com.au/wp-content/uploads/Mud-Crab-Grading-Scheme-V3-2016.pdf. Accessed 13 July 2023
9. Cheng, J.H., Sun, D.W., Zeng, X.A., Pu, H.: Bin: non-destructive and rapid determination of TVB-N content for freshness evaluation of grass carp (Ctenopharyngodon idella) by hyperspectral imaging. Innov. Food Sci. Emerg. Technol. **21**, 179–187 (2014). https://doi.org/10.1016/j.ifset.2013.10.013
10. Dixit, Y., Reis, M.M.: Hyperspectral imaging for assessment of total fat in salmon fillets: a comparison between benchtop and snapshot systems. J. Food Eng. **336** (2023). https://doi.org/10.1016/j.jfoodeng.2022.111212
11. He, H.J., Wu, D., Sun, D.W.: Nondestructive spectroscopic and imaging techniques for quality evaluation and assessment of fish and fish products. Crit. Rev. Food Sci. Nutr. **55**, 864–886 (2015). https://doi.org/10.1080/10408398.2012.746638
12. Xu, J.L., Riccioli, C., Sun, D.W.: Development of an alternative technique for rapid and accurate determination of fish caloric density based on hyperspectral imaging. J. Food Eng. **190**, 185–194 (2016). https://doi.org/10.1016/j.jfoodeng.2016.06.007
13. Kong, D., et al.: Hyperspectral imaging coupled with CNN: A powerful approach for quantitative identification of feather meal and fish by-product meal adulterated in marine fishmeal. Microchem. J. **180** (2022). https://doi.org/10.1016/j.microc.2022.107517
14. Yu, X., Wang, J., Wen, S., Yang, J., Zhang, F.: A deep learning based feature extraction method on hyperspectral images for nondestructive prediction of TVB-N content in Pacific white shrimp (Litopenaeus vannamei). Biosyst. Eng. **178**, 244–255 (2019). https://doi.org/10.1016/j.biosystemseng.2018.11.018
15. Siche, R., Vejarano, R., Aredo, V., Velasquez, L., Saldaña, E., Quevedo, R.: Evaluation of food quality and safety with Hyperspectral Imaging (HSI). Food Eng. Rev. **8**, 306–322 (2016). https://doi.org/10.1007/s12393-015-9137-8
16. Saha, D., Manickavasagan, A.: Machine learning techniques for analysis of hyperspectral images to determine quality of food products: a review. Curr. Res. Food Sci. **4**, 28–44 (2021). https://doi.org/10.1016/j.crfs.2021.01.002
17. Hassoun, A., Karoui, R.: Quality evaluation of fish and other seafood by traditional and nondestructive instrumental methods: advantages and limitations. Crit. Rev. Food Sci. Nutr. **57**, 1976–1998 (2017). https://doi.org/10.1080/10408398.2015.1047926
18. Esteves, E., Aníbal, J.: Sensory evaluation of seafood freshness using the quality index method: a meta-analysis. Int. J. Food Microbiol. **337** (2021). https://doi.org/10.1016/j.ijfoodmicro.2020.108934
19. Alasalvar, K., et al.: Freshness assessment of cultured sea bream (Sparus aurata) by chemical, physical and sensory methods. Food Chem. **72**, 33–40 (2001)
20. Wu, D., Sun, D.W.: Advanced applications of hyperspectral imaging technology for food quality and safety analysis and assessment: a review—part II: applications. Innov. Food Sci. Emerg. Technol. **19**, 15–28 (2013). https://doi.org/10.1016/j.ifset.2013.04.016
21. Rahman, A., Kondo, N., Ogawa, Y., Suzuki, T., Shirataki, Y., Wakita, Y.: Prediction of K value for fish flesh based on ultraviolet-visible spectroscopy of fish eye fluid using partial least squares regression. Comput. Electron. Agric. **117**, 149–153 (2015). https://doi.org/10.1016/j.compag.2015.07.018
22. Zhang, H., et al.: Non-destructive determination of fat and moisture contents in Salmon (Salmo salar) fillets using near-infrared hyperspectral imaging coupled with spectral and textural features. J. Food Compos. Anal. **92** (2020). https://doi.org/10.1016/j.jfca.2020.103567
23. Qu, J.H., Cheng, J.H., Sun, D.W., Pu, H., Wang, Q.J., Ma, J.: Discrimination of shelled shrimp (Metapenaeus ensis) among fresh, frozen-thawed and cold-stored by hyperspectral imaging technique. Lwt **62**, 202–209 (2015). https://doi.org/10.1016/j.lwt.2015.01.018

24. Shao, Y., Shi, Y., Wang, K., Li, F., Zhou, G., Xuan, G.: Detection of small yellow croaker freshness by hyperspectral imaging. J. Food Compos. Anal. **115** (2023). https://doi.org/10.1016/j.jfca.2022.104980

25. Tran, N.-T., Fukuzawa, M.: A portable spectrometric system for quantitative prediction of the soluble solids content of apples with a pre-calibrated multispectral sensor chipset. Sensors **20**, 5883 (2020). https://doi.org/10.3390/s20205883

26. Nguyen, C.-N., Phan, Q.-T., Tran, N.-T., Fukuzawa, M., Nguyen, P.-L., Nguyen, C.-N.: Precise sweetness grading of mangoes (Mangifera indica L.) based on random forest technique with low-cost multispectral sensors. IEEE Access **8**, 212371–212382 (2020). https://doi.org/10.1109/ACCESS.2020.3040062

27. Stevens, J.D., Murray, D., Diepeveen, D., Toohey, D.: Development and testing of an IoT spectroscopic nutrient monitoring system for use in micro indoor smart hydroponics. Horticulturae **9**, 185 (2023). https://doi.org/10.3390/horticulturae9020185

28. Shokrekhodaei, M., Cistola, D.P., Roberts, R.C., Quinones, S.: Non-invasive glucose monitoring using optical sensor and machine learning techniques for diabetes applications. IEEE Access **9**, 73029–73045 (2021). https://doi.org/10.1109/ACCESS.2021.3079182

29. Yang, B., Huang, X., Yan, X., Zhu, X., Guo, W.: A cost-effective on-site milk analyzer based on multispectral sensor. Comput. Electron. Agric. **179**, 105823 (2020). https://doi.org/10.1016/j.compag.2020.10582

Deep Learning for Detection and Classification of Nuclear Protein in Breast Cancer Tissue

Thuong-Cang Phan[1](✉) , Anh-Cang Phan[2] , Thi-My-Tien Le[2],
and Thanh-Ngoan Trieu[1]

[1] Can Tho University, Can Tho, Vietnam
{ptcang,ttngoan}@cit.ctu.edu.vn
[2] Vinh Long University of Technology Education, Vinh Long, Vietnam
{cangpa,tienltm}@vlute.edu.vn

Abstract. Breast cancer is a medical condition in which the cells of the mammary gland grow uncontrollably, creating malignant tumors, capable of dividing, and invading surrounding areas. Currently, the incidence of breast cancer is increasing and getting younger. Medical methods such as Ultrasound, Computed Tomography (CT), X-rays, and Magnetic Resonance Imaging (MRI) are the first options to diagnose breast cancer as well as other diseases in the mammary gland. However, these imaging methods are not enough to accurately determine the status of a breast tumor. To be able to accurately identify breast cancer and assess the extent of the disease, immunohistochemistry will be used to characterize the cells in the tissue sample, including breast cancer cells. In this work, we propose a method to detect and classify nuclear proteins in breast cancer cells on histological images using deep learning techniques. We take advantage of deep learning networks and the transfer learning approach to train network models on the SHIDCB-Ki-67 dataset of Shiraz University of Medical Sciences in Shiraz, Iran. DeepLabv3-MobileNet-V2, DeepLabv3-Xception, and DeepLabv3-DenseNet-121 are used to detect and classify nuclear proteins of breast cancer cells on histological images. Experimental results show that the proposed method with DeepLabv3-DenseNet-121 achieves higher accuracy (98.6%) than DeepLabv3-Xception and DeepLabv3-MobileNet-V2.

Keywords: SHIDCB-Ki-67 · DeepLabv3+ · MobileNet-v2 · Xception · DenseNet-121

1 Introduction

Breast cancer [7] is the most dangerous disease in women. The disease has been claiming the lives and health of thousands of people every day around the world. Like other cancers, breast cancer also needs to go through a long period of development before it begins to metastasize and spread to other parts of the body.

N. Thai-Nghe et al. (Eds.): ISDS 2023, CCIS 1950, pp. 15–28, 2024.
https://doi.org/10.1007/978-981-99-7666-9_2

In 2020, there were 2,261,419 new cases and 684,996 deaths from breast cancer worldwide. It was ranked first in terms of new incidence and fourth in mortality from cancer in both men and women. According to GLOBOCAN statistics in 2020 [3,10], Vietnam has 21,555 new cases of breast cancer, which is the most common cancer in women, accounting for 25.8% of all cancers in women. In the same year, more than 9,345 people died from the disease, ranking third after liver cancer and lung cancer [8]. Breast cancer not only has terrible health and psychological consequences but also negatively affects other aspects of a patient's life. In the past years, breast cancer was common in women aged 60–65 years old, now, women after 20 years old are at risk. Timely cancer detection and treatment is an important process to improve health status and prolong the life of patients. Breast cancer tumors have many different characteristics such as size, shape, density, percentage of stimulated cells, and ability to invade and spread [14]. The assessment and classification of these characteristics is important to decide whether treatment is appropriate. Diagnostic methods such as mammography and ultrasonography are the preferred methods of screening for breast cancer as well as for mammary diseases [12]. However, imaging methods such as ultrasound, computerized tomography (CT), X-ray, and magnetic resonance imaging (MRI) are not enough for accurate identification of breast tumor status [11,13]. Histology is a diagnostic method used in determining the type of breast cancer and assessing the extent of the disease. In breast cancer cells, Ki-67 is used to evaluate the extent of breast cancer tumors as well as determine the speed of cancer cell proliferation. Measuring the expression level of Ki-67 protein in breast cancer samples can help classify breast cancer tumors into groups with different properties and predict treatment outcomes. Besides, it has to be admitted that advanced information technology has led to better diagnosis and treatment for patients. Specifically, doctors are able to process medical images faster without missing out on any important features, as they are highlighted beforehand for review. Thus, advanced information technology is necessary to provide support to assess the extent of breast cancer for better treatment.

In recent years, there have been many studies on the application of deep learning based on the widely used convolutional neural networks (CNN) for the classification and detection of problems related to breast cancer. Agarap [1] presented a machine learning-based method to detect and diagnose breast cancer on mammography. The authors used the Wisconsin Diagnostic Breast Cancer (WDBC) dataset consisting of 569 breast tumor samples, of which 212 were benign and 357 were malignant. Each sample is an ultrasound image of a breast lump. The images were collected from various women between the ages of 25 and 90. Features including area, thickness, and standard deviation of pixels were extracted from ultrasound images. These features are used to train machine learning models such as Support Vector Machines (SVM), K-nearest neighbors (KNN), and Artificial Neural Networks (ANN). Dina A. Ragab et al. [19] presented an application of deep convolutional neural network to detect and classify breast cancer on mammographic images. The results showed that the combined method of CNN and SVM achieved an accuracy of 87.2% with an AUC equaling to 0.94 on the Curated Breast Imaging Subset of DDSM (CBIS-DDSM). Alqahtani et al. [2] proposed to use a multiscale channel recalibration model, msSE-ResNet, to address the challenge of automatic classification of breast cancer pathology images. The proposed method has been tested on the BreaKHis dataset, and compared with other classification methods. The network performed classification of benign/malignant breast pathology images collected at different magnifications with a classification accuracy of 88.87%. Alexander Rakhlin et al. [20] utilized several deep neural network architectures and gradient-boosted trees classifier to analyze breast cancer histology. They used the ICIAR 2018 (International Conference on Image Analysis and Recognition) dataset on microscopic images of breast cancer tissue stained with Hematoxylin and Eosin. The results showed that the proposed models has an accuracy of up to 93.8% on the testing dataset for the 2-class classification. Gnanasekaran et al. [9] presented a method of breast cancer classification using a deep convolutional neural networks on mammograms. The study was performed on the MIAS and the DDSM dataset. The test results showed that the proposed method has better classification accuracy than other traditional methods. The proposed model achieves accuracy of 92.54% and 96.47% on the two datasets. Spanhol et al. [22] proposed a method of classifying breast cancer on histopathology images using CNN. The authors evaluated the proposed system performance in the BreaKHis dataset containing 2,480 benign and 5,429 malignant samples. The results showed that the proposed CNN-based classifier performed better comparing with other machine learning models.

2   Background

2.1   Alteration of Nuclear Protein Ki-67 in Breast Cancer Cells

Breast cancer is a type of cancer that develops from cells in the mammary gland. Cancer cells have the ability to invade surrounding tissue and blood vessels, destroying tissue structures, and causing deterioration of normal cellular function. Ki-67 [5] is a protein found in cells that indicates the rate of cell division and growth. Ki-67-positive cells [24] are those in which the Ki-67 protein is found in their nucleus. A positive Ki-67 index is usually determined by counting the number of cells with the Ki-67 protein in a cell sample and comparing it to the total number of cells in that sample. Ki-67-positive cells are usually present in cells that are in the process of division. They are active cells that contribute to the generation of new cells. A Ki-67-positive cell usually has an increased division rate. Ki-67 negative cells [23] are those that do not have the presence of Ki-67 protein in their nucleus. This indicates that the cells are not in the process of division. The Ki-67 index is calculated as in Eq. 1.

$$ki - 67 score = \frac{ImmunoPositive}{ImmunoPositive + ImmunoNegative} \tag{1}$$

In the breast cancer histology report, the Ki-67 index is given as a percentage of breast cancer cells that are positive for this protein. The higher the level of Ki-67 protein compared to normal cells, the faster the abnormal cells are dividing and growing. This indicates that the cancer has a faster rate of division and a higher risk of recurrence. The level of Ki-67 is lower meaning that cancerous tumor tends to grow more slowly and has the potential to respond better to treatments [18]. In other words, the higher the Ki67-positive rate, the greater the risk of disease recurrence and the shorter the survival time [8,11,17]. Tumor cells can be suppressed by the body's own immune mechanisms. Tumor infiltrating lymphocytes (TILs) [6,16,17] were introduced as an immune component against tumor cell progression and have been shown to be beneficial in improving the outcomes of breast cancer patients.

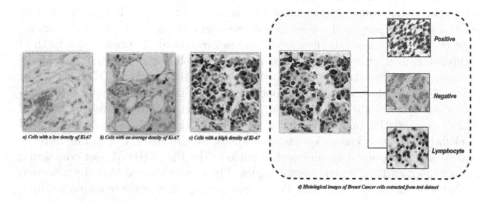

Fig. 1. Variation of nuclear protein Ki-67 in the assessment of proliferation index of breast cancer cells.

Figure 1 shows cells with low to high Ki-67 densities. Figure 1a the Ki-67 index is negative, which means that the percentage of Ki-67 is low (less than 30%). In Fig. 1b, the Ki-67 index is moderate. Figure 1c shows cells with a high density of Ki-67 (over 30% in). Fig 1d shows the detection and classification of Ki-67-positive tumor cells, Ki-67-negative tumor cells, and tumor cells with positive infiltrating lymphocytes.

The Ki-67 protein and tumor-infiltrating lymphocytes (TILs) [17] have been introduced as prognostic factors in predicting both tumor progression and probable response to chemotherapy. The estimation of the Ki-67 index and TILs is dependent on the observations of professional pathologists, and variations may also exist. Thus, automated methods using machine learning, namely Deep learning-based approaches have attracted public attention. However, deep learning methods need significantly annotated data. In this study, we propose to use SHIDC-BC-Ki-67 [17] as a dataset to estimate the Ki-67 index and determine TILs scores in breast cancer cells.

2.2 Deep Network Models

In recent years, convolutional neural networks have been widely used for the task of image classification. To detect and classify nuclear proteins in breast cancer cells, in this work, we used deep learning networks including DeepLabv3-MobileNet-V2, DeepLabv3-Xception, and DeepLabv3-DenseNet-121.

2.2.1 MobileNet-V2 MobileNet-V2 [21] is a significant improvement over MobileNet-V1 and enhances mobile image recognition capabilities including classification, object detection, and semantic segmentation. MobileNet-V2 is released as part of the TensorFlow-Slim Image Classifier Library. It builds on ideas from MobileNet-V1, using depth-separated convolution as efficient building blocks. However, this version introduces two new features to the architecture: linear bottlenecks and skip connection between bottlenecks. MobileNet-V2 models are faster with the same accuracy across the entire latency spectrum. In particular, it uses 2x fewer operations, needs 30% fewer parameters, and is about 30–40% faster on a Google Pixel phone than MobileNet-V1.

2.2.2 Xception Xception [4] is a deep convolutional neural network architecture consisting of depthwise separable convolutions developed by Google. Xception is an extension of the Inception architecture that replaces standard Inception modules with depthwise separable convolutions.

2.2.3 DenseNet-121 DenseNet [15] will be different from ResNet. The outputs of each mapping of the same length and width will be concatenated together into a depth block. Then to reduce the data dimension, a transition layer is used. This layer is a combination of a convolution layer to reduce the depth and a max pooling to reduce the length and width. In terms of accuracy, DenseNet has less than half the parameters but has higher accuracy than ResNet-50 on the ImageNet dataset. Besides, DenseNet is very effective in overfitting problems. It also applies BatchNormalization before convolution at the transition layers, thus reducing the vanishing gradient.

2.2.4 DeepLabv3+ DeepLabv3+ [25] was introduced by Google. The DeepLab family has appeared with versions from DeepLabv1, DeepLabv2, and DeepLabv3. DeepLabv3+ is a semantic segmentation architecture built on top of DeepLabv3 by adding a simple decoder module but effective to enhance segmentation results. The DeepLabv3+ model has an encryption phase and a decryption phase. The encoding phase extracts the required information from the image using a convolutional neural network while the decoding phase reconstructs an appropriately sized output based on the information obtained from the encoding phase. A decoder module has been added to give better segmentation results along object boundaries. DeepLab supports the following backbone networks: MobileNetv2, Xception, ResNet, PNASNet, and Auto-DeepLab.

2.2.5 Evaluation Metrics Loss function is determined by Eq. 2, where y is the actual value and p is the predicted value.

$$loss = -(ylog(p) + (1 - y)log(1 - p)) \tag{2}$$

Accuracy is the ratio between the number of correctly predicted data points and the total number of data points.

$$acc = \frac{TP + TN}{TP + TN + FP + FN} \tag{3}$$

3   Proposed Method

The proposed approach in detecting and classifying nuclear proteins of breast cancer cells is presented in Fig. 2. We build a system to detect and classify nuclear proteins of cancer cells on histological images using network architectures including DeepLabv3-MobileNet-V2, DeepLabv3-Xception, and DeepLabv3-DenseNet-121. The proposed approach consists of two phases, the training phase and the testing phase.

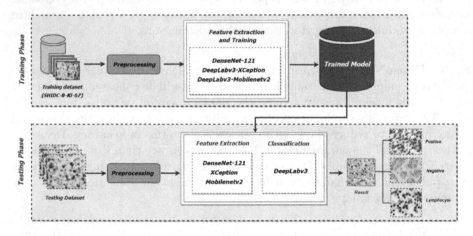

Fig. 2. The proposed approach in detecting and classifying nuclear proteins of breast cancer cells based on Deep Learning.

3.1   Training Phase

3.1.1   Preprocessing We use a number of image preprocessing methods to remove noise and enhance image quality while ensuring the image contrast so that the model can perform well. The transformation techniques are used such as histogram equalization, contrast adjustment, and noise removal to ensure that the images in the dataset are properly normalized. Additionally, to enhance the training dataset, we apply several techniques such as rotating, flipping, zooming, and changing the lighting. This helps the models learn from a variety of angles and lighting conditions.

3.1.2   Feature Extraction and Training Each image will be passed through feature extraction networks to extract the corresponding features. We propose to use the feature extraction networks such as MobileNet-V2, Xception, and DenseNet-121. At this stage, the dataset will be trained on the network models DeepLabv3-MobileNet-V2, DeepLabv3-Xception, and DeepLabv3-DenseNet-121. This helps to learn new features faster and shortens the training time. During the training process, when the loss value is not improved (not reduced), we will stop the training phase and move to the testing phase to compare and evaluate the models.

3.2   Testing Phase

In the testing phase, the input data will be images taken from the testing dataset for automatic detection and classification of Ki-67-positive tumor cells, Ki-67-negative tumor cells, and tumor cells with positive infiltrating lymphocytes.

4   Experiments

4.1   Scenarios and Training Parameters

In order to facilitate testing, comparison, and evaluation, we conduct experiments with three scenarios. In the training process, we change the parameters such as Batch size, Learning rate, Epoch, and Number of learning steps until reaching the optimal values to improve accuracy. The details of the training parameters are presented in Table 1.

Table 1. Proposed scenarios and training parameters

Scenarios	Training models	Feature extraction	Learning_rate	Batch_size	Num_step	Epochs	Num_classes
1	DeepLabv3	DenseNet-121	0.0001	128	64	200	3
2	DeepLabv3	Xception	0.0001	128	64	200	3
3	DeepLabv3	MobileNet-V2	0.0001	128	64	200	3

4.1.1   Data Description and Installation Environment The system is installed in Python language and runs on Google Colab Pro environment with the configuration of 25.4GB RAM and NVIDIA Tesla P100 GPU. The library to support training network models is TensorFlow 2.11.

The SHIDC-B-Ki-67 dataset (Table 2) consists of 2,357 images, which are divided into the training dataset and the testing dataset. Each image contains 69 cells on average and a total of 162,998 cells. To obtain a balanced set of training and testing, 70% (1,656) of each patient's images were randomly selected for the training phase and 30% (701) for the testing phase. Images were taken from slides prepared from breast mass tru-cut biopsies, which were further stained

for Ki-67 by immunohistochemistry method (IHC). These images are collected
in the RGB color space with 8 bits per color channel. The input image size is
256×256 pixels. Then, the Ki-67-stained images were labeled by the pathologists
with Ki-67-positive tumor cells, Ki-67-negative tumor cells, and tumor cells with
positive infiltrating lymphocytes. Labeling real world data is a challenging and
labor-intensive task.

Table 2. The SHIDC-B-Ki-67 dataset [17]

Type	Training - 1,656 IMG		Testing - 701 IMG	
	No. cells	Avg. cells/img	No. cells	Avg. cells/img
Ki-67-Positive	35,106	21.19	15,755	22.50
Ki-67-Negative	75,008	45.29	32,639	46.62
Positive-Lymphocyte	3,112	1.87	1,378	1.96

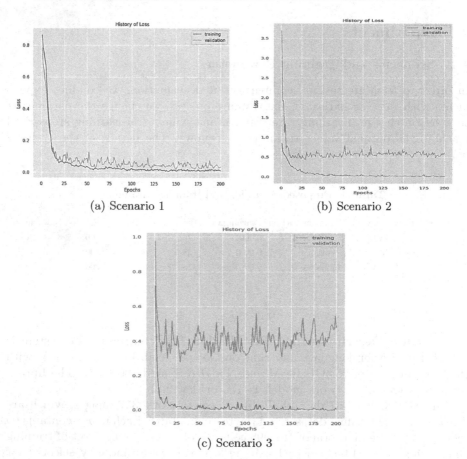

(a) Scenario 1   (b) Scenario 2

(c) Scenario 3

Fig. 3. Loss of three scenarios.

4.2 Training Results

Figure 3 shows the loss values of three scenarios. The measured loss of scenarios 1 to 3 is 0.02, 0.11, and 0.67, respectively. At the beginning of the training process below 25 epochs, in scenario 1, the loss value tends to decrease quickly. Then, the loss value decreases gradually until 200 epochs. This shows that after a long training time, effective features have been extracted and learned. We stop training at epoch 200th with the loss value of 0.02. In scenario 2, the loss value is 0.09 higher than that of scenario 1. It can be seen that from learning step 50th onwards the loss value no longer decreases. In scenario 3, the loss value decreases evenly at the learning steps above 25th. This shows that after a long time of training, the features have been extracted and learned more effectively. However, from learning step 50th onwards, the loss value tends to be unstable.

Figure 4 shows the accuracy of three scenarios. The accuracy of scenarios 1 to 3 are 0.98, 0.97, and 0.89, respectively. Scenario 1 has the highest accuracy compared to the two remaining scenarios. In general, scenario 1 extracts better

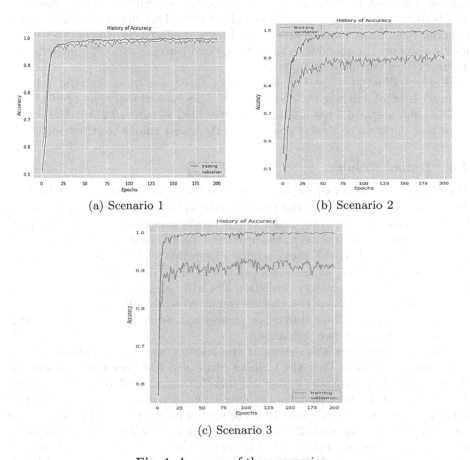

(a) Scenario 1   (b) Scenario 2

(c) Scenario 3

Fig. 4. Accuracy of three scenarios.

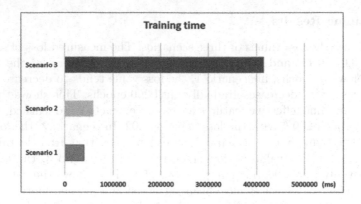

Fig. 5. Training time of three scenarios.

features than the other two scenarios. In scenario 2, the accuracy is 0.97. It can be seen that in learning steps below 25th, the accuracy rapidly increases and is quite stable from learning steps above 25th. This shows that after a long training time, features are extracted and learned effectively. Scenario 3 has the lowest accuracy compared with the others. From learning step 50th onwards, the accuracy does not increase anymore.

Figure 5 illustrates the training time of three scenarios. Scenario 1 has a training time of 110 min (4.58 h). Scenario 2 has a training time of 162.5 min (6.78 h) and scenario 3 has a training time of 1152.77 min (19.21 h). The results show that the training time of scenario 1 is the lowest among the three scenarios.

4.3   Testing Results

Table 3 presents the classification results from the experimental dataset. In these figures, red points represent the Ki-67-positive cells, green points represent the Ki-67-negative cells, and blue points represent the tumor cells with positive infiltrating lymphocytes. In case A, the prediction results of scenario 1 are superior to scenarios 2 and 3 with more accurate detection of Ki-67-positive cells. In case B, scenario 3 has an incorrect prediction of a Ki-67-positive cell. In case C, scenario 2 has an incorrect prediction of a Ki-67-negative cell, and scenario 3 has more cells that cannot be detected compared to the other two scenarios. In case D, scenarios 2 and 3 fail to detect cancer cells while scenario 1 gives better results than the other two scenarios.

Table 4 summarizes the results of the training and testing phases on the experimental dataset. After the training phase, the loss-accuracy values of the three scenarios 1 to 3 are 0.02–0.98, 0.11–0.97, and 0.67–0.89, respectively. Thereby, it can be seen that feature extraction in scenario 1 is more effective than in the other two scenarios. After the testing phase, the average accuracy of scenario 1 is 0.978, scenario 2 is 0.954, and scenario 3 is 0.887. As can be seen, scenario 1 is superior to the other 2 in terms of accuracy. Besides, the average prediction time in scenario 1 (8 s) is shorter than scenarios 2 (15 s) and 3 (20 s).

Table 3. Illustration of Ki-67 protein detection and classification.

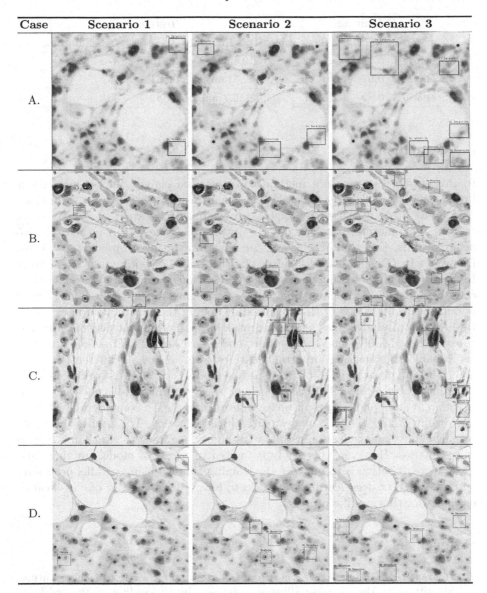

Table 4. Comparison of the proposed models

	Training			Testing	
	Accuracy	Loss_value	Training time	Accuracy	Detection time
Scenario 1	0.98	0.02	110 min	0.978	8 s
Scenario 2	0.97	0.11	162.5 min	0.954	15 s
Scenario 3	0.89	0.67	1,152.7 min	0.887	20 s

5 Conclusion

In this study, we propose a method to detect and classify nuclear proteins in breast cancer cells based on histological images using deep Learning techniques. The Ki-67 protein is used by experts to evaluate the extent of breast cancer tumors. This method not only detects but also classifies Ki-67-positive, Ki-67-negative cells, and tumor cells with positive infiltrating lymphocytes. We use DeepLabv3+ (DeepLabv3-MobileNet-v2, DeepLabv3-Xception and DeepLabv3-DenseNet-121) in detecting and classifying Ki-67 nuclear protein in breast cancer cells. DeepLabv3+ is capable of giving better segmentation results along object boundaries. This is one of the advanced models in the field of computer vision. Besides, DeepLabv3+ allows the integration of many different network architectures such as MobileNet-V2, Xception, and DenseNet-121. These architectures are capable of extracting important features from images at different levels for detecting and classifying Ki-67 protein. The model is capable of working in many different spatial scales. This is important in segmenting Ki-67 nuclear protein of different sizes and shapes in breast cancer images. We compare and evaluate the results of the network models with three scenarios. The results show that scenario 1 gives classification results with higher accuracy (0.98) than the other two scenarios. However, this study is limited in terms of model training time. The future direction is to improve the training time by using parallel processing and use different deep learning networks to have a broader view of classification methods for this specific problem.

References

1. Agarap, A.F.M.: On breast cancer detection: an application of machine learning algorithms on the wisconsin diagnostic dataset. In: Proceedings of the 2nd International Conference on Machine Learning and Soft Computing, pp. 5–9 (2018)
2. Alqahtani, Y., Mandawkar, U., Sharma, A., Hasan, M.N.S., Kulkarni, M.H., Sugumar, R.: Breast cancer pathological image classification based on the multiscale CNN squeeze model. Comput. Intell. Neurosci. (2022)
3. Bray, F., Ferlay, J., Soerjomataram, I., Siegel, R.L., Torre, L.A., Jemal, A.: Global cancer statistics 2018: globocan estimates of incidence and mortality worldwide for 36 cancers in 185 countries. CA Cancer J. Clin. **68**(6), 394–424 (2018)

4. Chollet, F.: Xception: deep learning with depthwise separable convolutions. In: Proceedings of the IEEE Conference on Vision and Pattern Recognition, pp. 1251–1258 (2017)
5. Davey, M.G., Hynes, S.O., Kerin, M.J., Miller, N., Lowery, A.J.: Ki-67 as a prognostic biomarker in invasive breast cancer. Cancers 13(17), 4455 (2021)
6. Denkert, C., et al.: Tumour-infiltrating lymphocytes and prognosis in different subtypes of breast cancer: a pooled analysis of 3771 patients treated with neoadjuvant therapy. Lancet Oncol. 19(1), 40–50 (2018)
7. DeSantis, C.E., Bray, F., Ferlay, J., Lortet-Tieulent, J., Anderson, B.O., Jemal, A.: International variation in female breast cancer incidence and mortality rates. Cancer Epidemiol. Biomarkers & prev. 24(10), 1495–1506 (2015)
8. Dubey, A.K., Gupta, U., Jain, S.: Breast cancer statistics and prediction methodology: a systematic review and analysis. Asian Pac. J. Cancer Prev. 16(10), 4237–4245 (2015)
9. Gnanasekaran, V.S., Joypaul, S., Meenakshi Sundaram, P., Chairman, D.D.: Deep learning algorithm for breast masses classification in mammograms. IET Image Proc. 14(12), 2860–2868 (2020)
10. Ha, L., et al.: Proportion and number of cancer cases and deaths attributable to behavioral risk factors in Vietnam. Int. J. Cancer (2023)
11. Harika, V., Kondi Vanitha, Y.H., Yamini, J., Sravani, T.: Diagnosis of cancer. Int. J. Res. Pharm. Chem. 5(3), 299–306 (2015)
12. Kajala, A., Jain, V.: Diagnosis of breast cancer using machine learning algorithms-a review. In: 2020 International Conference on Emerging Trends in Communication, Control and Computing (ICONC3), pp. 1–5. IEEE (2020)
13. Ke, D., Yang, R., Jing, L.: Combined diagnosis of breast cancer in the early stage by MRI and detection of gene expression. Exp. Ther. Med. 16(2), 467–472 (2018)
14. Kuttan, G.O., Elayidom, M.S.: Review on computer aided breast cancer detection and diagnosis using machine learning methods on mammogram image. Curr. Med. Imaging 19(12), 1361–1371 (2023)
15. Li, X., Shen, X., Zhou, Y., Wang, X., Li, T.Q.: Classification of breast cancer histopathological images using interleaved DenseNet with SENet (IDSNet). PLoS ONE 15(5), e0232127 (2020)
16. Mao, Y., Qu, Q., Chen, X., Huang, O., Wu, J., Shen, K.: The prognostic value of tumor-infiltrating lymphocytes in breast cancer: a systematic review and meta-analysis. PLoS ONE 11(4), e0152500 (2016)
17. Negahbani, F.: PathoNet introduced as a deep neural network backend for evaluation of Ki-67 and tumor-infiltrating lymphocytes in breast cancer. Sci. Rep. 11(1), 8489 (2021)
18. Petry, C., et al.: Evaluation of the potential of the Ki67 index to predict tumor evolution in patients with pituitary adenoma. Int. J. Clin. Exper. Pathol. 12(1), 320 (2019)
19. Ragab, D.A., Sharkas, M., Marshall, S., Ren, J.: Breast cancer detection using deep convolutional neural networks and support vector machines. PeerJ 7, e6201 (2019)
20. Rakhlin, A., Shvets, A., Iglovikov, V., Kalinin, A.A.: Deep convolutional neural networks for breast cancer histology image analysis. In: Campilho, A., Karray, F., ter Haar Romeny, B. (eds.) ICIAR 2018. LNCS, vol. 10882, pp. 737–744. Springer, Cham (2018). https://doi.org/10.1007/978-3-319-93000-8_83
21. Sandler, M., Howard, A., Zhu, M., Zhmoginov, A., Chen, L.C.: MobileNetv2: inverted residuals and linear bottlenecks. In: Proceedings of the IEEE Conference on Computer Vision and Pattern Recognition, pp. 4510–4520 (2018)

22. Spanhol, F.A., Oliveira, L.S., Petitjean, C., Heutte, L.: Breast cancer histopatho-
    logical image classification using convolutional neural networks. In: 2016 Interna-
    tional Joint Conference on Neural Networks (IJCNN), pp. 2560–2567. IEEE (2016)
23. Wolff, A.C., et al.: Recommendations for human epidermal growth factor receptor 2
    testing in breast cancer: American society of clinical oncology/college of American
    pathologists clinical practice guideline update. Arch. Pathol. Lab. Med. **138**(2),
    241–256 (2014)
24. Wolff, A.C., et al.: Human epidermal growth factor receptor 2 testing in breast can-
    cer: American society of clinical oncology/college of American pathologists clinical
    practice guideline focused update. Arch. Pathol. Lab. Med. **142**(11), 1364–1382
    (2018)
25. Zeng, H., Peng, S., Li, D.: Deeplabv3+ semantic segmentation model based on
    feature cross attention mechanism. In: Journal of Physics: Conference Series, vol.
    1678, p. 012106. IOP Publishing (2020)

Ensemble Learning with SVM
for High-Dimensional Gene Expression Data

Thanh-Nghi Do[1,2(✉)] and Minh-Thu Tran-Nguyen[1,2]

[1] College of Information Technology, Can Tho University, Campus II, 3/2 street, Can Tho City, Ninh Kieu District, Vietnam
{dtnghi,tnmthu}@ctu.edu.vn
[2] UMI UMMISCO 209 (IRD/UPMC), Paris, France

Abstract. The gene expression classification is the most important study in cancer diagnosis and drug discovery. Nevertheless, this task is very complicated to achieve accurate results because datasets have a very large number of dimensions and very few datapoints. In this paper, we propose the new ensemble learning algorithms with support vector machines (SVM) for efficiently handling the gene expression classification task. The Sherman-Morrison-Woodbury formula is used in the Newton SVM (NSVM) algorithm proposed by Mangasarian to make an extension of Newton SVM for dealing with datasets having a very large number of dimensions. Followed which, the ensemble learning trains the new extended Newton SVM for classifying gene expression datasets with simultaneously large number of datapoints and dimensions. The numerical test results on high-dimensional gene expression datasets show that our ensemble learning algorithms of Newton SVM are significantly faster and/or more accurate than the highly efficient standard SVM algorithm LibSVM.

Keywords: Gene expression classification · Newton Support Vector Machines · Ensemble learning

1 Introduction

The gene expression classification is the key study in cancer diagnosis and drug discovery. Gene expression datasets typically have a very few datapoints in a high-dimensional input space (genes). Therefore, the classification task is very complicated to achieve accurate results due to the curse of dimensionality phenomena. It is well-known as one of top 10 challenging problems in data mining research [25]. There have been many researches to adapt learning models for classification to these data [9, 11, 13, 14, 19, 20]. Support vector machines (SVM [23]) achieves the most accurate classification results.

The SVM algorithm is motivated by statistical learning theory. SVM algorithms use the idea of kernel substitution [1] for dealing with classification, regression and novelty detection tasks. Successful applications of SVMs have been reported for various fields such as face identification, text categorization and bio-informatics [12]. In spite of the prominent properties of SVM algorithms, they are not favorable to handle the challenge of large datasets. The training task of SVM leads to resolve the quadratic programming (QP), so

N. Thai-Nghe et al. (Eds.): ISDS 2023, CCIS 1950, pp. 29–40, 2024.
https://doi.org/10.1007/978-981-99-7666-9_3

that the computational cost of an SVM approach [17] is at least square of the number of training datapoints and the memory requirement. This makes SVM impractical for large datasets. The effective heuristics for scaling-up SVM learning task are to divide the original QP into series of small problems [2, 17], incremental learning [21] updating solutions in growing training set, parallel and distributed learning [18] on PC network or choosing interested datapoints subset [22] (active set) and boosting with SVM [7].

Our research purpose aims to scaling up SVM algorithms for dealing with high dimensional gene expression datasets (with simultaneously large number of datapoints and very high-dimensional input space). We propose a new extended SVM algorithm that is derived from the finite Newton method proposed by Mangasarian [16] for classification. The Newton SVM (NSVM) only requires solutions of linear equations instead of QP. If the dimensional input space is small enough (less than 10^3), even if there are millions datapoints, the NSVM algorithm is able to classify them in minutes on a personal computer (PC). For handling gene expression datasets with a very large number of dimensions and few training data, we propose to the new extended NSVM formulation using the Sherman-Morrison-Woodbury formula [10]. Followed which, we propose ensemble learning algorithms [3,4]) using the new extended NSVM that are able to handle gene expression datasets with simultaneously large number of datapoints and dimensions. Numerical test results on high-dimensional gene expression datasets [15] have shown that our ensemble learning algorithms with the new extended NSVM are fast and accurate compared with LibSVM [5].

The paper is organized as follows. Section 2 briefly presents the NSVM algorithm for classification tasks. In Sect. 3, we describe how to extend the NSVM to classify gene expression datasets. The experimental results are presented in Sect. 4. We then conclude in Sect. 5.

2   Newton Support Vector Machines

We start with a simple task for the linear binary classification, as shown in Fig. 1, with m datapoints x_i ($i = 1, \ldots, m$) in the n-dimensional input space R^n, having corresponding classes $y_i = \pm 1$. The SVM learning algorithm proposed by [23] aims to find the best separating hyper-plane (denoted by the normal vector $w \in R^n$ and the bias $b \in R$), i.e. furthest from both class $+1$ and class -1. This goal is accomplished through the maximization of the distance or margin between the supporting planes for each class ($x.w - b = +1$ for class $+1$, $x.w - b = -1$ for class -1). The margin between these supporting planes is $2/\|w\|$ (where $\|w\|$ is the 2-norm of the vector w). Any point x_i falling on the wrong side of its supporting plane is considered to be an error, denoted by z_i ($z_i \geq 0$). Therefore, SVM learning algorithms have to simultaneously maximize the margin and minimize errors. The standard SVMs pursue these goals with the quadratic programming of (1).

$$\min \ \Psi(w,b,z) = \frac{1}{2}\|w\|^2 + C\sum_{i=1}^{m} z_i \tag{1}$$
$$s.t. : y_i(w.x_i - b) + z_i \geq 1$$
$$z_i \geq 0$$

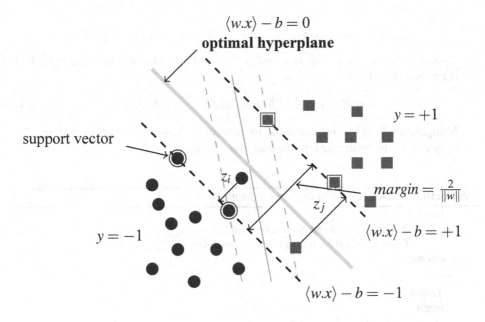

Fig. 1. Linear separation of the datapoints into two classes

where the positive constant C is used to tune errors and margin size.

The plane (w,b) is obtained by solving the quadratic programming (1). Then, the classification function of a new datapoint x based on the plane is:

$$predict(x) = sign(w.x - b) \tag{2}$$

SVM algorithms can use some other classification functions, including a polynomial function of degree d, a RBF (Radial Basis Function) or a sigmoid function. For changing from a linear to non-linear classifier, one must only substitute a kernel evaluation in (1) and (2) instead of the original dot product. More details about SVM and others kernel-based learning methods can be found in [6].

Unfortunately, Platt illustrates in [17] that the computational cost requirements of SVM solutions in (1) are at least $O(m^2)$, where m is the number of training datapoints, making classical SVM intractable for large datasets.

The Newton-SVM (NSVM) proposed by Mangasarian [16] reformulates the SVM problem (1). The new SVM formula achieves:

- the maximization of the margin by $min(1/2)\| w,b \|^2$
- the minimization of errors by $min(c/2)\|z\|^2$

Substituting for $z = [e - D(Aw - eb)]_+$ (where $(x)_+$ replaces negative components of a vector x by zeros, e is the column vector of 1) into the objective function Ψ of the quadratic programming (1) yields an unconstrained problem (3):

$$min\ \Psi(w,b) = (c/2)\|\ [e - D(Aw - eb)]_+\ \|^2 + (1/2)\|\ w,b\ \|^2 \tag{3}$$

By setting $[w_1, w_2, \ldots w_n, b]^T$ to u and $[A\quad -e]$ to E, then the unconstrained problem (3) is rewritten by (4):

$$min\ \Psi(u) = (c/2)\|\ (e - DEu)_+\ \|^2 + (1/2)u^T u \tag{4}$$

Mangasarian [16] has proposed the finite stepless Newton method for solving the strongly convex unconstrained minimization problem (4).

Algorithm 1: Newton SVM learning algorithm

---

input :

    Training dataset represented by A and D

    Constant $c > 0$ for tuning errors, margin size

output:

    (w, b)

Training:

begin

    1. Create matrix $E = [A\quad -e]$

    2. Starting with $u_0 \in R^{n+1}$ and $i = 0$

    3. **repeat**

        3.1. The gradient of Ψ at u_i is:

$$\nabla\Psi(u_i) = c(-DE)^T(e - DEu_i)_+ + u_i \tag{5}$$

        3.2. The generalized Hessian of Ψ at u_i is:

$$\partial^2\Psi(u_i) = c(-DE)^T diag([e - DEu_i]_*)(-DE) + I \tag{6}$$

        with $diag([e - DEu_i]_*)$ denotes the $(n+1) \times (n+1)$ diagonal matrix whose j^{th} diagonal entry is sub-gradient of the step function $(e - DEu_i)_+$ and I is the identity matrix of size $(n+1) \times (n+1)$.

        3.3. Updating

$$u_{i+1} = u_i - \partial^2\Psi(u_i)^{-1}\nabla\Psi(u_i) \tag{7}$$

        3.4. Increment $i = i + 1$

    until $\nabla\Psi(u_i) < tol$;

    4. Optimal plane (w, b): $(w_1, w_2, \ldots, w_n, b)$ via u_i

end

Mangasarian has illustrated that the sequence u_i of Algorithm 1 terminates at the global minimum solution (with a number of iterations varying between 5 and 8). The NSVM algorithm requires thus only solutions of linear equations (7) of $(n+1)$ variables (w_1, w_2, ..., w_n, b) instead of the quadratic programming (1). If the dimensional input space is small enough (less than 10^3), even if there are millions datapoints, these algorithms are able to classify them in some minutes on a PC (being much faster than the standard LibSVM [5] while giving competitive test correctness).

3   Newton Support Vector Machines for Gene Expression Datasets

3.1   Newton Support Vector Machines for Classifying Large Number of Dimensions

Gene expression classification tasks handle datasets with a very large number of dimensions (many ten thousands of dimensions) and few training datapoints (hundreds of datapoints). Thus, the $(n+1) \times (n+1)$ matrix $\partial^2 \Psi(u_i)$ of (6) in the NSVM algorithm is too large and the inverse matrix $\partial^2 \Psi(u_i)^{-1}$ in (7) has a high computational cost. Therefore, the original NSVM algorithm is not suited for classifying gene expression datasets.

To overcome these problems, we propose to extend the NSVM algorithm by applying the Sherman-Morrison-Woodbury formula [10] to $\partial^2 \Psi(u_i)^{-1}$. Thus, it leads to obtain a new dual algorithm that only depends the inverse matrix of $(m) \times (m)$ (where m datapoints $\ll n$ dimensions).

$$(A + UV^T)^{-1} = A^{-1} - A^{-1}U(I + V^T A^{-1} U)^{-1} V^T A^{-1} \tag{8}$$

By setting $Q = diag(\sqrt{c[e - DEu_i]_*})$ and $P = Q(-DE)$, we can re-write the inverse matrix $\partial^2 \Psi(u_i)^{-1}$ in Eq. (7) as follows:

$$\partial^2 \Psi(u_i)^{-1} = (I + P^T P)^{-1} \tag{9}$$

Thus, the Sherman-Morrison-Woodbury formula (8) is applied to the right part of (9), the inverse matrix $\partial^2 f(u_i)^{-1}$ is as (10):

$$\Rightarrow \partial^2 \Psi(u_i)^{-1} = I - P^T (I + PP^T)^{-1} P \tag{10}$$

The new extended NSVM formula uses $\partial^2 \Psi(u_i)^{-1}$ formula in Eq. (10) to update u_i. And then, the algorithmic complexity of the new extended NSVM only depends on the inversion of the $(m) \times (m)$ matrix $(I + PP^T)$. Therefore, this new extended NSVM formulation described in Algorithm 2 can handle datasets with very large number of dimensions and few training data because the cost of storage and computation scale on the number of training datapoints.

3.2   Ensemble Learning with Newton Support Vector Machines for Classifying Large Amounts of High Dimensional Datapoints

For classification of massive datasets with simultaneously large number (at least 10^4) of datapoints and dimensions, there are at least two problems: the learning time increases dramatically with the training data size and the memory requirement increases according to data size. The NSVM algorithms need to store and invert a matrix with size $(m) \times (m)$ (the new extended NSVM) or $(n+1) \times (n+1)$ (the original NSVM). This requires too much main memory and very high computational time.

For scaling-up the NSVM to large datasets, we propose to apply the ensemble approach, including Bagging [4] and Arc-x4 [3] to NSVM algorithms. The proposed

---

Algorithm 2: Extended Newton SVM algorithm for large number of dimensions

---

    input :

        Training dataset represented by A and D

        Constant $c > 0$ for tuning errors, margin size

    output:

        (w, b)

    Training:

    begin

        1. Create matrix $E = [A \quad -e]$

        2. Starting with $u_0 \in R^{n+1}$ and $i = 0$

        3. **repeat**

            3.1. The gradient of Ψ at u_i is:

$$\nabla \Psi(u_i) = c(-DE)^T (e - DEu_i)_+ + u_i \qquad (11)$$

            3.2. The generalized Hessian of Ψ at u_i is:

$$\partial^2 \Psi(u_i) = I - P^T (I + PP^T)^{-1} P \qquad (12)$$

            with $Q = diag(\sqrt{c[e - DEu_i]_+})$ and $P = Q(-DE)$ and I is the identity matrix of size $(n+1) \times (n+1)$.

            3.3. Updating

$$u_{i+1} = u_i - \partial^2 \Psi(u_i)^{-1} \nabla \Psi(u_i) \qquad (13)$$

            3.4. Increment $i = i + 1$

        until $\nabla \Psi(u_i) < tol$;

        4. Optimal plane (w, b): $(w_1, w_2, \ldots, w_n, b)$ via u_i

    end

---

ensemble learning brings out two advantages. The first one is to be able to overcome the large scale problem and the second one is to maintain the classification accuracy.

The ensemble learning with NSVM is described as in Algorithm 3. Ensemble learning algorithms call repeatedly NSVM learning algorithms *NumIt* times to classify datasets. Here, NSVM algorithms are used to train weak classifiers in ensemble learning algorithms.

With large number of datapoints in dimensional input space being small enough, the original NSVM described in Algorithm 1 is called in ensemble learning algorithms for training weak classifiers.

For dealing with a very large number of dimensions and few datapoints or simultaneously large number of datapoints and dimensions, ensemble algorithms use the new extended NSVM described in Algorithm 2 to train weak classifiers.

At each training step for a weak classifier, ensemble learning algorithms sample a subset of datapoints from the training dataset according to the distribution weights over the training datapoints. For the Arc-x4 mode, it needs increasing the weights of incorrectly classified datapoints in last iterations so that the next weak learner is forced to focus on the hard datapoints in the training dataset.

Algorithm 3: Ensemble learning with Newton SVM for large amounts of high dimensional datapoints

---

input :

    Training dataset with m datapoints:

    $\{x_i, y_i\}_{i=1,m}$, $x_i \in R^n$ and $y_i \in \pm 1$

    Constant $c > 0$ for tuning errors, margin size

    Number of iterations $NumIt$

output:

    (w, b)

Training:

begin

    1. Initial distribution of m datapoints: $p_1(i) = 1/m$

    2. **for** $t \leftarrow 1$ **to** $NumIt$ **do**

        2.1. Sampling S_t of datapoints using p_t

        2.2. Learning $NSVM_t$ from S_t: $h_t = NSVM_t(S_t, c)$

        2.3. **if** *Arc-x4* **then**

            2.3.1. Computing predicting error for each datapoint x_i with previous classifiers h_t:

            $\varepsilon_i = \sum_t h_t(x_i) \neq y_i$

            2.3.2. Updating distribution of m datapoints:

            for $i \leftarrow 1$ **to** m **do**

                $p_{t+1}(i) = \frac{1+\varepsilon_i^4}{Z_t}$ (where Z_t is the normalization factor)

            end

        end

    end

    3. Optimal plane (w, b) is obtained by aggregating models h_t

end

---

4   Evaluation

Table 1. Description of Gene expression datasets

ID	Datasets	Classes	Points	Dimensions	Protocol
1	ALL-AML Leukemia	2	72	7129	38 trn - 34 tst
2	Breast Cancer	2	97	24481	78 trn - 19 tst
3	Ovarian Cancer	2	253	15154	leave-1-out
4	Lung Cancer	2	181	12533	32 trn - 149 tst
5	Prostate Cancer	2	136	12600	102 trn - 34 tst
6	Ovarian Cancer NCI-QStar	2	216	373410	leave-1-out
7	Translation Initiation Sites	2	13375	927	3-fold

We are interested in the evaluation of the performances of our ensemble learning algorithms with NSVM in terms of the learning time, the accuracy on large datasets. To pursue this goal, we implement ensemble learning algorithms with NSVM in C/C++, using the high performance linear algebra library, ATLAS/Lapack [8,24]. We also use the highly efficient standard SVM algorithm LibSVM [5] in the performance evaluation of ensemble learning algorithms, including Arc-x4-NSVM and Bag-NSVM. All tests were run under a machine Linux Fedora 32, Intel(R) Core i7-4790 CPU, 3.6 GHz, 32 GB RAM.

The experiment uses 7 high-dimensional gene expression datasets [15] described in Table 1. All datasets have large number of dimensions. Especially, Translation Initiation Sites dataset has simultaneously large number of datapoints and dimensions. The test protocols are presented in the last column of Table 1. With datasets having training set (trn) and testing set (tst) available, we used the training data to tune the parameters of the algorithms for obtaining a good accuracy in the learning phase. Arc-x4-NSVM and Bag-NSVM train 50 weak NSVM classifiers. For LibSVM, NSVM, we tuned the positive constant c for trade-off of errors and the margin size. Then the obtained model is evaluated on the test set. If the training set and testing set are not available then we used cross-validation protocols to evaluate the performance. With datasets having less than three hundred datapoints, the test protocol is leave-one-out cross-validation (loo). It involves using a single datapoint from the dataset as the testing data and the remaining datapoints as the training data. This is repeated so many that each datapoint in the dataset is used once as the testing data. With dataset having more than three hundred datapoints, 3-fold cross-validation is used to evaluate the performance. The dataset is partitioned into 3 folds. A single fold is retained as the validation set, and the remaining 2 folds are used as training data. The cross-validation process is then repeated 3 times (folds). The results from the 3 folds are then averaged to produce the final result.

We obtain classification results in terms of the training time showed in Table 2 and Fig. 2, in terms of the classification correctness showed in Table 3 and Fig. 3. Best results and second ones are in bold and italic.

Table 2. Classification results in terms of training time (sec)

ID	Datasets	Training time (sec)		
		LibSVM	Arc-x4-NSVM	Bag-NSVM
1	ALL-AML Leukemia	9.14	*5.01*	**1.82**
2	Breast Cancer	269.66	*75.43*	**8.16**
3	Ovarian Cancer	403.60	*10.13*	**8.11**
4	Lung Cancer	20.80	*3.51*	**0.47**
5	Prostate Cancer	1.6	*0.7*	**0.51**
6	Ovarian Cancer NCI-QStar	158.95	*20.13*	**9.46**
7	Translation Initiation Sites	314.00	*50.27*	**35.72**

Table 3. Classification results in terms of accuracy (%)

ID	Datasets	Accuracy(%)		
		LibSVM	Arc-x4-NSVM	Bag-NSVM
1	ALL-AML Leukemia	**97.06**	**97.06**	**97.06**
2	Breast Cancer	*73.68*	**84.21**	*73.68*
3	Ovarian Cancer	**100**	**100**	**100**
4	Lung Cancer	**98.66**	*98.00*	97.32
5	Prostate Cancer	73.53	**97.06**	**97.06**
6	Ovarian Cancer NCI-QStar	**97.69**	*97.22*	*97.22*
7	Translation Initiation Sites	**92.41**	*92.08*	91.41

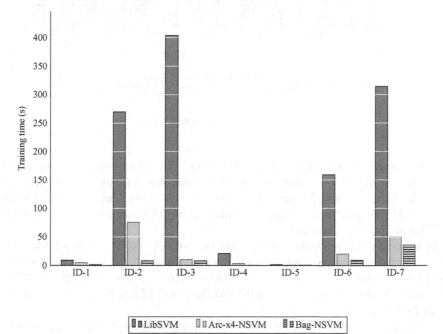

Fig. 2. Comparison of training time (sec)

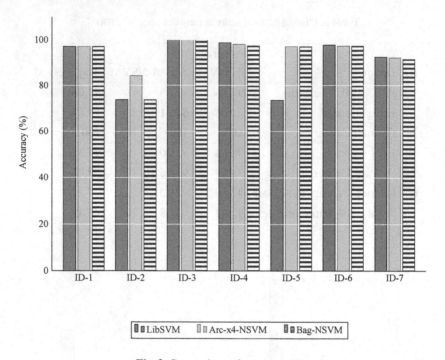

Fig. 3. Comparison of accuracy (%)

The average training time of LibSVM, Arc-x4-NSVM and Bag-SVM are 168.25 sec, 23.60 sec and 9.18 sec, respectively. The comparison in terms of the training time shows that LibSVM is 7.13 times and 18.33 times slower than our Arc-x4-NSVM and Bag-NSVM. Our ensemble learning algorithms with NSVM are always faster than Lib-SVM for performing all datasets.

In terms of the classification correctness, our Arc-x4-NSVM and Bag-NSVM give very competitive accuracy compared to LibSVM. LibSVM, Arc-x4-NSVM and Bag-NSVM achieve the average accuracy of 90.43%, 95.09% and 93.39%. Arc-x4-NSVM has 2 wins, 2 ties, 3 losses, against LibSVM. Bag-NSVM has 1 win, 3 ties, 3 losses, versus LibSVM.

These results show that our ensemble learning algorithm with NSVM are favorable to deal with very high-dimensional gene expression datasets but also with simultaneously very large number of datapoints and dimensions, e.g. the Translation Initiation Sites dataset. They can learn accurate classification models in short training time.

5   Conclusion and Future Works

We have presented ensemble learning algorithms with NSVM for dealing with high-dimensional gene expression datasets. The Sherman-Morrison-Woodbury formula [10] is used in NSVM to make the extended NSVM for very large number of dimensions and few training datapoints. The ensemble learning [3,4]) trains the new extended

NSVM to classify gene expression datasets with simultaneously large number of data-points and dimensions. The numerical test results on high-dimensional gene expression datasets [15] show that our Arc-x4-NSVM, Bag-NSVM improve the training speed while achieving good accuracy compared with LibSVM.

In the future, we intend to provide more empirical tests on the large benchmark of gene expression datasets. A forthcoming improvement will be to extend these algorithms for multi-class classification problems.

Acknowledgments. This work has received support from the College of Information Technology, Can Tho University. We would like to thank very much the Big Data and Mobile Computing Laboratory.

References

1. Bennett, K., Campbell, C.: Support vector machines: hype or hallelujah ?. In: SIGKDD Explorations, pp. 1–13 (2000)
2. Boser, B., Guyon, I., Vapnik, V.: An training algorithm for optimal margin classifiers. In: proc. of 5th ACM Annual Workshop on Computational Learning Theory of 5th ACM Annual Workshop on Computational Learning Theory, pp. 144–152. ACM (1992)
3. Breiman, L.: Arcing classifiers. Annals Stat. **26**(3), 801–849 (1998)
4. Breiman, L.: Bagging predictors. Mach. Learn. **24**(2), 123–140 (1996)
5. Chang, C.C., Lin, C.J.: LIBSVM: a library for support vector machines. ACM Trans. Intell. Syst. Technol. **2**(3), 1–27 (2011), software available at http://www.csie.ntu.edu.tw/cjlin/libsvm
6. Cristianini, N., Shawe-Taylor, J.: An Introduction to Support Vector Machines and Other Kernel-based Learning Methods. Cambridge University Press (2000)
7. Do, T., Poulet, F.: Towards high dimensional data mining with boosting of PSVM and visualization tools. In: ICEIS 2004, Proceedings of the 6th International Conference on Enterprise Information Systems, Porto, Portugal, 14–17 April 2004, pp. 36–41 (2004)
8. Dongarra, J., Pozo, R., Walker, D.: LAPACK++: a design overview of object-oriented extensions for high performance linear algebra. In: Proceedings of Supercomputing, pp. 162–171 (1993)
9. Furey, T.S., Cristianini, N., Duffy, N., Bednarski, D.W., Schummer, M., Haussler, D.: Support vector machine classification and validation of cancer tissue samples using microarray expression data. Bioinform. **16**(10), 906–914 (2000)
10. Golub, G., Loan, C.V.: Matrix Computations, 3rd edn. John Hopkins University Press, Baltimore, Maryland (1996)
11. Golub, T.R., et al.: Molecular classification of cancer: class discovery and class prediction by gene expression monitoring. Science **286**(5439), 531–537 (1999)
12. Guyon, I.: Web page on svm applications (1999). http://www.clopinet.com/isabelle/Projects/SVM/app-list.html
13. Hastie, T., Tibshirani, R., Friedman, J.: The Elements of Statistical Learning. SSS, Springer, New York (2009). https://doi.org/10.1007/978-0-387-84858-7
14. Huynh, P., Nguyen, V.H., Do, T.: Novel hybrid DCNN-SVM model for classifying rna-sequencing gene expression data. J. Inf. Telecommun. **3**(4), 533–547 (2019)
15. Jinyan, L., Huiqing, L.: Kent ridge bio-medical dataset repository. Nanyang Technological University, School of Computer Engineering (2004)
16. Mangasarian, O.: A finite newton method for classification problems. Technical Report 01–11, Data Mining Institute, Computer Sciences Department, University of Wisconsin (2001)

17. Platt, J.: Fast training of support vector machines using sequential minimal optimization. In: Schölkopf, B., Burges, C., Smola, A. (eds.) Advances in Kernel Methods - Support Vector Learning, pp. 185–208 (1999)
18. Poulet, F., Do, T.N.: Mining very large datasets with support vector machine algorithms. In: Camp, V.O., Filipe, J., Hammoudi, S., Piattini, M. (eds.) Enterprise Information Systems, pp. 177–184 (2004)
19. Shinmura, S.: High-dimensional Microarray Data Analysis. Springer, Singapore (2019). https://doi.org/10.1007/978-981-13-5998-9
20. Statnikov, A.R., Wang, L., Aliferis, C.F.: A comprehensive comparison of random forests and support vector machines for microarray-based cancer classification. BMC Bioinform. **9** (2008)
21. Syed, N., Liu, H., Sung, K.: Incremental learning with support vector machines. In: Proceedings of the ACM SIGKDD International Conference on KDD. ACM (1999)
22. Tong, S., Koller, D.: Support vector machine active learning with applications to text classification. In: proc. of the 17th International Conference on Machine Learning, pp. 999–1006. ACM (2000)
23. Vapnik, V.: The Nature of Statistical Learning Theory. Springer-Verlag (1995). https://doi.org/10.1007/978-1-4757-3264-1
24. Whaley, R.C., Dongarra, J.: Automatically tuned linear algebra software. In: Ninth SIAM Conference on Parallel Processing for Scientific Computing (1999), cD-ROM Proceedings
25. Yang, Q., Wu, X.: 10 challenging problems in data mining research. Int. J. Inf. Technol. Decis. Mak. **5**(4), 597–604 (2006)

A New Integrated Medical-Image Processing System with High Clinical Applicability for Effective Dataset Preparation in ML-Based Diagnosis

Kotori Harada[1], Takahiro Yoshimoto[1], Nam Phong Duong[1], My N. Nguyen[1,2] (ID),
Yoshihiro Sowa[3(✉)], and Masayuki Fukuzawa[1(✉)] (ID)

[1] Graduate School of Science and Technology, Kyoto Institute of Technology, Matsugasaki, Sakyo-Ku, Kyoto 606-8585, Japan
`fukuzawa@kit.ac.jp`
[2] College of Information and Communication Technology, Can Tho University, 3-2 Street, Ninh Kieu District, Can Tho, Vietnam
[3] Department of Plastic Surgery, Jichi Medical University, Yakushiji, Shimotsuke-Shi, Tochigi 329-0498, Japan
`ysowawan@gmail.com`

Abstract. We have developed a new medical-image processing system with high clinical applicability for effective dataset preparation in computer-aided diagnosis (CAD) based on machine learning (ML) techniques. Despite the wide range of application of ML-based CAD, it remains a challenge to apply this technique to clinical diagnosis of specific diseases or evaluation of specific medical practices. This is due to the lack of an effective framework for preparing a sufficient number of datasets. Consequently, there is absence of a guideline or standard procedure for processes such as image acquisition, anonymization, annotation, preprocessing and feature extraction. To address this ongoing issue, we proposed a system that was designed to integrate the special functions such as incremental anonymization, annotation assistance, and hybrid process flow for preprocessing and feature extraction. The incremental anonymization aimed to enable batch processing after image acquisition, reducing the daily workload of medical specialists in their institutions. Cross annotation and error correction were also supported even outside the medical institution by cogeneration of annotation sheets with anonymized images and by its OCR-based data-collection process. A hybrid process flow combining a simple manual operation and complemental automation algorithm was adopted to accelerate preprocessing and feature extraction. The system prototype successfully streamlined the dataset preparation process, including the generation of 3D breast-mesh closures and their associated geometric features. This substantial enhancement in efficiency demonstrates the system's high clinical applicability and its potential to significantly contribute to the field of breast reconstruction evaluation.

Keywords: Computer-Aided Diagnosis · Machine Learning · Image Processing · Dataset Preparation · Clinical Applicability

N. Thai-Nghe et al. (Eds.): ISDS 2023, CCIS 1950, pp. 41–50, 2024.
https://doi.org/10.1007/978-981-99-7666-9_4

1 Introduction

Machine learning (ML) technique has undergone significant evolution and has been successfully adopted in various application domains in recent years [1, 2]. Especially in the field of computer-aided diagnosis (CAD) with medical images, several ML models have been developed, demonstrating excellent diagnostic performance comparable to that of medical specialists for certain diseases given sufficient training datasets readily available for use. However, it is required to prepare new datasets for the clinical diagnosis of specific diseases or the evaluation of specific medical practices by repeating various hard-to-automate processes such as image acquisition, anonymization, annotation, preprocessing and feature extraction. The image acquisition and annotation processes are exclusive to medical specialists, and anonymization should be completed within the medical institutions where the images were acquired to protect patient privacy during collaborations with other medical specialists or research institutions. Consequently, these critical procedures demand additional time and effort from medical specialists, in addition to their daily clinical practices, leading to difficulties in dataset preparation.

Numerous attempts have been made to assist image analysis and dataset preparation processes. In the specific context of plastic surgery, the software BCCT.core was developed to assist in the objective evaluation of esthetic outcome in breast reconstruction surgery and has shown a significant correlation with a subjective measure of the Harris scale [3–5]. However, it was difficult for collective evaluation of a large number of cases due to the requirements of interactive operation in almost every procedure such as marking nipple-areola complex, scars, and so on. Sowa et al. proposed a framework of image analysis system aimed at evaluating the esthetic outcome in breast reconstruction surgery to cover image anonymization, annotation collection, and ROI extraction [6]. The initial prototype system demonstrated its usefulness for collecting the case in the medical institutions. However, it did not include all the required processes of dataset preparation, thus limiting its clinical applicability.

In this study, we have developed a new image-processing system with high clinical applicability for effective dataset preparation in ML-based CAD. In order to support the dataset preparation, the developed system was designed to integrate special functions such as incremental anonymization and filtering, annotation assistance, and hybrid functions combining manual operation and automatic algorithm for preprocessing and feature extraction.

The remainder of this paper is as follows. The typical process flow of dataset preparation for ML-based CAD is introduced in Sect. 2. Section 3 presents the proposed system prototype targeted at esthetic outcome evaluation in breast reconstruction surgery. In Sect. 4, a dataset preparation is examined using the developed system prototype to demonstrate its performance and clinical applicability. Finally, Sect. 5 summarizes the achievements and concludes this paper.

2 Process Flow of Dataset Preparation for ML-Based Medical Image Diagnosis

In the context of ML-based CAD, the effectiveness of collaboration between medical specialists and information scientists is widely recognized. Moreover, multiple medical institutions frequently collaborate in preparing a large number of training datasets. In this study, we assumed a typical collaborative process where medical institutions, collaborative medical specialists, and research institutions incrementally prepare new training datasets while exchanging the intermediate data safely. Figure 1 shows typical process flow of dataset preparation for ML-based CAD. A final dataset consists of anonymized and preprocessed clinical images, their extracted feature primitives, and an annotation list that includes diagnosis and findings of all the cases. The function of each section and its corresponding process flow are as follows.

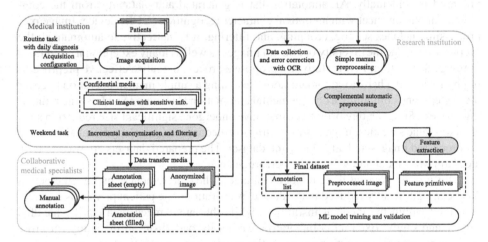

Fig. 1. Typical process flow of dataset preparation for ML-based CAD

2.1 Medical Institution

Medical institutions are responsible for image acquisition and its anonymization. Although the clinical image is acquired interactively and stored in a confidential media for each patient as routine task with daily diagnosis, a pre-defined acquisition configuration is favorable to be applied commonly. On the other hand, anonymization can be performed for many patients at once. On a similar note, it is advantageous to perform annotation process intensively for multiple patients. Since collaborative medical specialists commonly perform the annotation process in parallel to reduce deviation between evaluators, it is effective to generate reusable annotation sheets in advance. Opting for an optical answer sheet or equivalence is also preferable to streamline subsequent data collection and facilitate error correction.

2.2   Collaborative Medical Specialists

Collaborative medical specialists, with the access admission to the data transfer media, are responsible for observing anonymized images and recording their diagnosis and findings on the annotation sheets. These specialists are not necessarily members of the medical institution in which the image acquisition was performed. Hence, it is vital and effective to exchange anonymized images and annotation sheets through the data transfer media. Each clinical case is presumed to be evaluated multiple times by different specialists. Thus, the final annotation list should be designed to link multiple annotations to a corresponding image.

2.3   Research Institution

As research institutions receive the data transfer media on a regular basis, datasets are prepared incrementally. An annotation list is generated automatically from the completed annotation sheet with an optical character recognition (OCR) device. Image preprocessing includes scaling, cropping, and filtering. It is ineffective to automate all the preprocessing due the diversity of clinical images as well as many exceptional conditions in image acquisition. In this study, we propose a process to realize effective preprocessing by identifying hard-to-automate steps, designing a simple manual operation to cover these steps, and combining a complemental automation algorithm to address the remaining process. Some feature primitives that have undergone significant size reduction such as pixel-value statistics, fingerprints, or transformed images, are also extracted from the preprocessed images and added into the dataset. The prepared datasets are used to train a certain ML model and validate its performance.

 This process flow has three major advantages. The requirement of human interaction has been significantly reduced in the dataset preparation. The patient privacy is securely protected within the medical institution, while the anonymized data can be handled timely and safely. External collaborators can easily participate in the study for cross annotation to reduce deviation of annotation between evaluators.

3   System Prototyping for Esthetic Outcome Evaluation in Breast Reconstruction Surgery

Based on the process flow shown in Fig. 1, we developed a prototype of an integrated image-processing system for effective dataset preparation by integrating the functions of incremental anonymization and filtering, annotation assistance, preprocessing and feature extraction. It was targeted at the esthetic outcome evaluation in breast reconstruction surgery in plastic surgery.

3.1   Breast Reconstruction Surgery and Esthetic Outcome Evaluation

Figure 2(a) illustrates an example of breast that has undergone reconstruction surgery. Breast reconstruction is a plastic surgery operation aiming to restore the shape of the breast which has been affected and deformed during breast cancer resection. The esthetic outcome of the reconstructed breast, which is the degree to be restored, is important as a measure of quality of reconstruction surgery because it greatly affects the patient's quality of life (QOL). The esthetic outcome of the reconstructed breast is subjectively evaluated by plastic surgeon specialists.

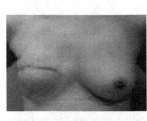

Evaluation viewpoint	Evaluation score (Scoring criteria)
Breast volume	2 (equivalent), 1 (slightly different), 0 (considerably different)
Breast shape	2 (equivalent), 1 (slightly different), 0 (considerably different)
Scars (Non-symmetric)	2 (inconspicuous), 1 (slightly inconspicuous), 0 (conspicuous)
Breast softness (Non-symmetric)	2 (soft), 1 (slightly hard), 0 (hard)
Nipple-Areola Complex (NAC) color	1 (equivalent), 0 (different)
NAC size and shape	1 (equivalent), 0 (different)
Nipple location (nipple-to-sternal notch distance)	1 (less than 2cm), 0 (2cm and more)
Lowest inframammary point (height)	1 (less than 2cm), 0 (2cm and more)
Total score	12-11(excellent), 10-8(good), 7-5(fair), 4-0(poor)

(a) Example of breast reconstruction surgery   (b) Evaluation viewpoints of esthetic outcome recognized among Japanese plastic surgeons

Fig. 2. (a) Example of breast reconstruction surgery, and (b) evaluation viewpoints of esthetic outcome recognized among Japanese plastic surgeons [7]

Figure 2(b) shows the evaluation viewpoints of esthetic outcome that are well-recognized among Japanese plastic surgeons [7]. In general, the evaluation of a reconstructed breast esthetic outcome involves comparing it with the healthy breast on the opposite side, rather than comparing it with the breast before reconstruction. As a result, the evaluation focuses on discerning the disparities between the left and right breasts, except for scars and breast softness. Evaluation scores are designed so that a significant weight is attributed to viewpoints with a substantial impact on visual impressions. The evaluation viewpoints can be classified into three groups: the shape, appearance, and softness. However, for this study, the softness-related viewpoint was excluded because characterizing it accurately from the images proved to be difficult.

3.2 Breast Image Acquisition

Among the various viewpoints related to the esthetic outcome, the appearance-related aspects are expected to be adequately represented through 2D images acquired using a conventional digital camera. However, the shape-related ones may not be as effectively captured using 2D images alone. To address this issue, we have collected both 2D and 3D images during the postoperative follow-up of breast reconstruction surgery.

As a pretreatment, the plastic surgeon places two types of markers on the patient's body surface before image acquisition. The first type consists of two cross marks, one at the center of the clavicle and the other 25 cm below it along the body axis, used for scaling purposes. The second type involves outlining the breast region, which the plastic surgeon places directly on the patient's body surface.

The 2D images include a patient's upper body and are standardized to be acquired from six different angles using a conventional digital camera. At each angle, the chest is centered, and the face is excluded. These 2D images also include supplemental images used to provide patient identification or additional information.

On the other hand, 3D images represent an approximated curved surface constructed from numerous colored triangular meshes that share edges. These 3D images are obtained through hand-scanning with a depth camera (Intel Realsense L515 [8]) and by combining the captured 3D information from multiple viewpoints using an acquisition software (Imfusion RecFusion [9]). Subsequently, the face region is manually removed from the 3D images before recording them.

3.3   Incremental Anonymization and Filtering

Figure 3(a) shows process flow of incremental anonymization and filtering in the developed system prototype. The patient's privacy information may be present in the file names, directory names, or file headers, which may pose a challenge when attempting to remove such sensitive data each time. This system is designed to mitigate the challenges associated with privacy concerns by automating the anonymization process for multiple images. This can be carried out by a scheduled interval, such as on a weekly or monthly basis.

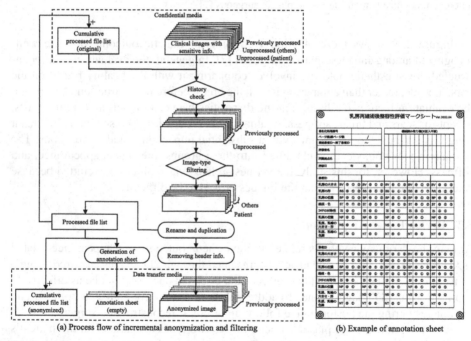

(a) Process flow of incremental anonymization and filtering   (b) Example of annotation sheet

Fig. 3. (a) Process flow of incremental anonymization and filtering, and (b) example of annotation sheet

In order to ensure the seamless transfer of unprocessed images into the data transfer media, an incremental processing approach is necessary for the anonymization of clinical images. This approach skips the previously processed images, focusing solely

on anonymizing the newly added images, facilitating their cumulative storage. Therefore, a cumulative processed file list is generated in the confidential media to record the
information of each processed image.

The anonymization process begins with a history-check to identify unprocessed
images by comparing all the clinical images in the confidential media and the cumulative
processed file list. These unprocessed images are then duplicated into the data transfer
media while maintaining the original directory structure, however their file names as well
as directory names needs to be anonymized (anonymous-IDs). The header information in
the image such as the exchangeable image file format (EXIF) is also deleted except for its
essential parts such as timestamp and resolution to avoid retrospective identification of
the patient or the medical institution. The information of the processed images is added
into the cumulative processed file list in the confidential media and the data transfer
media in original and anonymous form, respectively. The empty annotation sheets are
also generated during the anonymization process. We adopted a style of optical answer
sheet with several annotation fields for 10 cases per page by assuming intensive and
retrospective annotation by medical specialists quarterly. The anonymous-IDs of each
case are filled in the sheet in advance to avoid human error. An example of the sheet is
shown in Fig. 3(b). The collaborative medical specialists and the research institutions
can safely perform the cross annotation, preprocessing and feature extraction using only
the data transfer media containing all the required data.

3.4 Preprocessing and Feature Extraction

In the esthetic outcome evaluation in breast reconstruction surgery, extracting breast
region is compulsory in preprocessing since it is considered to be effective to focus
solely on the breast region instead of the entire chest. As for the 3D images, we designed
a hybrid process flow to extract 3D breast mesh and its feature primitives. This process
flow is displayed in Fig. 4.

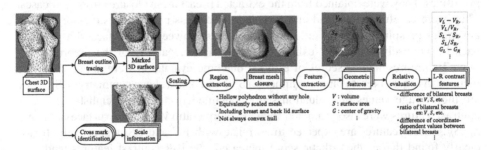

Fig. 4. Hybrid process flow to extract 3D breast mesh and its feature primitives

The scaling process is to normalize the 3D mesh dimension based on the coordinates
of the two pretreatment cross marks. The coordinates of nipples are also manually
identified if it is visually recognized, and recorded as feature primitive intrinsic in the
mesh. The region extraction step is to automatically extract a voluminous closure of
the 3D breast mesh using the manually traced breast outline. This process includes two

steps: first, extracting the surface skin using the breast outline, and then generating a curved surface as a back-lid with an assumed smoothness.

The feature extraction step is to automatically extract various geometric feature primitives from each breast mesh closure. These features encompass size-related measures, such as volume and surface area, as well as position-related characteristics, such as the center of gravity. In addition, alternative sets of feature primitives are extracted from a circumscribed cuboid or a cylinder of the breast mesh closure to examine the validity of the approximated alternatives of feature primitive. The relative evaluation step involves obtaining derivative primitives, such as differences or ratios, by comparing the geometric features between both breasts. These derivative primitives are named as L-R contrast features. The process flow for 2D images is still currently under development.

4   Experimental Results

A dataset preparation was conducted using the developed system prototype. The breast image acquisition was performed from May 2019 to December 2022 at the previous affiliated medical institutions (Kyoto Prefectural University of Medicine and Kyoto University) of the author (Sowa, Y.) and 177 cases of clinical images were obtained under the approval of their ethics committee. The incremental anonymization process was performed four times at the medical institutions, with each iteration involving approximately 40 cases. A comparison between the data transfer media and the cumulative processed file list confirmed the successful cumulative duplication of anonymized images and incremental generation of annotation sheets. Cross annotation was performed by four collaborative medical specialists while exchanging the data transfer media. The preprocessing and feature extraction procedures were carried out at a research institution (Kyoto Institute of Technology), which also handled the data collection of cross annotation results.

Figure 5 shows typical examples of extracted breast mesh closures and their feature primitives. They were obtained from the extracted breast mesh closures from four cases. They were carefully examined and revealed a seamless polyhedral structure without any holes or splits and with all intersection lines between the surface skin and the back-lid precisely connected, exhibiting a proficient capability of the preprocessing flow. Two representative scores of esthetic outcome evaluation were noted above the pictures in Fig. 5: breast volume symmetry (BV) and breast shape symmetry (BS). They are the average values of evaluation scores obtained from four specialists. Two L-R contrast features were also examined such as volume ratio V_L/V_R and surface-area ratio S_L/S_R. These features are expected to correlate with BV and BS, respectively. It was clearly found that as the esthetic score increased, the L-R contrast features tended to approach 1.0, suggesting a strong correlation between the extracted features and the esthetic outcome. Therefore, the results confirmed the appropriate extraction of feature primitives.

The time performance of this system was also examined. The breast image acquisition including pretreatment required approximately 5 min for each case, which was performed without significantly hindering daily practices. The incremental anonymization process required approximately 3 min for each iteration. Since this process is performed as a

Case		A	B	C	D
Esthetic outcome evaluation score	BV*	2.00	1.00	0.00	0.00
	BS**	2.00	0.75	0.00	0.00
3D breast mesh closure	Left	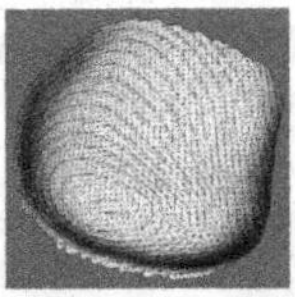		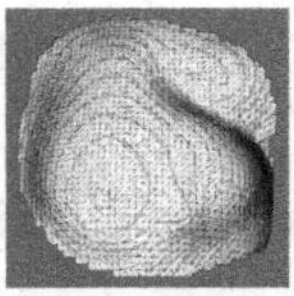	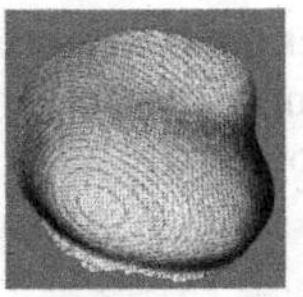
	Right	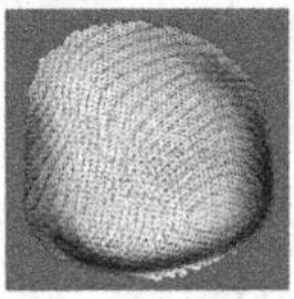		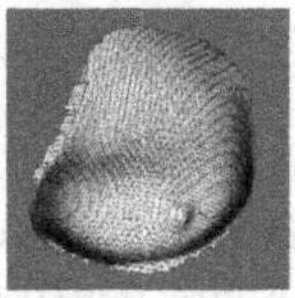	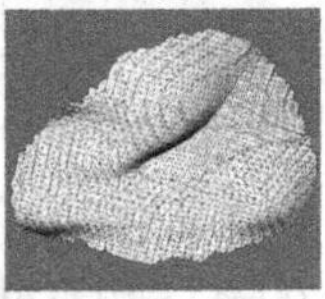
L-R contrast features	V_L/V_R	1.0	1.1	0.5	13.1
	S_L/S_R	1.0	1.2	0.6	2.4

*Breast volume, **Breast Shape

Fig. 5. Typical examples of extracted breast mesh closures and their feature primitives

batch, it was realized with little additional time and effort from medical specialists. It took approximately 16 min for each case to perform the preprocessing and feature extraction, most of which were the manual operation such as identification of cross marks and visible nipples, and tracing of breast outline. Considering that the clinical image acquisition frequency is typically less than 10 cases per day, the overall duration is acceptable and unlikely to cause a bottleneck in dataset preparation. Since this system covers the entire processes of dataset preparation and achieved enough time performance mentioned above, the developed system is considered to have high clinical applicability.

5 Conclusion

In this paper, we have introduced a novel medical-image processing system tailored for effective dataset preparation in ML-based CAD, focusing on esthetic outcome evaluation in breast reconstruction surgery. The development of a system prototype integrated specialized functions including incremental anonymization, annotation assistance, and hybrid process flow for preprocessing and feature extraction.

The incremental anonymization and filtering capabilities allowed for batch processing after acquiring a certain number of cases, reducing the daily workload of medical specialists within their institutions. Furthermore, the cross annotation process and its error correction by external specialists were significantly facilitated through the cogeneration of annotation sheets with anonymized images and the application of OCR-based collection methods.

A key characteristic of the system was the implementation of a hybrid process flow, seamlessly combining manual operations with complementary automation algorithms for effective preprocessing and feature extraction. This approach led to a substantial reduction in effort required to prepare datasets, including 3D breast-mesh closures and their associated geometric features, thereby demonstrating the high clinical applicability of the developed prototype.

Acknowledgement. The authors extend their gratitude to Mr. Kazuya Koyanagi for his invaluable contribution during the initial development stages of this study and the creation of the system prototype.

This work was supported by JSPS Core-to-Core Program (grant number: JPJSCCB20230005).

References

1. Litjens, G., et al.: A survey on deep learning in medical image analysis. Med. Image Anal. **42**, 60–88 (2017). https://doi.org/10.1016/j.media.2017.07.005
2. Anwar, S.M., Majid, M., Qayyum, A., Awais, M., Alnowami, M.R., Khan, M.K.: Medical image analysis using convolutional neural networks: a review. J. Med. Syst. **42** (2018). https://doi.org/10.1007/s10916-018-1088-1
3. Preuss, J., Lester, L., Saunders, C.: BCCT.core – Can a computer program be used for the assessment of aesthetic outcome after breast reconstructive surgery? The Breast **21**, 597–600 (2012). https://doi.org/10.1016/j.breast.2012.05.012
4. Cardoso, M.J., et al.: Turning subjective into objective: the BCCT.core software for evaluation of cosmetic results in breast cancer conservative treatment. The Breast **16**, 456–461 (2007). https://doi.org/10.1016/j.breast.2007.05.002
5. Heil, J., et al.: Objective assessment of aesthetic outcome after breast conserving therapy: Subjective third party panel rating and objective BCCT.core software evaluation. The Breast **21**, 61–65 (2012). https://doi.org/10.1016/j.breast.2011.07.013
6. Sowa, Y., Nguyen Ngoc, M., Fukuzawa, M.: Toward a system for esthetic outcome evaluation with machine-learning. PEPARS **166**, 27–34 (2020). [in Japanese]
7. Yano, K.: Cosmetic results after breast reconstruction. Japan. J. Clin. Med. **65**(6), 465–468 (2007) [in Japanese]
8. Intel Realsense L515. https://www.intelrealsense.com/lidar-camera-l515. Accessed 6 July 2023
9. Imfusion Recfusion. https://www.recfusion.net/. Accessed 6 July 2023

An Intelligent Polarimetric System for High-Throughput and High-Sensitivity Quality Assessment of Commercial Semiconductor Wafers

Shun Setomura, Nao Arai, Yuta Kimura, and Masayuki Fukuzawa(✉) ⓘ

Graduate School of Science and Technology, Kyoto Institute of Technology, Matsugasaki, Sakyo-Ku, Kyoto 606-8585, Japan
fukuzawa@kit.ac.jp

Abstract. An intelligent polarimetric system has been developed for high-throughput quality assessment of commercial semiconductor wafers by measuring a small amount of residual strains while scanning the wafer at high speed. Residual strain is an important quality measure in semiconductor wafers and computer-controlled polarimetric system is widely used to characterize its distribution. In order to handle the novel semiconductors such as silicon carbide (SiC) and gallium nitride (GaN), more high-throughput and high-sensitivity characterization is expected, but it was difficult due to the limitation of the conventional system both in sensitivity and scanning speed. In order to achieve both high sensitivity and fast scanning, the developed system adopted a small-scale SoC including a control law accelerator (CLA) with ultra-low interrupt latency. A special firmware of distributed processing was originally developed where the CLA performed fast polarization control and its synchronized photometry, while the CPU covered overall processes and communication to the host. The developed system exhibited 5 times faster scanning speed and 10 times higher photometric sensitivity compared with the scanning-type conventional system, which revealed the potential for the world class performance in the quality assessment of the commercial semiconductor wafers and implied the applicability for further study on advanced strain analysis based on measurement informatics.

Keywords: Intelligent Polarimetric System · Measurement Informatics · Residual Strain · Semiconductor Wafer · Real-Time Control · Low Latency

1 Introduction

Polarimetric system is aimed at measuring the birefringence in a sample by analyzing the transmitted polarized light incident on the sample at different polarization direction. It has been widely used to measure strain-induced birefringence in metals and transmission materials because of its advantages of measurement throughput for large-area samples compared with other strain assessment techniques such as X-ray topography and Raman spectroscopy [1, 2]. In the case of semiconductor wafers manufactured as substrates for LSIs and optical devices, quantitative assessment of residual strain distribution is very important because inhomogeneous strain distribution may cause various issues in device fabrication even if the strain amount is small. Therefore, various intelligent-type polarimetric systems have been developed to increase the sensitivity as well as the assessment throughput.

Yamada developed a high-sensitivity computer-controlled polariscope to characterize the residual strain in commercial thin wafers [3]. Fukuzawa and Yamada evaluated various commercial semiconductor wafers for strain-induced problems by developing an improved version of the previous polariscope, named as scanning infrared polariscope (SIRP) [4–6]. Figure 1 shows typical 2D maps of residual strain measured in commercial indium phosphide (InP) wafers with SIRP to demonstrate the variety of strain distribution. The measured value $|S_r - S_t|$ is the absolute difference of in-plane strain components between radial and tangential directions of the wafer, which corresponds to so-called 'principal strain'. The measurement values of SIRP are the phase retardation δ and the principal direction ψ of birefringence, and in the case of InP, $|S_r - S_t|$ can be evaluated from the following equation:

$$|S_r - S_t| = \frac{\lambda\delta}{\pi d n_0^3}\sqrt{\left(\frac{\cos 2\psi}{p_{11} - p_{12}}\right)^2 + \left(\frac{\sin 2\psi}{p_{44}}\right)^2} \qquad (1)$$

where d, n_0 and p_{ij} are the thickness, refractive index and photoelastic coefficients in the sample, respectively.

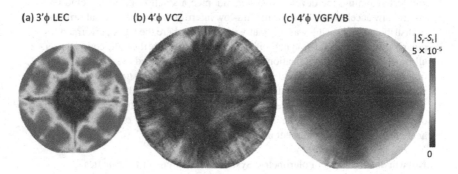

(a) 3'φ LEC   (b) 4'φ VCZ   (c) 4'φ VGF/VB

$|S_r\text{-}S_t|$
5×10^{-5}

0

Fig. 1. Typical 2D maps of residual strain examined with SIRP

The SIRP had the highest level of sensitivity in the world at the time of development and keeps still high sensitivity to be applicable for characterizing the commercial wafer of defect-free silicon (Si) crystal for LSIs [7]. However, improvement in measurement throughput is desired with the recent increase in diameter of commercial wafers including novel semiconductors such as silicon carbide (SiC) and gallium nitride (GaN) for power electronic devices. It is also expected to achieve further sensitivity because it enables us to develop advanced techniques based on measurement informatics such as a separated analysis of S_r and S_t. The imaging-type polarimeters (xIPs) were developed to improve the characterization throughput [7], but the sensitivity was inferior to that of SIRP. Geiler et al. developed an alternative polarimetric system (SIRD) based on faster r-θ scan [8], but quantitative evaluation of $|S_r - S_t|$ was difficult.

This study aims to develop an intelligent polarimetric system with both high scanning speed and high sensitivity for characterizing the residual strain in semiconductor wafers, which enables us to handle a large-diameter commercial wafers of SiC and GaN as well as to develop advanced strain analysis based on measurement informatics.

The remainder of this paper is as follows. Section 2 explains the design and implementation of a high sensitivity and fast scanning infrared polariscope (HSF-SIRP) developed in this study. Section 3 shows the experimental results and discusses the expected performance of the system. The concluding remarks are described in Sect. 4.

2 High-Sensitivity and Fast Scanning Infrared Polariscope (HSF-SIRP)

2.1 Functional System Design

Figure 2 shows a functional system block diagram of HSF-SIRP developed in this study. This system consists of an optical system, a series of drivers and converters, a real-time controller, and a host PC with a GUI-based hosting software. The optical system is similar to a linear polariscope, and a light source, motorized polarizer and analyzer, a motorized XY stage, and an optical sensor are aligned with a common optical axis. They are directly driven by the real-time controller through their own drivers or converters. This controller includes a light controller to adjust and calibrate the intensity of the light source, a polarization and scan controller to bring a certain polarization state of the optical system during scanning the wafer with XY stage, a photometer to acquire the transmitted light intensity, an overall controller to manage the photometry and scanning, and a host interface to transfer the photometry data as well as to communicate commands with the host PC.

In the developed system, δ and ψ of birefringence are obtained by analyzing the transmitted light intensities under various polarization states. Assume the direction of polarizer and analyzer ϕ and the transmitted light intensities $I^{\|}(\phi)$ and $I^{\perp}(\phi)$ under the parallel and crossed Nichol conditions, respectively. The transmission ratio to be acquired by the system is obtained by the following equation.

$$I_r(\phi) \cong \frac{I^{\perp}(\phi)}{I^{\|}(\phi)} \tag{2}$$

δ and ψ are expressed as the following equations by using discrete sine and cosine transform:

$$\delta = 2\arcsin\left[16\left(I_{\sin}^2 + I_{\cos}^2\right)\right]^{\frac{1}{4}}, \psi = \frac{1}{4}\arctan\frac{I_{\sin}}{I_{\cos}} \tag{3}$$

$$I_{\sin} \equiv \frac{1}{J}\sum_{j=0}^{J-1} I_r(\phi_j)\sin4\phi_j = -\frac{1}{4}\sin4\psi\sin^2\frac{\delta}{2} \tag{4}$$

$$I_{\cos} \equiv \frac{1}{J}\sum_{j=0}^{J-1} I_r(\phi_j)\cos4\phi_j = -\frac{1}{4}\cos4\psi\sin^2\frac{\delta}{2} \tag{5}$$

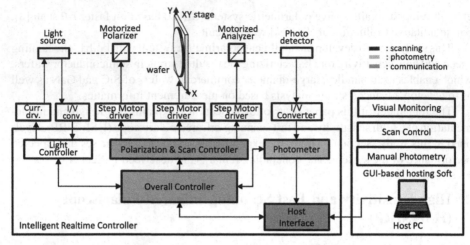

Fig. 2. Functional system block diagram of HSF-SIRP

$$\phi_j = \frac{2\pi j}{J}, (j = 0, 1, \ldots, J - 1) \tag{6}$$

where J is the division number of ϕ when $I_r(\phi)$ is measured.

From Eqs. (1) to (6), the sensitivity of $|S_r - S_t|$ depends on the order of detectable $I_r(\phi)$. It was about 10^{-7} in SIRP, which means that $I^\perp(\phi) \ll I^\parallel(\phi)$. Since the photometer has a limited dynamic range in general, it is necessary to optimize the light source intensity and the detector gain separately in the measurements of $I^\parallel(\phi)$ and $I^\perp(\phi)$. Assume the light source intensities $I_0^\parallel$ and $I_0^\perp$, the photometer gains $G^\parallel$ and $G^\perp$, and the photometer value $P^\parallel(\phi)$ and $P^\perp(\phi)$ in the parallel and crossed Nichols conditions, respectively. $I_r(\phi)$ can also be expressed as Eq. (7).

$$I_r(\phi) \cong \frac{I^\perp(\phi)}{I^\parallel(\phi)} = \frac{P^\perp(\phi)}{P^\parallel(\phi)} \times \frac{G^\parallel}{G^\perp} \times \frac{I_0^\parallel}{I_0^\perp} \tag{7}$$

According to Eq. (7), it is desirable for high sensitivity that $G^\parallel/G^\perp \ll 1$ and $I_0^\parallel/I_0^\perp \ll 1$ are satisfied and their product is configurable at least down to 10^{-7} so as to keep the precision of $P^\perp(\phi)$ and $P^\parallel(\phi)$ even near the minimum $I_r(\phi)$.

A performance bottleneck in the wafer scanning is the rotation speed of the polarizer and analyzer and its synchronizing photometry. Fast rotation requires acceleration/deceleration rotation of the polarizer and analyzer, and its synchronized photometry should have variable interval. In order to implement it, the polarization and scan controller requires two special functions. The former is an intra-cycle update function that generates the period of next drive pulse within the present pulse period [9], and the latter is a low-latency triggering function that sends a trigger signal to the photometer at the beginning of each pulse period. Such timing-critical and complicated process was difficult to be implemented in software such as interrupt handlers of conventional microcontroller. In this study, we designed and implemented a real-time controller by adopting a small-scale SoC (C2000 real-time microcontroller by Texas Instruments) equipped with a control law accelerator (CLA) with ultra-low interrupt latency in order to meet the requirements for high sensitivity and to avoid the complexity of overall control in photometry and scanning [10].

2.2 Implementation of Real-Time Controller

Figure 3 shows a block diagram of real-time controller implemented in this study. In this controller, the CLA handles timing-critical processes such as intra-cycle update and conditional branch based on the recent status such as the number of drive pulses, the direct memory access controller (DMAC) transfers the photometric data and communication commands asynchronously, while the CPU focuses on overall control and command interpretation from the host PC. Such role partitioning enabled both flexibility and low latency while minimizing time dependencies of each task in the software.

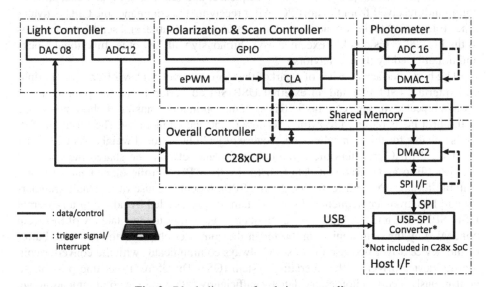

Fig. 3. Block diagram of real-time controller

The polarization and scan controller consists of a CLA, an enhanced pulse width modulator (ePWM), a general purpose IO (GPIO) and a part of a shared memory. The drive pulse is generated by CLA altering the GPIO output at appropriate timing for each motor. The drive pulse period is counted by ePWM and notified to the CLA. The CLA sends a trigger signal to the photometer block at the beginning of each pulse period, and then updates the next period within the cycle according to an acceleration data table in the shared memory. The CLA also manages the total pulse number including acceleration and deceleration by counting the notification, and automatically finishes a series of pulse generation after sending an interrupt to the CPU when the counts meet a preconfigured number. The advantage of this block is that CPU is not involved in driving the polarizer, analyzer or XY stage.

The photometer consists of a 16-bit analog-to-digital converter (ADC16), a DMAC and a part of the shared memory. The ADC16 begins to digitize the transmitted light intensity just after receiving the trigger signal from CLA and issues the end-of-conversion trigger to DMAC. The DMAC transfers the converted data to the shared memory. In the same manner as the polarization controller, the photometer block does not involve the CPU at all, except for initial settings.

The light controller consists of a digital-to-analog converter (DAC) and ADC12. The DAC sends an analog voltage to control the driving current of the light source, while the ADC12 digitizes the light source intensity. Since light control is static, this block is handled by the CPU when a corresponding command is received. It should be noted that a certain control voltage does not always result in the same light intensity due to the temperature characteristics of the light source and current driver. In order to calibrate the light intensity, we adopted a structure that can monitor the actual light intensity directly.

The overall controller consists of C28x CPU. The essential functions are initialization of various peripherals just after booting sequence, and interpretation and execution of commands received from the host PC. After receiving the command, the CPU activates other functional blocks, updates an internal configuration table, and sends return values to the host PC. This block is executed asynchronously with any other functional blocks and not affected by their behaviors.

The host interface consists of a part of the shared memory, a DMAC, a serial peripheral interface (SPI I/F) and an external USB-SPI converter. Since the photometry is always repeated during scanning in this system, continuous transfer of photometry data to the host PC is required to avoid overflow of the shared memory. Therefore, the SPI I/F is suitable for this controller side because of its light protocol weight. As with other blocks, a DMAC transfers the communication data between the shared memory and the SPI I/F to avoid frequent data-transfer by the CPU. On the other hand, it is not suitable for host PC to directly handle the SPI protocol because of its small granularity and high transfer frequency. In this system, it was avoided by adopting an external USB-SPI converter. The high transfer frequency and requested low latency of SPI were concealed by the communication buffer in the converter chip and the software buffer of the device driver because the host PC always communicates with the converter chip via the device driver of the operating system (OS). Therefore, it became possible to continuously receive photometry data by sufficiently following the communication rate of SPI even in a GUI application.

3   Experimental Results

A prototype of the proposed system was developed using the C2000 real-time microcontroller (TMS320F28379d). The real-time controller firmware was originally developed from scratch and both the CPU and the CLA were clocked at 200 MHz. In order to accurately observe the behavior of the polarization and scan controller, we added a debug-support function in the CLA firmware. It was to generate the task duration signals by altering some GPIO outputs at the start and end of the CLA tasks for 1) pulse generation, 2) photometer triggering, and 3) pulse-period update. A high level of the task duration signal indicates that the task is running. Since it takes only a few instructions for the CLA to alter the GPIO output, the probe effect of this modification on performance evaluation is negligible.

3.1 Waveform Analysis for Realtime Performance of HSF-SIRP

In order to evaluate the essential performance of HSF-SIRP, we performed waveform analysis of the task duration signals. Figure 4 shows the typical waveforms of the task duration signals and the drive pulse. It was obtained by acquiring the output signal of the real-time controller with an oscilloscope when the drive pulse reached the maximum frequency of 50 kHz during accelerated rotation of the polarizer and analyzer.

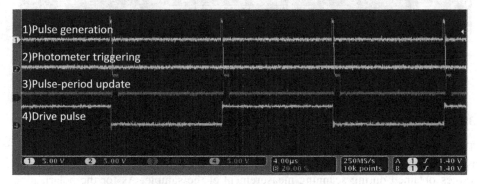

Fig. 4. Typical waveforms of the task duration signals and the drive pulse

The waveform revealed that the three CLA tasks were executed in sequence with a very small duration compared with the period of drive pulse, while the drive-pulse signal was altered at exact intervals of 20 μs compared with the fluctuation of the task execution. Therefore, it was confirmed that the real-time controller succeeded in generating the drive pulses and synchronous triggering the photometer. Figure 5 shows the same waveforms as shown in Fig. 4 with a short time-axis scale to observe them in detail. The t_{d1} represents the time from the beginning of the pulse generation task to the alteration edge of the drive pulse, t_{d2} does that from the alteration edge of the drive pulse to the end of the photometry triggering task, t_{w3} does the duration of the pulse-period update task, and t_{wtot} does the duration of all the CLA tasks for a cycle of drive pulse generation, which are noted as the white arrows in Fig. 5.

t_{d1} was around 120 ns and remained almost constant over all the cycles of drive pulse generation with the deviation of a few nanoseconds, whose small and constant delay was enough for exact control of the pulse period of 20 μs. t_{d2} was around 90 ns and was minimum duration among the CLA tasks, which demonstrated that the low-latency photometer triggering was succeeded. t_{w3} was around 520 ns and was maximum in the CLA tasks. Although this task includes intra-cycle update of the drive pulse period and conditional branch to manage acceleration and deceleration, the duration is much smaller than the drive pulse period. Therefore, it was confirmed that the intra-cycle update was also succeeded with sufficient time margin. Furthermore, t_{wtot} was around

720 ns and corresponded to only 4% of the drive pulse period, which implies a plenty of room for additional CLA tasks for further development.

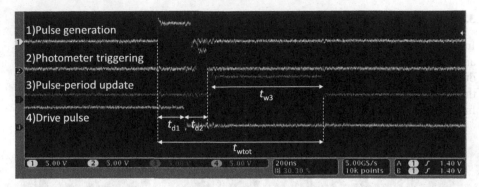

Fig. 5. The waveforms same as shown in Fig. 4 with a short time-axis scale.

3.2 Comparison of System Performance Between SIRP and HSF-SIRP

Table 1 shows essential performances of SIRP and HSF-SIRP. Among these, $I_0^{\parallel}/I_0^{\perp}$ is the average value of 10 measurements, and the other items represent the maximum values obtained during scanning measurement of the sample. As for the polarization control and scanning, the drive pulse frequency of HSF-SIRP achieved 5 times as fast as that of SIRP while keeping the synchronous photometry. Since that frequency limits the rotation speed of the polarizer and analyzer, it also exhibited that the HFS-SIRP achieved scanning speed up to 5 times as fast as SIRP if the scan pitch is the same. The gain ratio $G^{\parallel}/G^{\perp}$ and the light-intensity ratio $I_0^{\parallel}/I_0^{\perp}$ of HSF-SIRP became around 3 and 5 times as wide as those of SIRP, respectively. According to the Eq. (7), this result exhibited that the order of detectable $I_r(\phi)$ was improved more than ten times. The data transfer rate of HSF-SIRP was 5 MB/s when a delay of 100 ms was allowed, which was sufficient for the maximum transfer rate that could occur with scanning.

Table 1. Essential performances of SIRP and HSF-SIRP

Function	Item	SIRP	HSF-SIRP
Polarization control and scanning	Drive pulse frequency [kHz]	10	50
Light control	$I_0^{\parallel}/I_0^{\perp}$	2/7	100/2048
Photometry	$G^{\parallel}/G^{\perp}$	1/1000	1/3000
Communication	Data transfer rate [MB/s]	1	5

The HSF-SIRP revealed 5 times faster scanning speed and 10 times higher photometric sensitivity compared with the SIRP. Since the performance of SIRP is still enough for characterizing the conventional semiconductor wafers, the HSF-SIRP has a potential for the world class performance in the quality assessment of novel semiconductor wafers.

Additionally, improved sensitivity of HSF-SIRP enables us to measure very low-strain samples that could not be measured with conventional methods, which leads drastic increase of measurement data. Therefore, further research based on measurement informatics will be possible, such as the development of new machine learning models for advanced strain analysis trained by the measurement data of HSF-SIRP.

4 Conclusion

An intelligent polarimetric system has been developed for high-throughput quality assessment of commercial semiconductor wafers by measuring a small amount of residual strains while scanning the wafer at high speed. A small-scale SoC including a CLA with ultra-low interrupt latency was adopted to construct a real-time controller, and a special firmware of distributed processing was originally developed where the CLA performed fast polarization control and its synchronized photometry, the DMAC transferred photometry data to the host, and the CPU covered overall processes and communication to the host. The developed system exhibited 5 times faster scanning speed and 10 times higher photometric sensitivity compared with the scanning-type conventional system, which revealed the potential for the world class performance in the quality assessment of the commercial semiconductor wafers and implied the applicability for further study on advanced strain analysis based on measurement informatics.

Acknowledgement. This work was partly supported by JSPS KAKENHI Grant Number JP23K04600 and by JSPS Core-to-Core Program (grant number: JPJSCCB20230005).

References

1. Runyan, W.R., Shaffner, T.J.: Semiconductor Measurements and Instrumentation, 2nd edn. The McGraw-Hill Companies Inc, United States of America (1998)
2. Schroder, D.K.: Semiconductor Material and Device Characterization, 3rd edn. Wiley, United States of America (2006)
3. Yamada, M.: High-sensitivity computer-controlled infrared polariscope. Rev. Sic. Instrum. **64**(7), 1815–1821 (1993)
4. Fukuzawa, M., Yamada, M.: Proceedings of IPRM2008 (2008)
5. Fukuzawa, M., Yamada, M.: J. Cryst. Growth **229**, 22–25 (2001)
6. Wada, O., Hasegawa, H.: InP-Based Materials and Devices Physics and Technology. Wiley, United States of America (1999)
7. Fukuzawa, M., Yamada, M.: Phys. Stat. sol. (c) **5**(9), 2941–2943 (2008)
8. Geiler, H.D., et al.: Mater. Sci. Semicond. Process. **9**, 345–350 (2006)
9. Texas Instruments Application Note. https://www.ti.com/lit/an/spracn0f/spracn0f.pdf. Accessed 13 July 2023
10. Texas Instruments Reference Manual. https://www.ti.com/lit/ug/spruhm8i/spruhm8i.pdf. Accessed 13 July 2023

Big Data, IoT, and Cloud Computing

Strengthening Information Security Through Zero Trust Architecture: A Case Study in South Korea

H. H. Nguyen[1], Yeram Lim[2], Minhui Seo[2], Yunyoung Jung[2], Minji Kim[2], and Wonhyung Park[2(✉)]

[1] College of Information and Communication Technology, Can Tho University, Can Tho, Vietnam
nhhoa@ctu.edy.vn

[2] Department of Convergence Security, Sungshin Women's University, Seoul, South Korea
{20200952,20211066,20211097,20212565,wonhyung}@sungshin.ac.kr

Abstract. This paper explores the enhancement of Information Security Management Systems (ISMS) through the adoption of Zero Trust architecture, focusing on its implementation in South Korea. The study examines the limitations of traditional ISMS models, particularly the Korean Information Security Management System, and posits that integrating Zero Trust principles can address these shortcomings. Utilizing a case-study approach, this paper provides an in-depth analysis of the Zero Trust model, with specific emphasis on Multi-Factor Authentication and Decentralized Identifiers. It presents real-world implementations and discusses policy frameworks that South Korea has adopted to transition towards Zero Trust. This research aims to contribute to both the theoretical and practical understanding of integrating Zero Trust architecture in ISMS protocols, offering insights into overcoming challenges and future directions.

Keywords: Zero Trust · Information Security Management System (ISMS) · Multi-Factor Authentication (MFA) · Decentralized Identifiers (DID)

1 Introduction

In the current era, where cyber threats are omnipresent and constantly evolving, organizations worldwide are grappling with the daunting task of securing their information assets. The stakes are particularly high in South Korea, a nation renowned for its technological advancements but also susceptible to a range of cybersecurity threats. The established approach to cybersecurity, largely rooted in perimeter-based models, has shown its limitations in the face of advanced threats, including insider attacks. One glaring example is the incident at LG U+ in South Korea, where an insider threat led to a significant security breach. Such instances call for a comprehensive rethink of existing Information Security Management Systems (ISMS).

The focus of this paper is to explore how Zero Trust Architecture can serve as a paradigm shift in how organizations view and implement information security, addressing the limitations of traditional perimeter-based models. Unlike traditional approaches,

N. Thai-Nghe et al. (Eds.): ISDS 2023, CCIS 1950, pp. 63–77, 2024.
https://doi.org/10.1007/978-981-99-7666-9_6

Zero Trust eliminates the notion of inherent trust for any entity within an organization's network, instead requiring continuous verification. The 'never trust, always verify' philosophy of Zero Trust is increasingly pertinent in a world characterized by remote work, cloud computing, and an escalating complexity of cyber threats [10].

Specifically, we delve into the Korean ISMS (K-ISMS), South Korea's national standard for information security. K-ISMS, while robust in many aspects, follows a traditional, perimeter-based security model [7]. We argue that integrating Zero Trust principles into K-ISMS can offer a more dynamic and adaptive cybersecurity posture. This is particularly vital for a country like South Korea, which faces unique regulatory landscapes and geopolitical considerations in its cybersecurity strategy.

The paper is structured as follows: Sect. 2 offers a review of the literature, laying the groundwork by discussing Zero Trust philosophy, traditional security models, and the structure and limitations of K-ISMS. Section 3 outlines the limitations of current ISMS frameworks in general, followed by Sect. 4, which proposes an updated ISMS checklist that incorporates Zero Trust principles. In Sect. 5, we present case studies to validate the effectiveness of a Zero Trust approach, with a particular focus on its implementation in the South Korean context. Finally, Sect. 6 summarizes our findings and outlines future directions for research and implementation.

2   Global Cybersecurity Standards

2.1   Paradigm Shift: Zero Trust Architecture

The Zero Trust model represents a paradigm shift in network security, fundamentally eliminating the concept of trust from the network architecture [1]. It mandates continuous verification of all users and systems, regardless of their origin, thereby providing a robust defense against both internal and external threats. This approach has gained considerable traction, particularly in cloud computing environments, where it is increasingly seen as the standard.

The concept of Zero Trust was first formulated by John Kindervag in 2010 during his tenure as a principal analyst at Forrester Research [12]. Contrary to traditional cybersecurity models that focus on establishing a secure perimeter, Kindervag's groundbreaking idea was to orient network security from the inside out. This approach was a departure from conventional methods that trusted anything within the perimeter and sought to block external threats.

Several fundamental principles underlie the Zero Trust architecture. The principle of "Least-Privilege Access" ensures that users and systems have only the minimum levels of access or permissions required to perform their tasks [2]. In addition, the network is segmented into smaller, more easily managed sections, known as "micro-segments." This strategy aims to limit the damage an attacker can do if they gain access to one part of the network, essentially confining them to that specific area and preventing free lateral movements. Another cornerstone is "Continuous Verification," which mandates ongoing, dynamic assessment of the security status of both users and systems to adjust access controls accordingly [5].

Zero Trust has found particular resonance in the era of cloud computing. Traditional perimeter-based defenses are not well-suited for cloud environments, which often lack a clearly defined perimeter. This makes Zero Trust especially effective, as it doesn't rely on distinguishing between a secure internal network and an untrusted external one. The model has also gained traction due to the rise of remote work. Unlike traditional security frameworks that often falter when applied to remote working scenarios, Zero Trust's assumption that threats can originate from anywhere makes it inherently better at securing remote work environments.

In the age of the Internet of Things (IoT) and Bring Your Own Device (BYOD) policies, Zero Trust is even more relevant. The influx of various new devices into organizational networks has further eroded the boundaries of what used to be considered a secure internal perimeter. Zero Trust's granular access controls and segmentation strategies offer a more targeted and effective defense mechanism than broad, perimeter-based approaches.

The efficacy of Zero Trust has been demonstrated in multiple sectors through various case studies. Google's BeyondCorp initiative shows how large organizations with complex networks can maintain security without relying on a traditional internal network. Likewise, the United States Department of Defense has adopted Zero Trust principles, showcasing the model's applicability even in the most stringent and sensitive of environments [9].

However, Zero Trust is not without its detractors. Critics often point to the complexity and expense associated with implementing this model, particularly in organizations that rely on legacy systems. Moreover, there can be resistance from employees who are accustomed to less restrictive network access, as well as a steep learning curve to overcome.

Looking forward, advancements in machine learning and artificial intelligence hold the promise of automated, real-time decision-making that could further enforce Zero Trust principles. Research is also underway to explore how emerging technologies like blockchain could enhance the inherently trustless and decentralized aspects of Zero Trust architectures. Overall, Zero Trust provides a more holistic, adaptive, and future-proof framework for cybersecurity. Its core philosophy of 'never trust, always verify' aligns well with the evolving complexities of modern network environments, which are increasingly dynamic and lack clear perimeters [10].

2.2 Limitations of Traditional Security Architectures

Traditional security models have their roots in an era when organizations primarily operated closed, on-premise networks. The focus of these early frameworks was to construct a secure boundary around the internal network, safeguarding it from external threats. This led to the development of technology solutions like firewalls and intrusion detection systems, essentially forming what is often termed as a "perimeter-based" or "castle-and-moat" security model [13].

These models were built on several core principles. One of the main tenets was the concept of "Trust by Default," meaning that once users or systems were authenticated and passed through the external defenses into the internal network, they were generally trusted. In addition, while not as granular as Zero Trust models, traditional security models did often include some form of network segmentation. This meant dividing the internal network into various zones, each with different levels of trust and access controls.

These models also primarily relied on signature-based threat detection to identify and counter known malware and attack vectors. However, this method proved less effective against novel or highly customized attacks.

A considerable part of the traditional model's architecture is designed to fend off external threats. A typical strategy involves deploying various security measures at the network's edge, like firewalls, Virtual Private Networks (VPNs), and intrusion detection and prevention systems (IDS/IPS). While effective in blocking a good deal of external threats, traditional models fall short in countering internal threats effectively. The inherent trust bestowed upon entities already inside the perimeter makes these models susceptible to insider attacks. Whether by malice or mistake, insiders can often bypass internal security measures due to their privileged network positions.

This vulnerability has been laid bare in numerous high-profile security breaches. Cases such as the 2014 Target breach and the 2017 Equifax hack serve as sobering examples. Once the attackers gained initial access to the network, their movements within the system were hardly restricted, leading to disastrous consequences. Another pressing issue is the compatibility of traditional models with modern technologies like cloud services, mobile computing, and Internet of Things (IoT) devices. These technologies often defy the concept of a fixed, clearly defined security perimeter, rendering traditional models less effective.

Adopting a modern security stance over a traditional one poses its own set of organizational challenges. These can range from resistance to change from network administrators and end-users, to budget constraints and the complexity of retrofitting or phasing out legacy systems. Nevertheless, as the cybersecurity landscape evolves, the strategies to mitigate threats are changing as well. Traditional models are increasingly seen as outdated and are expected to play a diminishing role in the future of network security.

In summary, while traditional security models have served organizations well for a time, they are increasingly ill-equipped to handle the contemporary threat landscape. With threats becoming more sophisticated and originating not just from outside but also from within organizational boundaries, reliance solely on traditional models is increasingly seen as a liability.

2.3 Standardization in Security: Information Security Management Systems

ISMSs are comprehensive frameworks that involve a blend of policies, procedures, and both technical and physical controls aimed at managing an organization's risk surrounding its information assets. This system adopts a holistic approach that integrates people, processes, and technology. Among ISMS frameworks, the ISO/IEC 27001 standard has gained global recognition and serves as a template for various national security standards [14].

The primary aim of any ISMS is to safeguard an organization's information assets from unauthorized access, damage, or theft. This involves not only technological measures but also an alignment with external laws and regulations to ensure compliance. Beyond that, ISMS frameworks are designed to identify and prioritize information risks and apply appropriate countermeasures. Ensuring the continuity of critical business operations, especially in the face of security incidents, is another crucial objective.

An effective ISMS revolves around several core components. The first step is the development of overarching security policies that articulate the organization's stance on security. Following this is a rigorous risk assessment phase where the existing risks are identified and evaluated. The subsequent step involves implementing various control measures like firewalls, encryption protocols, and multi-factor authentication to mitigate these risks. Continuous monitoring and review mechanisms are also put in place to evaluate the effectiveness of the ISMS. Lastly, specific procedures are established for incident response, to ensure that any security incidents are appropriately managed and mitigated.

ISMS frameworks have found worldwide adoption across diverse sectors including healthcare, finance, and government organizations. They have come to serve as a common language for discussing and managing security postures and have often become a prerequisite for both business contracts and regulatory compliance.

3 Challenges in Existing Frameworks

Current ISMSs carry a range of limitations that can compromise the overall security of an organization [14]. One of the most glaring issues is their susceptibility to insider threats. While most ISMS approaches are designed to ward off external attackers, they often implicitly trust traffic and activities that originate within the network. This is an alarming oversight, as demonstrated by the recent incident at LG U+ in South Korea, where an insider threat led to a significant security breach. Not only do such models lack the granularity to identify suspicious internal activities, but they also usually have inadequate logging and alerting capabilities to flag such events in real-time. The internal actors, who might be disgruntled employees, contractors, or even business partners with privileged access, can manipulate their position within the organization to bypass security controls easily. This glaring blind spot often leaves organizations vulnerable to data theft, sabotage, or even more devastating forms of cyber-espionage.

Another limitation lies in the adaptability of traditional ISMS frameworks to emerging cybersecurity threats [14]. Most of these systems are static by nature, built around predefined rules and known threat signatures. Consequently, they often struggle to identify and adapt to new forms of attack vectors, such as zero-day vulnerabilities, advanced persistent threats (APT), and polymorphic or metamorphic malware. As cybercriminal tactics become increasingly sophisticated, traditional ISMS require frequent and complex updates to remain effective, which can be both time-consuming and resource-intensive. Even when updates are available, there may be delays in deploying them due to testing requirements, during which time the organization remains vulnerable.

Furthermore, the legacy nature of some ISMS solutions makes them less compatible with newer technologies such as cloud-based services, the Internet of Things (IoT), and decentralized blockchain applications [13]. As businesses increasingly move towards digital transformation, the inability to seamlessly integrate security controls in mixed or evolving environments creates additional risks. This often results in a patchwork of solutions that lack coherence and can create security gaps.

Moreover, traditional ISMS are typically ill-equipped to handle the dynamism and scalability demands of today's enterprise environments. With organizations rapidly adopting multi-cloud infrastructures, edge computing, and mobile workforces, the fixed and perimeter-centric nature of many ISMS solutions becomes a bottleneck. These systems often don't scale well and can't provide the real-time, context-aware security decisions needed in a fast-paced, distributed architecture.

Compliance is another area where existing ISMS approaches often fall short. The landscape of regulatory requirements is constantly evolving, with laws like the General Data Protection Regulation (GDPR) in Europe and the California Consumer Privacy Act (CCPA) in the United States mandating stringent data protection measures. Many traditional ISMS are not agile enough to adapt to these ever-changing compliance landscapes, leaving organizations at risk of legal repercussions.

To sum up, while traditional ISMSs have served a purpose, they are increasingly showing their age and limitations. From their vulnerabilities to insider threats and lack of adaptability to new kinds of cyber risks, to their struggles with modern technological environments and compliance demands, it is becoming evident that new approaches to information security are not just preferable but essential.

4 Proposed Zero Trust K-ISMS Framework

4.1 Overview of Korean ISMSs

The Korean Information Security Management System, commonly referred to as K-ISMS, is South Korea's national standard for information security [15, 16]. It was introduced by the Korean government as part of its strategy to strengthen the nation's cybersecurity posture. Although initially aimed at public sector organizations and critical infrastructure, the scope of K-ISMS has expanded to include private corporations, especially those operating in sectors like finance and information technology [16].

K-ISMS encompasses a wide range of areas from IT governance and policy management to physical security controls. The framework is mostly in alignment with international ISMS standards, yet it is tailored to comply with specific South Korean laws and regulations. One of its most distinguishing features is its certification process, a vital requirement for business-to-business and government contracts in South Korea. The system also includes rigorous auditing processes to ensure compliance, typically involving both internal and external audits. Furthermore, K-ISMS often comes under the direct purview of governmental agencies, which ensures a high level of scrutiny.

However, K-ISMS is not without its limitations. It largely adheres to traditional, perimeter-based security models, which are effective against external threats but less so against internal ones. Once an entity is within the "trusted" network perimeter, K-ISMS does not generally enforce stringent access controls or ongoing monitoring.

Additionally, the heavy emphasis on certification and compliance can sometimes distract from the actual effectiveness of security measures.

Various case studies illustrate both the strengths and weaknesses of K-ISMS. On the positive side, companies adopting the K-ISMS standard have generally experienced a decline in external security incidents. Conversely, incidents involving internal staff members leaking data indicate a need for the framework to focus more on internal threats.

The Korean government and cybersecurity community are actively engaged in updating K-ISMS to address its limitations. The focus is gradually shifting towards incorporating principles from Zero Trust architectures and behavior-based analytics as future directions to evolve K-ISMS into a more comprehensive and effective security framework.

In conclusion, K-ISMS offers a robust and nationally tailored approach to information security in South Korea. However, its reliance on traditional perimeter-based security models means that it requires further evolution to adequately handle modern, sophisticated threats. These threats are increasingly originating not just from outside organizational perimeters but also from within them, underscoring the need for a more nuanced and comprehensive approach to information security.

4.2 ISMS Checklist with Zero Trust

The transition to a Zero Trust architecture necessitates a comprehensive reevaluation and, in many cases, a complete overhaul of existing ISMS protocols. This shift is not a mere addition to existing measures but a transformation that impacts every layer of an organization's security framework. A fundamental element of this transformation is a checklist that provides guidelines for integrating Zero Trust principles into multiple areas of ISMS, from network access to database management.

For network access, traditional ISMS solutions often depend on IP management, terminal authentication, and segmenting the network into zones based on business requirements and perceived importance, such as DMZs, server farms, and development areas. However, in a Zero Trust model, the validation process for insiders with network access would be significantly more stringent. Rather than relying solely on IDs or passwords, a Multi-Factor Authentication (MFA) would be used to ensure that insiders are precisely who they claim to be. Additionally, the network would be micro-segmented, giving each user differentiated levels of access to various network segments based on the principle of minimum necessary privileges. This granular approach would substantially mitigate risks, even in cases where a user's credentials are compromised.

In terms of information system access, current ISMS protocols typically outline who is permitted to access different information systems, along with the methods and secure means for doing so. Zero Trust adds several supplementary layers to these existing authentication and control measures. For instance, the device or location from which a user is attempting to access the system would be verified. Also, these systems would be programmed to automatically lock after periods of inactivity and to demand reauthentication at regular intervals. These extra layers would significantly tighten security, making unauthorized access increasingly challenging.

When considering application access, traditional methods usually restrict this based on user roles and the sensitivity of the accessed information. Zero Trust amplifies these controls by incorporating MFA to authenticate the user's identity more robustly before allowing application access. Continuous verification methods would also be employed to maintain the user's identity and the connection's security posture through the session.

Concerning database access, traditional ISMS practices involve identifying the type of information stored in databases and establishing access controls based on this information and user roles. Zero Trust, however, advocates for a more granular approach. It suggests that databases should be segmented based on the sensitivity and importance of the stored information. Access control policies would then be meticulously crafted to differentiate between types of users and applications, granting only the most minimal, essential access.

4.3 Use-Case: Decentralized Identifiers

The integration of Decentralized Identifiers (DIDs) into a Zero Trust framework offers an intriguing yet substantial upgrade in the realm of Multi-Factor Authentication (MFA) [17]. Unlike traditional centralized identity databases, which have several vulnerability points, DIDs employ blockchain technology to provide a non-centralized method for securely managing identities. This deviation from centralized systems significantly enhances the integrity and resilience of identity management operations.

Firstly, the technology behind DIDs provides several advantages over traditional systems. Since blockchain operates on a decentralized model, it eliminates single points of failure that could be targeted in centralized databases. Each identity is verified through a network of nodes, making it extremely difficult for malicious actors to alter or fake identity data. Additionally, the cryptography that underpins blockchain provides a secure and private way for users to validate their identity without exposing sensitive information.

The immutability of blockchain technology further fortifies the robustness of DIDs. Once identity data is added to the blockchain, it cannot be changed without a consensus among network participants. This makes it a powerful deterrent against types of cyberattacks that aim to manipulate or steal identity data. Consequently, DIDs provide not just a second or third factor for authentication but a fundamentally more secure architecture for identity verification.

DIDs also offer enhanced user privacy. In traditional identity management systems, users are often required to disclose a large amount of personal information to a central authority, increasing the risk of data breaches. With DIDs, users have more control over their personal data and can choose what to share and with whom, which aligns well with data minimization principles of modern data protection regulations like GDPR.

Moreover, DIDs can be integrated into various aspects of an organization's operations, from secure employee login to customer interactions, without requiring a complete overhaul of existing systems. Their modular nature makes it feasible to integrate them incrementally, allowing organizations to phase in blockchain-based identity verification at a pace that suits their needs and capabilities.

From an operational standpoint, DIDs can reduce the administrative burden of managing identities. Traditional systems often involve cumbersome processes of password resets, account recoveries, and verification. With DIDs, many of these processes can be automated or streamlined, saving time and resources.

The scalability of blockchain technology also means that DIDs can easily be adapted to handle a large number of identities. As organizations grow, it's crucial for their identity management systems to scale accordingly. DIDs are inherently scalable, providing a sustainable solution for expanding businesses.

Finally, using DIDs can enhance customer trust. The transparent yet secure nature of blockchain allows customers to see exactly how their data is being used, which can increase their trust in an organization's ability to safeguard their information.

In summary, the incorporation of DIDs into a Zero Trust framework offers numerous advantages. These range from enhanced security and user privacy to operational efficiencies and scalability. DIDs represent not just an addition to MFA but a paradigm shift in how identities can be managed securely and efficiently in a digitally connected world. This makes them an essential component of a modern, robust, and forward-looking ISMS, fully aligned with the principles of Zero Trust.

4.4 In-Depth Analysis of Use-Cases and Examples

Reinforcing Access Control Mechanisms Through Zero Trust Principles
The advent of Zero Trust architecture has revolutionized traditional approaches to cybersecurity by introducing rigorous user authentication and authorization procedures applicable to both internal and external actors. This is particularly critical in the realm of access control, a vital part of cybersecurity that determines who can interact with specific data and systems. The Zero Trust framework subjects every user request to meticulous verification, leveraging multiple technologies rooted in Zero Trust.

The typical workflow of Zero Trust-based access control is illustrated in Fig. 1, and each phase of this process integrates vital security technologies:

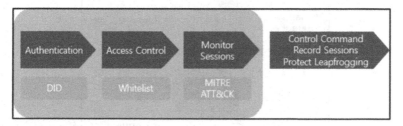

Fig. 1. In-Depth Workflow of Zero Trust Access Control

The authentication phase of this workflow begins with the use of DIDs. Typically based on blockchain technology, DIDs offer robust protection against both tampering and impersonation attempts. They stand out as more secure compared to traditional username/password authentication schemes and serve as a formidable foundation for implementing Zero Trust.

Following successful authentication via DID, the user's credentials are verified against a whitelist. This list comprises users who have previously been vetted and are approved to access designated resources. In contrast to a blacklist, which excludes known harmful entities, a whitelist permits access only to recognized and trusted individuals. This method considerably narrows the threat vector by eliminating unauthorized access at the outset.

Upon securing access, the next step involves continuous monitoring of user activities. This is often accomplished through security frameworks like MITRE ATT&CK. These frameworks are designed to recognize unusual patterns that may signify malicious activities. If any anomalies are detected, they trigger immediate alerts and can lead to restrictive actions, thereby maintaining the system's overall integrity.

Lastly, each command or query initiated by the user is subject to rigorous control and is meticulously logged. In addition to this, there are protective measures set up to counter 'leapfrogging,' a common tactic employed by attackers who exploit one compromised system as a stepping stone to infiltrate others. These measures ensure that even if a user gains unauthorized access to one part of the system, they cannot use it as a launching pad to compromise other areas, further bolstering the security architecture.

By integrating these diverse yet complementary components, the Zero Trust model offers a layered and highly effective approach to access control.

Implementing DID for User Authentication Within Zero Trust

Multi-Factor Authentication (MFA) serves as a foundational element in Zero Trust frameworks, particularly when augmented by traditional username and password (ID/PW) systems. Known as a default-deny strategy, this approach allows entry only to users who have navigated through a series of rigorous authentication mechanisms. The significance of pre-defined policies is paramount in this context, as they facilitate the real-time enforcement of security measures.

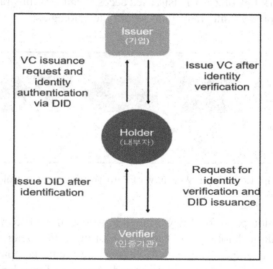

Fig. 2. Comprehensive Structure of DID-based Authentication

Various methods contribute to the MFA process. The traditional ID/PW remains the elementary step, usually acting as the first factor in this multi-tiered authentication sequence. One Time Pad (OTP) serves as an additional layer, commonly delivered as a temporary code to a user's mobile device. Biometric verification, which could include fingerprint scanning or facial recognition, adds yet another dimension. DIDs represent a more modern innovation in this lineup. Built on blockchain technology, DIDs offer heightened security through a decentralized mode of identity verification.

As illustrated in Fig. 2, the authentication journey initiates with users obtaining their DIDs from trustworthy verification agencies. Once equipped with these DIDs, users are in a position to request a Verifiable Credential (VC) from their organization, establishing their status as an insider. After the organization confirms the user's identity, it issues the VC, enabling the user to access specified organizational resources.

The process culminates when users, attempting to access internal systems, must furnish a Verifiable Presentation (VP) rooted in their pre-issued VC. This VP encompasses only the indispensable data requisite for the present authentication effort. The organization cross-references this VP with its established policy guidelines, ensuring that users gain the least-privilege access, thus maintaining high-security standards.

By adopting DID-based authentication in this intricate manner, organizations not only elevate their security protocols but also achieve finer control over resource accessibility. This method markedly surpasses the security and efficiency offered by traditional ID/PW authentication systems.

5   Empirical Analysis in South Korea

5.1   Zero Trust in the Public Sector

The implementation of Zero Trust models in the public sector has garnered significant attention, primarily due to the critical responsibilities that governmental agencies hold in governance and public safety. In the United States, the Biden administration has made it a priority to improve the nation's cybersecurity posture [18]. This effort aims to replace the traditional perimeter-based security models with Zero Trust architectures, a move motivated by a wave of sophisticated cyber-attacks that have exposed vulnerabilities in government systems.

The Cybersecurity and Infrastructure Security Agency (CISA), a key agency in the United States, has been pivotal in this transition. Not only has it published comprehensive guidelines, but it has also spearheaded pilot projects to demonstrate the efficacy of Zero Trust principles. The approach is multifaceted, involving not just strong authentication processes like multi-factor authentication but also real-time monitoring and continuous verification mechanisms. The latter are designed to confirm identities dynamically and to scrutinize data transfers and user behaviors for any anomalies.

State and local governments are not far behind in adopting this security model. For example, the state of California has initiated projects to evaluate how Zero Trust can protect citizen data and critical infrastructure like power grids and water supplies [18]. These pilot projects are vital in understanding the complexities and challenges in applying Zero Trust models at different levels of government.

Internationally, countries like the United Kingdom and Canada have also been proactive in embracing Zero Trust principles [19]. The U.K.'s National Cyber Security Centre (NCSC) has endorsed Zero Trust as a viable strategy for public sector organizations and has provided its own set of guidelines and best practices. Canada has a similar approach and has been focusing on integrating Zero Trust measures in healthcare, a sector often targeted by cybercriminals. These international cases provide valuable perspectives on how Zero Trust can be tailored to meet country-specific requirements and threat landscapes.

5.2   Zero Trust in South Korea: A Comprehensive Approach

South Korea's emergence as a technological juggernaut has not only led to innovations but also to new kinds of vulnerabilities. As part of its national cybersecurity strategy, South Korea is making strides in developing and adopting Zero Trust security models. The Ministry of Science and ICT, in collaboration with the Korea Internet & Security Agency (KISA), has been spearheading these initiatives, taking a multi-pronged approach to fortify the nation's digital infrastructure.

Policy Forum: A Catalyst for Change

One of the cornerstone initiatives has been the establishment of the Policy Forum on Zero Trust and Supply Chain Security. Unlike conventional forums that serve merely as discussion platforms, this forum aims to enact tangible change. Comprising cybersecurity experts, policymakers, and industry leaders, the forum has been instrumental in setting the strategic direction for Zero Trust adoption in the country. Through roundtable discussions and white papers, the forum outlines the roadmap for transitioning to a Zero Trust architecture, setting key performance indicators (KPIs) and timelines.

Sector-Specific Pilot Programs

The forum's recommendations have been swiftly acted upon, with the launch of several pilot programs targeting sectors that are vital for national security and public welfare. In the energy sector, the focus has been on safeguarding the smart grid systems that control electricity distribution. In healthcare, the attention is on protecting patient records and critical medical systems. Public transportation systems are also being looked at, particularly the control systems for subways and rail networks. These pilots are comprehensive studies to evaluate the scalability, effectiveness, and return on investment (ROI) of implementing Zero Trust principles.

Leveraging Emerging Technologies

South Korea is also on the cutting edge when it comes to synergizing Zero Trust models with emerging technologies like Artificial Intelligence (AI) and blockchain. The government is exploring the potential of AI-powered analytics to identify anomalies in network traffic, enabling real-time mitigation of cyber threats. AI's predictive capabilities can also help preemptively identify vulnerabilities in the system, allowing for proactive rather than reactive security measures. Blockchain technology is being tested for its applicability in creating immutable and secure identity verification processes. By storing identity data on a decentralized ledger, South Korea aims to negate the risks associated with centralized identity databases, which are often the targets of cyber attacks.

Public-Private Partnerships and Incentives

Realizing that a resilient cybersecurity posture is a collective responsibility, the South Korean government has incentivized private corporations to adopt Zero Trust architectures. A range of incentives, including but not limited to, tax benefits, grants, and public recognition awards have been established to encourage companies to invest in new security technologies. Through these public-private partnerships, South Korea intends to create a robust cybersecurity ecosystem that can withstand the challenges posed by an increasingly complex threat landscape.

Geopolitical Context and International Collaboration

South Korea's focus on cybersecurity also has a geopolitical dimension. Given its proximity to nations with varying cybersecurity postures, South Korea aims to set a benchmark for cybersecurity in the region. To that end, the government is actively participating in international cybersecurity forums and bilateral discussions, sharing its Zero Trust frameworks and learning from the experiences of other nations.

In summary, South Korea's approach to implementing Zero Trust models is both exhaustive and flexible, adapting to the needs of different sectors and types of organizations. With a forward-thinking policy forum, sector-specific pilot programs, a focus on technological innovation, and a robust public-private partnership model, South Korea aims to build a cybersecurity infrastructure that is resilient, agile, and capable of meeting the challenges of today and tomorrow.

In Summary

The initiatives undertaken in the public sector of the United States and the comprehensive strategies in South Korea exemplify the depth and breadth of Zero Trust implementation across different organizational and geopolitical contexts. These case studies demonstrate the adaptability of Zero Trust models in addressing a range of security challenges, from safeguarding national critical infrastructures to protecting sensitive data. Both nations showcase how Zero Trust can be more than a theoretical concept, proving its practical efficacy in enhancing cybersecurity postures significantly. These real-world implementations provide a robust foundation of lessons learned and best practices that can guide other organizations, public or private, considering the leap to a more secure, Zero Trust environment.

6 Conclusion and Future Directions

The adoption of Zero Trust Architecture represents a revolutionary shift in the field of information security management, offering a nuanced and adaptive strategy against evolving cyber threats. This paper has demonstrated that K-ISMS, South Korea's national standard for information security, could be significantly fortified through the integration of Zero Trust principles. In particular, we have highlighted the benefits of incorporating advanced authentication mechanisms like Multi-Factor Authentication (MFA) and Decentralized Identifiers (DID) into existing ISMS protocols.

While our focus has been on South Korea, the implications of our findings have global relevance, especially for nations seeking to update their cybersecurity frameworks to address both current and emerging threats. Moreover, we believe that the deployment of Zero Trust can be highly beneficial across various sectors, not only in technology but also in areas such as healthcare, finance, and critical national infrastructure.

By sharing our insights, we aim to contribute to the ongoing discourse on how to better protect our digital ecosystems in an increasingly interconnected yet perilous world.

Future research could explore the economic feasibility of transitioning to Zero Trust, as well as develop metrics for evaluating its effectiveness post-implementation. Longitudinal studies can offer more definitive conclusions on the impact of Zero Trust on cybersecurity resilience over time. Furthermore, the interplay between Zero Trust and data protection regulations like GDPR can be a rich avenue for future investigation.

Due to constraints related to the length of this paper as well as the need to maintain the confidentiality of sensitive information from various organizations, we were unable to include detailed numerical data or conduct in-depth statistical analyses pertaining to specific systems, companies, or agencies. However, we recognize the importance of empirical validation for our findings. As a result, our future research endeavors will focus on thoroughly exploring the real-world implications of adopting a zero-trust framework in digital systems. We aim to collaborate with relevant stakeholders to collect more comprehensive data and provide a more nuanced understanding of the challenges and benefits associated with implementing zero-trust architectures.

References

1. Migeon, J.-H., Bobbert, Y.: Leveraging zero trust security strategy to facilitate compliance to data protection regulations. In: Arai, K. (ed.) Intelligent Computing: Proceedings of the 2022 Computing Conference, Volume 3, pp. 847–863. Springer International Publishing, Cham (2022). https://doi.org/10.1007/978-3-031-10467-1_52
2. García-Teodoro, P., Camacho, J., Maciá-Fernández, G., Gómez-Hernández, J.A., López-Marín, V.J.: A novel zero-trust network access control scheme based on the security profile of devices and users. Comput. Netw. **212**, 109068 (2022)
3. KISA [Internet]. https://isms.kisa.or.kr/main/csap/intro/
4. FedRAMP [Internet]. https://www.fedramp.gov/cloud-service-providers/
5. Lee, S., Park, M., Lee, S., Park, W.: Security enhancement through comparison of domestic and overseas cloud security policies. In: Proceedings of the Korea Information and Communication Engineering General Academic Conference, pp. 268–270. Gunsan (2021)
6. Kang, B., Kim, S.J.: Study on security grade classification of financial company documents. J. Korea Inst. Inform. Secur. Cryptol. **24**(6), 1319–1328 (2014)
7. Kim, Y.-D., Kim, J.-S., Lee, K.-H.: Major issues of the national cyber security system in south Korea, and its future direction. Korean J. Defense Anal. 435–455 (2013)
8. ICOO Developing and Using Security Classification Guides [Internet]. https://www.nsa.gov/portals/
9. Act on the development of Cloud Computing and Protection of Its users (2023). https://law.go.kr/LSW/admRulInfoP.do?admRulSeq=2100000218804
10. DOD Releases Path to Cyber Security Through Zero Trust Architecture [Internet]. https://www.defense.gov/News/News-Stories/Article/Article/3229211/dod-releases-path-to-cyber-security-through-zero-trust-architecture/
11. Buck, C., Olenberger, C., Schweizer, A., Völter, F., Eymann, T.: Never trust, always verify: a multivocal literature review on current knowledge and research gaps of zero-trust. Comput. Secur. **110**, 02436, ISSN 0167-4048 (2021)
12. Zaid, B., Sayeed, A., Bala, P., Alshehri, A., Alanazi, A.M., Zubair, S.: Toward secure and resilient networks: a zero-trust security framework with quantum fingerprinting for devices accessing network. Mathematics **11**, 2653 (2023)
13. Dumitru, I.-A.: Zero trust security. In: Proceedings of the International Conference on Cybersecurity and Cybercrime-2022. Asociatia Romana pentru Asigurarea Securitatii Informatiei (2022)

14. Giovanna, C., et al.: The ISO/IEC 27001 information security management standard: literature review and theory-based research agenda. TQM J. **33.7**, 76–105 (2021)
15. Hong, S., Park, S., Park, L.W., Jeon, M., Chang, H.: An analysis of security systems for electronic information for establishing secure internet of things environments: focusing on research trends in the security field in South Korea. Futur. Gener. Comput. Syst. **82**, 769–782 (2018)
16. Kim, H., Lee, K., Lim, J.: A study on the impact analysis of security flaws between security controls: an empirical analysis of K-ISMS using case-control study. KSII Trans. Internet Inform. Syst. **11**(9) (2017)
17. Kortesniemi, Y., Lagutin, D., Elo, T., Fotiou, N.: Improving the privacy of IoT with decentralised identifiers (dids). J. Comput. Networks Commun. (2019)
18. Casey, B.M.: Cybersecurity Policies of the United States: An Analysis of Political and National Defense Motivations influencing Its Constant Development (Doctoral dissertation, Utica University) (2022)
19. Brett, M.: Zero trust computing through the application of information asset registers. Cyber Secur. Peer-Review. J. **5**(1), 80–94 (2021)

An AIoT Device for Raising Awareness About Trash Classification at Source

Ngoc-Sang Vo[1,2], Ngoc-Thanh-Xuan Nguyen[1,2], Gia-Phat Le[1,2],
Lam-Tam-Nhu Nguyen[1,2], Ho Tri Khang[1,2], Tien-Phat Tran[1,2],
and Hoang-Anh Pham[1,2(✉)] (iD)

[1] Ho Chi Minh City University of Technology (HCMUT), 268 Ly Thuong Kiet
Street, Ward 14, District 10, Ho Chi Minh City, Vietnam
anhpham@hcmut.edu.vn
[2] Vietnam National University Ho Chi Minh City (VNU-HCM), Linh Trung Ward,
Ho Chi Minh City, Vietnam

Abstract. Waste segregation is a critical issue for environmental protection and sustainable growth. In Vietnam, public awareness and action on waste separation at source remain limited, highlighting the importance of engaging individuals, particularly students, in transforming waste disposal practices. Modern technologies, including the Internet of Things (IoT) and Artificial Intelligence (AI), have revolutionized various aspects of our lives and offer promising solutions to raise public awareness on this issue. This paper proposes an IoT device named BEG (BACHKHOA Eco-friendly Guide) integrating AI-based Computer Vision technology to classify waste via a camera. Unlike existing smart trash cans, which classify and dispose of the trash automatically, our device provides information about the waste type to guide users on proper disposal, thus reinforcing awareness of garbage classification at source. We also introduce the BEGNet, a Convolutional Neural Network (CNN) employing RegNetY120 as its backbone, which demonstrates superior performance in accuracy compared to other approaches on both the Trashnet dataset and our custom dataset - BKTrashImage. The proposed BEG device will improve knowledge about waste segregation, reduce improperly disposed waste, and foster a thriving circular economy.

Keywords: Artificial Intelligence · Computer Vision · Internet of Things · Human Awareness · Waste Segregation

1 Introduction

In the modern era, the utilization of science and technology, including artificial intelligence (AI), has become indispensable across various aspects of life. Deep learning, a subfield of artificial intelligence (AI), has witnessed remarkable advancements and emerged as a powerful tool in multiple applications. Its ability to automatically learn and extract complex patterns from large datasets has led to breakthroughs in many domains, including computer vision, natural language processing, and speech recognition.

Promoting sustainable practices and mitigating the environmental impact of improper waste management requires individuals worldwide to be aware of waste sorting at source. Vietnam, however, still lacks awareness and action in this area. In [14], the survey results conducted on 1513 students at the Vietnam Maritime University showed that 95.85% of the students have an understanding of the harmful effects of single-use plastic, but only 57% of the students could identify groups of waste sorting. The student's concern and sense of the meaning of plastic codes/labels are still limited. Based on this survey results, most students are aware of the harms of plastic waste, but this awareness is still insufficient. Therefore, it is essential to implement propaganda, dissemination, and education to raise students' awareness about waste sorting, reduction, reuse, recycling, and treatment.

In that scenario, the integration of cutting-edge technologies holds significant promise in promoting sustainable practices. This study introduces a revolutionary prototype called BEG (Bachkhoa Eco-friendly Guide) that utilizes computer vision and IoT to educate people about environmental issues and improve waste management. This study makes the following significant contributions. Firstly, we develop the BEGNet, an efficient trash image classification model. Secondly, we propose an IoT-based architecture for effectively implementing and deploying the trash bin system. Thirdly, we build a BKTrashImage dataset containing 6205 images of five categories: paper cups, aluminum cans, milk boxes, PET bottles, and foam boxes. Finally, we conduct experiments to demonstrate the proposed model's effectiveness compared to state-of-the-art trash classification approaches on an existing Trashnet dataset and our BKTrashImage dataset.

2 Related Works

2.1 Trash Image Classification on the Trashnet Dataset

Several studies have focused on developing robust Convolutional Neural Network (CNN) models for trash image classification using the Trashnet dataset [17], which includes six categories: glass, paper, cardboard, plastic, metal, and trash. The dataset was captured using various mobile devices, such as the Apple iPhone 7 Plus, Apple iPhone 5S, and Apple iPhone SE.

Aral et al. [2] explored transfer learning models based on popular CNN architectures, including Densenet121, DenseNet169, InceptionResnetV2, MobileNet, and Xception, for trash classification on the Trashnet dataset. They split the dataset into 70% for training, 13% for validation, and 17% for testing, using a batch size of 8 and an input size of 224 × 224. The Densenet121 model achieved the highest accuracy of 95%.

Ruiz et al. [12] investigated the effectiveness of various CNN models, such as VGG, Inception, and ResNet, for automatic trash categorization. They used 80% of the Trashnet dataset for training, 10% for validation, and the remaining 10%

for testing. The Inception-ResNet combination achieved an average accuracy of 88.66% on the Trashnet dataset.

Vo et al. [15] proposed a robust Trash Classification model (DNN-TC) using Deep Neural Networks. They modified the original ResNext-101 model by adding two fully connected layers after the global average pooling layer. The authors used 60% of the Trashnet dataset for training, 20% for validation, and the remaining 20% for testing. DNN-TC achieved an accuracy of 94%.

Lam et al. [10] used computer vision technology to transform a conventional trash bin into an intelligent one. Their method, MobileNetV2, achieved 85.59% and 71.00% accuracy on Dataset 1 and Dataset 2, respectively.

In addition to the above-mentioned methods for trash classification, several well-known CNN models, such as RegNet [4,11], ResNext [16], ImageNet [9], VGG [13], ResNet [5], and DenseNet [7], which were originally developed for image classification, can also be used as baseline models for trash classification.

2.2   Previous Studies on Smart Trash Cans

Previous studies on smart waste management systems have highlighted the potential of integrating advanced technologies, such as the Internet of Things (IoT) and Artificial Intelligence (AI), to modernize waste management practices. Lam et al. [10] proposed a trash bin system architecture that utilizes a camera to capture waste images and a central processing unit to analyze them and determine which bin is appropriate for disposal. Ali et al. [1] introduced an IoT-based smart waste bin monitoring and municipal solid waste management system that addresses the increasing challenges of waste generation in developing countries due to population growth and urbanization. Their system can detect fires within the bins using ultrasonic sensors (HC-SR04) and temperature-humidity sensors (HW-505), potentially saving lives and reducing economic losses. Kanade et al. [8] developed an intelligent garbage monitoring system that uses an ultrasonic sensor to track waste levels, with data uploaded to the Cloud via the Firebase platform, allowing users to monitor the fill status of garbage bins remotely.

In general, substantial progress has been made in trash image classification, with various studies developing robust CNN models tailored for the Trashnet dataset [17]. These studies have achieved high accuracies, indicating the potential for successful application in smart waste management systems. However, despite significant advancements in smart waste management research [1,8,10], current systems still lack features that could enhance users' awareness of upstream trash classification, an essential aspect of waste management.

3   The Proposed BEG and Its Implementation

3.1   The Proposed Trash Image Classification Model (BEGNet)

RegNetX and RegNetY, introduced in [11], offer a network design space aimed at decreasing computation and the number of epochs needed for training. This design space has the added benefit of being more easily interpretable. RegNetY specifically refers to a convolutional network design space that generates simple and regular models with parameters including depth (d), initial width ($w_0 > 0$), and slope ($w_a > 0$), and produces a unique block width for each block. The RegNet model is subject to a key constraint, which is a linear parameterization of block widths, as specified in Eq. 1.

$$u_j = w_0 + w_a \cdot j \qquad (1)$$

In addition to RegNetX, a modification for RegNetY that involves incorporating Squeeze-and-Excitation blocks was also introduced in [6]. These blocks are a type of architectural unit designed to enhance the representational capabilities of a network by allowing for dynamic channel-wise feature recalibration. The structure of the Squeeze-and-Excitation block is illustrated in Fig. 1, and its operation can be described as follows:

- A convolutional block is taken as input.
- The input channels are "squeezed" into a single scalar value using average pooling.
- A dense layer, followed by a ReLU activation function, is applied to add non-linearity, and the output channel complexity is reduced by a certain ratio.
- Another dense layer, followed by a sigmoid activation function, produces a smooth gating function for each channel.
- Finally, the feature maps of the convolutional block are weighted based on the output of the side network, which is referred to as the "excitation".

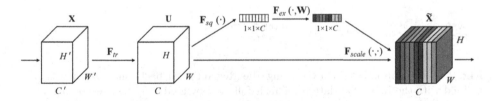

Fig. 1. A Squeeze-and-Excitation block.

To classify images, a robust framework for trash classification, namely BACHKHOA Eco-friendly Guide Network (BEGNet), which utilizes Deep Neural Networks, is introduced. The BEGNet employs the RegNetY120 pre-trained on the ImageNet dataset [3] as its backbone, followed by a Global Average Pooling layer, a Dropout layer, a Dense layer indicating the number of classes, and finally a customized activation function that is inspired by the Sigmoid activation function and defined by Eq. 2.

$$Customized\ Activation\ Function = \frac{1}{1 + e^{-\frac{1}{5}x}} \qquad (2)$$

The selection of the Dropout Rate and Customized Activation Function configurations is based on empirical evaluation, as discussed in Sect. 4. Figure 2 provides further clarification on our proposed method, where N_class represents the number of trash labels.

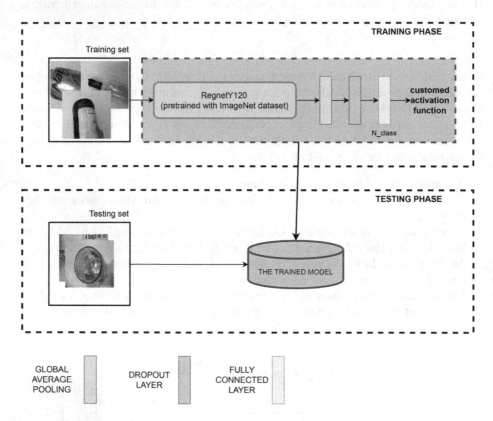

Fig. 2. BEG's approach involves using the RegNetY120 backbone, which was pretrained with the ImageNet dataset. This is followed by a Global Average Pooling layer, a Dropout layer, and a Sigmoid activation function.

3.2   The Proposed IoT-Based Architecture for Prototyping BEG Device

Figure 3 depicts the workflow of our proposed BEG system that utilizes an IoT-based architecture. The system comprises three interconnected components: smart trash bins, a cloud-hosted database, and a mobile application, all communicating in real-time via the Internet.

The smart trash bins include three separate bins for each type of waste, as well as the following primary modules: a Jetson Nano embedded computing board,

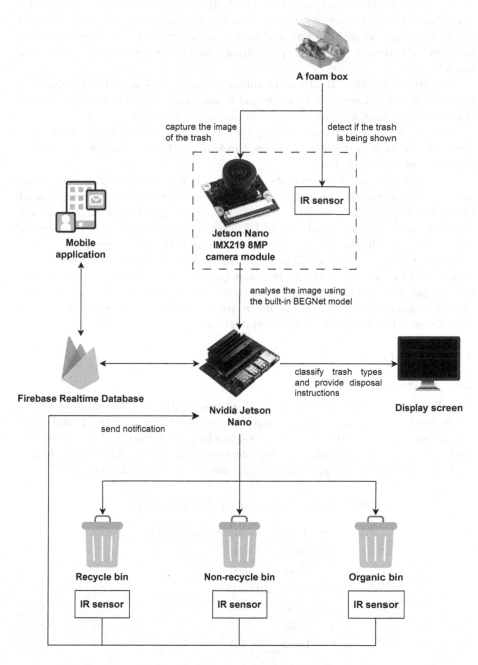

Fig. 3. The workflow of BEG.

a Jetson Nano IMX219 8MP camera module, a 1024×600 10.1-inch capacitive touch screen LCD display, and four HC-SR04 ultrasonic sensors. Each of the three bins has one ultrasonic sensor to measure waste levels based on the distance inside the bins, while an additional sensor is attached to the camera module to detect the presence of trash, preventing constant image feeding into the AI models.

We utilize Firebase Realtime Database as the cloud-hosted database. This platform enables developers to store and synchronize data among users in real-time, ensuring seamless communication between system components.

For the mobile application, we employ Flutter as the development framework due to its compatibility with the chosen Firebase platform, providing a consistent and efficient user experience.

The product workflow consists of several key steps. First, the IR sensor detects the presence of trash in front of the camera, triggering the camera module to capture an image and temporarily store it in the Jetson Nano's storage. The Jetson Nano then employs our BEGNet model to analyze the image and obtain the classification result. Upon determining the trash type, the system displays guidance for proper disposal on the screen. Additionally, an IR sensor is installed in each trash bin to measure the volume of waste present. Once a bin reaches its capacity, a signal is transmitted to the real-time database, which updates the mobile application to notify users immediately. Figure 3 provides a visual representation of the entire architecture.

4   Evaluation

4.1   Datasets

The Trashnet dataset consists of 2527 images captured using mobile devices, categorized into six classes: glass, paper, cardboard, plastic, metal, and trash. The images showcase the objects against a white background, illuminated with sunlight and/or room lighting. Table 1 displays the image statistics for each class, while Fig. 4 exhibits sample images from each class within the dataset.

Table 1. The statistic of the Trashnet dataset.

No	Classes	Number of images
1	Glass	501
2	Paper	594
3	Cardboard	403
4	Plastic	482
5	Metal	410
6	Trash	137

In order to tackle the issue of increasing students' understanding and awareness of source-based garbage classification, our focus is on developing a comprehensive dataset that includes classes representing trash commonly disposed of by students. To accomplish this, we combine the Trashnet dataset [17] with additional images sourced from the Internet and manually captured images of trash items. The resulting dataset, named BKTrashImage, comprises six classes: Paper Cup, Aluminum Can, Milk Box, PET Bottle, and Foam Box, totalling 6205 images. We strive to maintain a balanced distribution of classes in the dataset. Table 2 illustrates the distribution of the BKTrashImage dataset.

Fig. 4. Samples of Trashnet dataset.

To have more insight into how BKTrashImage looks, Fig. 5 shows some samples of each class of the dataset.

Table 2. The statistic of the BKTrashImage dataset.

No	Classes	Number of images
1	Paper Cup	1556
2	Aluminum Can	1531
3	Milk box	1624
4	PET bottle	1499
5	Foam box	1395

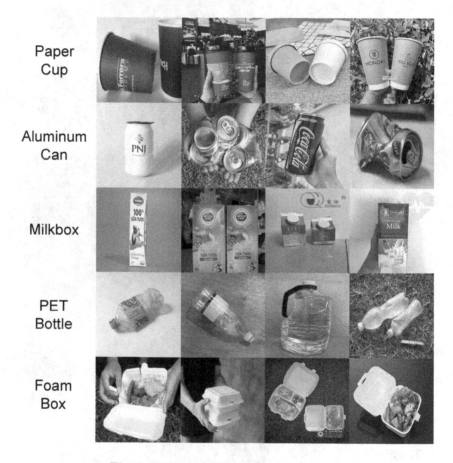

Fig. 5. Samples of BKTrashImage dataset.

4.2 Experimental Setting

In this section, we present a quantitative assessment of the proposed image classification approach (BEGNet) and other methods, using two experimental datasets, Trashnet and BKTrashImage, as introduced in the previous section.

We reserve 80% of each dataset for training and the remaining 20% for testing purposes.

The experimental methods were implemented using Python 3.10.11 and executed on the TensorFlow framework, a free software library for deep learning in the Python programming language. The operating system utilized was Ubuntu 20.04.5 LTS, with a GPU NVIDIA A100-SXM4-40GB and 83.48 GB of RAM.

We employ state-of-the-art methods for trash classification, as referenced in Sect. 2, including Densenet121_Aral [2], Inception-ResNet_Ruiz [12], MobileNetV2_Lam [10], and DNN-TC_Vo [15]. We compare these approaches with our proposed method for trash image classification. During the preprocessing phase preceding training, the images are resized to 224×224 and normalized by dividing each pixel value by 255, facilitating more convenient computation. Stochastic Gradient Descent (SGD) is used as the optimization algorithm for the competing methods, with a learning rate α set at 0.0001. We also utilize a batch size of 128 and train models over 20 epochs, evaluating their performance on the testing set after each epoch. During the testing phase, we compare the experimental methods based on their classification performance on the testing set.

4.3 Experimental Results

BEGNet Model Parameter Selection. To begin with, we present experimental results regarding the selection of the Dropout Rate and activation function for the BEGNet model. The performance of BEGNet, in terms of accuracy on the Trashnet dataset, was evaluated based on the joint settings of these two variables. We tested four discrete values of Dropout Rate: 0.0, 0.2, 0.5, and 0.8. Additionally, we examined four activation functions: Softmax, Sigmoid, Hard Sigmoid, and a self-customized activation function. The initial experiment indicated that BEGNet achieved optimal performance with a 0.2 Dropout Rate and the self-customized activation function. Table 3 presents the accuracy of BEGNet for each joint setting.

Table 3. The accuracy of BEGNet on each joint setting of Dropout Rate values and activation functions.

	Dropout Rate			
	0.0	0.2	0.5	0.8
Sigmoid	94.26	94.46	93.86	93.86
Softmax	93.47	94.06	94.06	93.47
Hard Sigmoid	89.31	90.50	88.51	91.41
Customized Activation Function	94.65	**95.45**	94.85	93.86

The accuracy of the proposed BEGNet model, as well as that of other experimental methods, when evaluated on the Trashnet dataset, is displayed in

Table 4. Our proposed approach outperforms the competing methods, with an accuracy of 95.45%. In comparison, Densenet121_Aral, Inception-ResNet_Ruiz, DNN-TC_Vo, and MobileNetV2_Lam obtained accuracies of 93.66%, 93.27%, 93.07%, and 88.12%, respectively.

Table 4. The accuracy of the experimental methods on the Trashnet dataset.

No	Methods	Accuracy on the testing set (%)
1	Densenet121_Aral [2]	93.66
2	Inception-ResNet_Ruiz [12]	93.27
3	MobileNetV2_Lam [10]	88.12
4	DNN-TC_Vo [15]	93.07
5	BEGNet	**95.45**

To further demonstrate the effectiveness of our proposed approach, we evaluate the experimental methods on the BKTrashImage dataset, which includes types of trash commonly discarded by students. As shown in Table 5, our method achieved the highest accuracy of 97.17% compared to all other experimental methods. The next best performing models were Inception-ResNet_Ruiz (96.45%) and Densenet121_Aral (96.32%). Although MobileNetV2_Lam and DNN-TC_Vo achieved high accuracies of 95.73% and 95.79%, respectively, they still fell short by over 1% compared to our proposed model.

Table 5. The accuracy of the experimental methods on the BKTrashImage dataset.

No	Methods	Accuracy on the testing set (%)
1	Densenet121_Aral [2]	96.32
2	Inception-ResNet_Ruiz [12]	96.45
3	MobileNetV2_Lam [10]	95.73
4	DNN-TC_Vo [15]	95.79
5	BEGNet	**97.17**

4.4 Discussion

To demonstrate the effectiveness of our approach, we compared it with state-of-the-art methods on both the Trashnet and BKTrashImage datasets. Trashnet is a smaller dataset in which most images contain a single object. Meanwhile, the BKTrashImage dataset focuses on trash items frequently discarded by students and follows the same format. We empirically evaluated BEGNet on both datasets and achieved notable results.

For the Trashnet dataset, our approach achieved an accuracy rate of 95.45%, outperforming other approaches. For the BKTrashImage dataset, our approach achieved the highest accuracy of 97.17%. These results demonstrate our proposed approach's effectiveness and potential for practical applications.

5   Conclusion

This paper presents an AIoT device that integrates AI-based Computer Vision technology for waste classification. The most significant contribution is to develop the BEGNet model by modifying the RegNetY120 architecture to enhance classification performance. The proposed BEGNet surpasses other state-of-the-art methods in accuracy on both the Trashnet and our BKTrashImage datasets. We also design an IoT-based architecture for the practical implementation and deployment of the trash bin system.

Distinct from existing smart waste management systems that automatically classify and dispose trash, our approach classifies waste and provides users with guidance on proper disposal, raising their awareness of source-based garbage classification. Therefore, the proposed solution, BACHKHOA Eco-friendly Guide (BEG), could offer significant environmental, societal, and economic benefits. The proposed system promotes behavioral change by raising students' awareness of trash classification at the source, positioning students to play a critical role in transforming society's waste disposal practices.

In conclusion, BEG is actively addressing waste segregation challenges by fostering environmental awareness among students, but continuous efforts are needed for further optimization. The critical areas for improvement include expanding the datasets to cover a broader range of waste categories for a more comprehensive approach, modifying the Deep Neural Network to enable the detection of multiple types of trash at once instead of the current item-by-item process, and refining the mobile application to enhance stakeholder interaction in downstream waste management, which are critical steps towards advancing this initiative.

Acknowledgement. This research is funded by the Office for International Study Programs (OISP) under grant number SVOISP-2023-KH&KTMT-46. We acknowledge Ho Chi Minh City University of Technology (HCMUT), VNU-HCM for supporting this study.

References

1. Ali, T., Irfan, M., Alwadie, A.S., Glowacz, A.: IoT-based smart waste bin monitoring and municipal solid waste management system for smart cities. Arab. J. Sci. Eng. **45**, 10185–10198 (2020)
2. Aral, R.A., Keskin, Ş.R., Kaya, M., Hacıömeroğlu, M.: Classification of TrashNet dataset based on deep learning models. In: 2018 IEEE International Conference on Big Data (Big Data), pp. 2058–2062. IEEE (2018)

3. Deng, J., Dong, W., Socher, R., Li, L.J., Li, K., Fei-Fei, L.: ImageNet: a large-scale hierarchical image database. In: 2009 IEEE Conference on Computer Vision and Pattern Recognition, pp. 248–255. IEEE (2009)
4. Dollár, P., Singh, M., Girshick, R.: Fast and accurate model scaling. In: Proceedings of the IEEE/CVF Conference on Computer Vision and Pattern Recognition, pp. 924–932 (2021)
5. He, K., Zhang, X., Ren, S., Sun, J.: Deep residual learning for image recognition. In: Proceedings of the IEEE Conference on Computer Vision and Pattern Recognition, pp. 770–778 (2016)
6. Hu, J., Shen, L., Sun, G.: Squeeze-and-excitation networks. In: Proceedings of the IEEE Conference on Computer Vision and Pattern Recognition, pp. 7132–7141 (2018)
7. Huang, G., Liu, Z., Van Der Maaten, L., Weinberger, K.Q.: Densely connected convolutional networks. In: Proceedings of the IEEE Conference on Computer Vision and Pattern Recognition, pp. 4700–4708 (2017)
8. Kanade, P., Alva, P., Prasad, J.P., Kanade, S.: Smart garbage monitoring system using internet of things (IoT). In: 2021 5th International Conference on Computing Methodologies and Communication (ICCMC), pp. 330–335. IEEE (2021)
9. Krizhevsky, A., Sutskever, I., Hinton, G.E.: ImageNet classification with deep convolutional neural networks. Commun. ACM **60**(6), 84–90 (2017)
10. Lam, K.N., et al.: Using artificial intelligence and IoT for constructing a smart trash bin. In: Dang, T.K., Küng, J., Chung, T.M., Takizawa, M. (eds.) FDSE 2021. CCIS, vol. 1500, pp. 427–435. Springer, Singapore (2021). https://doi.org/10.1007/978-981-16-8062-5_29
11. Radosavovic, I., Kosaraju, R.P., Girshick, R., He, K., Dollár, P.: Designing network design spaces. In: Proceedings of the IEEE/CVF Conference on Computer Vision and Pattern Recognition, pp. 10428–10436 (2020)
12. Ruiz, V., Sánchez, Á., Vélez, J.F., Raducanu, B.: Automatic image-based waste classification. In: Ferrández Vicente, J.M., Álvarez-Sánchez, J.R., de la Paz López, F., Toledo Moreo, J., Adeli, H. (eds.) IWINAC 2019. LNCS, vol. 11487, pp. 422–431. Springer, Cham (2019). https://doi.org/10.1007/978-3-030-19651-6_41
13. Simonyan, K., Zisserman, A.: Very deep convolutional networks for large-scale image recognition. arXiv preprint: arXiv:1409.1556 (2014)
14. Thi, D.P., Thuy, H.D.T.: Survey and assessment of the students' awareness and behaviors about waste sorting and single use plastic consumption habits. https://jmst.vimaru.edu.vn/index.php/tckhcnhh/article/view/21
15. Vo, A.H., Vo, M.T., Le, T., et al.: A novel framework for trash classification using deep transfer learning. IEEE Access **7**, 178631–178639 (2019)
16. Xie, S., Girshick, R., Dollár, P., Tu, Z., He, K.: Aggregated residual transformations for deep neural networks. In: Proceedings of the IEEE Conference on Computer Vision and Pattern Recognition, pp. 1492–1500 (2017)
17. Yang, M., Thung, G.: Classification of trash for recyclability status. CS229 Proj. Rep. **2016**(1), 3 (2016)

A Big Data Approach for Customer Behavior Analysis in Telecommunication Industry

Hong-Phuc Vo[1,2], Khoa-Gia-Cat Nguyen[1,2], Kim-Loc Nguyen[1,2], and Thanh-Van Le[1,2(✉)]

[1] Ho Chi Minh City University of Technology (HCMUT), Ho Chi Minh, Vietnam
`ltvan@hcmut.edu.vn`
[2] Vietnam National University Ho Chi Minh City (VNU-HCM),
Ho Chi Minh, Vietnam

Abstract. The evolution of telecommunications has led to a profound transformation in the realm of communication, revolutionizing how this industry mines customer behavior for their business outcomes. The analysis of user historical activities promoted paramount importance in driving strategic decision-making to enhance customer experiences and recommend ways to attract customers more effectively. While the demand is growing, some telecom data analytics either use small datasets or provide a high abstract level of analysis result. When the number of customers increases significantly, it becomes impractical to customize service for each customer under the same approach. This paper provides a comprehensive examination of the challenges, needs, and solutions associated with the analysis of user data within the telecom domain. We focus on three key user data analysis problems: user clustering, user classification, and revenue prediction derived from user insights. With Florus - our proposed big data framework, we have carried out the telecom customer behavior analysis with a large dataset. The experiment result demonstrates the promising performance and its potential for long-term use.

Keywords: Customer behavior analysis · big data framework · regression · clustering

1 Introduction

The contemporary telecommunications sector is currently experiencing an unprecedented increase in user-generated data, which shows no signs of abating. This rapid expansion of telecom data introduces a substantial degree of intricacy, presenting a spectrum of challenges and prospects for stakeholders within the industry. The sheer volume and complexity of the data being generated are fundamentally reshaping the way businesses operate and engage with their customers. Every interaction represented by voice calls, text messages, internet usage, or location tracking, contributes to the vast reservoir of data that is being amassed. What adds further complexity is the diverse range of data formats, spanning from structured, semi-structured to unstructured, further augmenting the intricate nature of the telecom data landscape.

N. Thai-Nghe et al. (Eds.): ISDS 2023, CCIS 1950, pp. 91–105, 2024.
https://doi.org/10.1007/978-981-99-7666-9_8

When seeking a business enhancement strategy, the corporations found that customer behavior analysis is the key to maintaining and developing their relationship with customers, who bring income. Undoubtedly, the e-commerce industry has a long-time experience in mining customer spending. In [11], the customer spending score predictions can be inferred through the regressions method. This work utilizes of customer spending budget and increases sale value. However, the sample of the research is only 200 customers, a small number compared to the large scale of the telecom company. Despite the simplicity, this technique may not be applicable when the volume and complexity are significantly different.

Considering the customer behavior analysis in the aforementioned domain, to win a larger market share than competitors, telecom research is mainly focused on the churn proportion, these analysis [3,9,15] expose reasons and alert the provider about the tendency of the user to terminate service, require action on services to maintain the loyalty in this sector [2], not all the potential clients are alike [11]. Relative to the focused study sample, based on telecom customers' usage, scientists can predict their demographic information [4]. In this paper, we will concentrate directly on the revenue of users from 3 primary services - service subscription, phone call, and internet usage - that most telecom company offer.

Another related research about telecom user revenue is Forecast and Analyze the Telecom Income based on ARIMA Model [14]. This research was conducted on the revenue of the provinces in China, in the format of a time series of Yen income for the company. This study represents the total revenue in general, but can not provide the granular forecast or specify the marketing strategy for each consumer due to the high abstraction level of data aggregation.

Under the large scale of dataset, the telecom industry demands a detailed, low bias, low error solution for every customer. Not only about providing insights to enhance the customer experience but also providing a general conclusion for a higher level of customer management on strategy motivation. The study about telecom customer behavior has not yet focused on their spending, which mainly contributes to the revenue of the company, for every of millions of users. In this paper, we propose a method for predicting future charged amounts for each user based on their usages. To tackle the imbalance, the regression analysis will be conducted on groups of segments, where the segment is a cluster of users. In application, to feature the decision support system, we provided a classification model, with its outcome will invoke the corresponding revenue prediction model.

As the processing data amount raising, the traditional implementation for machine learning suffers the bottleneck or exceeds the upper bound of the memory. In our experiment, the algorithm of scikit-learn can only handle a table with 2 million rows $\times$ 4 columns, much smaller than the size of this industry data. Three solutions to keep working with this approach are sampling the dataset to the lower number of records, or scaling up the system, or scaling out the system. While sampling risks the value of unmanipulated information, scaling up faces the infeasibility of physical computing resources. Therefore, scaling out the system and using the big data framework to construct a stronger cluster seems the best approach. Specifically, we will set up an Apache Spark cluster and use its algorithm library API to support our analysis process. The implementation of these setups is part of the Florus - a framework for handling large datasets.

The paper is organized in the following way. Section 2 describes the background of the Apache Spark framework. Section 3 proposes a pipeline for analysis of telecom user behavior based on big data techniques. Section 4 represents the obtained model and chosen parameters for each stage. In particular, Sect. 4.3 again evaluates the application process and testing on the unknown dataset. Finally, Sect. 5 summarizes all the conducted results and describes future work.

2 Apache Spark Framework

2.1 Architecture

Apache Spark is a multi-language engine for executing data engineering, data science, and machine learning on a single-node machine or a cluster. It is an open-source architecture that implements the MapReduce model along with Hadoop Distributed File System (HDFS). While HDFS stores the result of each MapReduce phase in the disk, Spark maintains it in memory when possible [10].

A spark application architecture is also designed to follow the Master/Slave concept where the master is called driver and the slave is called executor. When an application is started, the driver first creates a Spark Context, which acts like a gateway to access all functionalities of Spark, to connect to its cluster manager such as Yarn, Meros or Kubernetes depending on how a Spark cluster is deployed. Then, it will request the cluster manager resources and allocate some executors in the worker node [10,12].

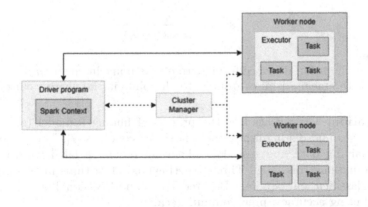

Fig. 1. Architecture of a Spark application

Figure 1 above shows the architecture of a Spark application where the dashed line shows the process to request resources, allocate the executor, and the solid lines show the process of passing data through the driver and executor as tasks.

2.2   Spark MLlib

MLlib, Spark's open-source distributed machine learning library, provides efficient functionality for a wide range of learning settings and includes several underlying statistical, optimization, and linear algebra primitives. Shipped with Spark, MLlib supports several languages and provides a high-level API that leverages Spark's rich ecosystem to simplify the development of end-to-end machine learning pipelines [8]. The MLLib provides 3 types of model algorithm API, namely: Classification, Regression, and Clustering.

Cluster Analysis. Cluster analysis or simply clustering is the process of partitioning a set of data objects (or observations) into subsets [5]. Each of these subsets contains similar objects, whose similarities are different from the other groups. A K-means model or any other clustering analysis can be evaluated by the most common metrics, which are the Silhouette score and Inertia score:

– **Inertia**: Inertia measures how internally coherent clusters are.

$$Inertia = \sum_{k=1}^{K} \sum_{x_i \in C_k} distance(x_i, c_k)^2 \tag{1}$$

Where:
  - C_k: is the Kth cluster
  - x_i: is the ith point in the C_k
  - c_k: is the centroid of C_k
– **Silhouette**: Silhouette score is used to evaluate the quality of clusters created using clustering algorithms in terms of how well samples are clustered with other samples that are similar to each other.

$$s(x_i) = \frac{b_i - a_i}{max(a_i, b_i)} \tag{2}$$

Where:
  - a_i is the mean distance from x_i to others point in cluster x_i.
  - b_i is the mean distance from x_i to all points in the nearest cluster of x_i

Classification. Classification is the process of finding a model (or function) that describes and distinguishes data classes or concepts [7]. The prerequisites include training data and test data. They all require a label feature to predict when missing this label. Three experimented algorithms in this paper are Logistic Classification, Gradient-Boosted Trees, and Decision Tree, and they are evaluated using accuracy, precision, and recall.

Regression. Regression analysis is a statistical methodology that is most often used for numeric prediction, hence the two terms tend to be used synonymously [6]. This method visualizes the distribution trends of data. The specified Linear regression or Gradient-boosted tree regression APIs can be evaluated for the application in regression.

3   Proposed Approach

User behavior analysis is vital for the management of the telecommunication industry, being considered one of the most important factors contributing to business success. The *revenue* represents the charged amounts on phone calls and internet usage of customers, two main services of this industry. This study aims to predict the spending of each user based on their recorded activities in the nearest 2 months period. Even though high benefits would come for customer relationship management, inaccurate analysis can be adverse to the marketing and sale strategy. Especially when businesses witness the number potential of customers with high revenue contributions much lower than the others. Consequently, the recommendations for this group do not satisfy these customers, not even enhance the profits but cost money and effort to lose customers.

We propose serial steps from customer clustering to revenue prediction so that the enhancement could be suitable for both the high revenue and high quantity customer groups. The approach will have 2 flows: the training process for building appropriate models and predicting process in the Decision Support System (DSS). Figure 2 illustrates the stages inside and the differences between them.

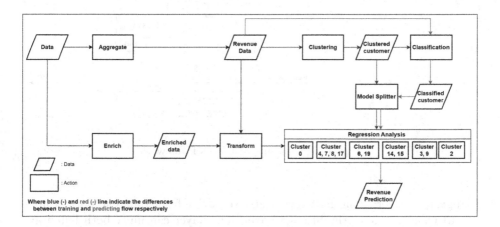

Fig. 2. Data flow of the proposed approach

In our experiment, we use over 550 million anonymized and daily aggregated records from telecom customer usages. The main focus of this dataset is package subscriptions, phone calls, internet charged fees, and top-up amounts in 3 months. There is no demographics data included to reveal user identities, as well as any raw format of user activities. Due to the large volume, we will conduct the process with our Florus framework for all relevant tasks in collecting and analyzing data.

3.1   Florus - A Framework for Handling Large Dataset

Florus is a Lakehouse architecture framework designed for handling large datasets. We design this system to support end users to ingest data from multiple sources, then process and visualize the stored data into graphics. For machine learning demand, Florus provides the interface to read data, train models, and save the result in our environment.

The system includes the user interface, a set of micro-services, and the infrastructure. This framework can apply to any dataset and does not require redesigning the architecture of Fig. 3. In general, there are 5 layers, each holds a specific function in the system.

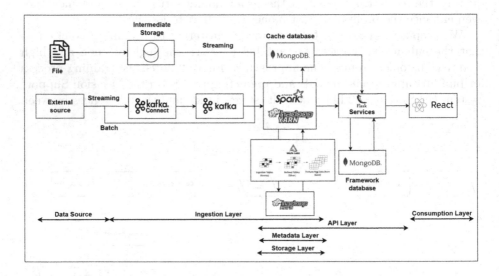

Fig. 3. Architecture of Florus framework

- Ingestion layer: This first layer retrieves data from different data sources and transport it to the Storage Layer. This layer can unify both batch and streaming processing.
  - Batch: Scheduling an event to read from the database.
  - Streaming: Connecting to the source database to stream newly appended data by Kafka and Kafka Connector. Later, data will be proceeded to streaming into HDFS by Spark Structured Streaming. Users can also upload files to our intermediate storage before starting streaming files.
- Storage layer: The Storage layer is made up of two distinct components that belong to two separate purposes:
  - Using HDFS to store data received from the Ingestion layer. The Medallion architecture supports data management in this layer by 3 levels of cleanliness: Bronze (raw data), Silver (validated, enriched), and Gold (refined, aggregated data).

- Using MongoDB to store data about the framework, including user information, project details, metadata of data sources, tables, and other framework related metadata.
- Metadata layer: This layer provides Metadata of the HDFS and furnishes management features:
  - ACID Transaction: Delta Lake framework ensures consistency as multiple parties concurrently read or write data.
  - Caching: MongoDB is used for caching to store the content of the analyzed result to speed up the query on HDFS.
- API layer: The core of this layer is the Flask component, which hosts the API endpoints. The processing can associate Spark and SparkML to read/write on HDFS or to operate the machine learning model.
- Consumption layer: ReactJS is used to provide the user interface for their interaction with the system such as setting data sources, processing data files, training machine learning models, ... In addition, Apache ECharts will help the user represent visualization for aggregated result.

3.2   Cluster Settings

Besides, the implementation under the Spark application in the above framework also requires setting up the infrastructure and configuring resources. In our experiments, we conducted the work on a Spark cluster with **one master node and three slave nodes** by the instruction of [1]. Each computing node has 32GB RAM, 1TB storage and is running the Ubuntu 22.04.1 LTS version of the operating system. Table 1 lists the important parameters in our Spark nodes.

Table 1. Configuration for Spark application

Configuration	Key	Value
Number of executors	spark.executor.instances	14
Number of cores for drivers	spark.driver.cores	3
Number of cores for each executors	spark.executor.cores	3
Size of memory used for driver	spark.driver.memory	3 GB
Size of memory used for each executors	spark.executor.memory	3 GB
Number of partitions in RDDs returned by transformations	spark.default.parallelism	126
Number of partitions to use when shuffling data for join or aggregation	spark.sql.shuffle.partition	126

3.3   Customer Behavior Analysis

For tackling our objectives, we have carried out multiple steps to develop a set of models and their correspondences to give a throughout the analysis. In order to enhance the accuracy and handle the analysis on a large number of users, the following tasks will be integrated into the pipeline:

1. Preparation:

   The user's revenue is first aggregated from data-using, calling using log by month and by week. In addition, the user's revenue is affected by their behavior, which is usually related to their subscriptions and many other relevant uses. Each plan allows users to purchase a fixed amount of non-charge services within a specific period. Customer segments represent a set of users with similar charge fees, counted calls, and internet services,... By generating multiple features, the dataset then be enriched. Additionally, the enriched dataset also requires transformation before entering the final stage by splitting into 3 subsets as follows:

   - Train set: Use for training model
   - Validate set: Use for model selection and tunning hyperparameter
   - Test set: Use for testing each step of the pipeline and the whole pipeline

2. Clustering:

   Clustering user datasets can bring numerous benefits to businesses. Firstly, grouping customers with similar purchasing behaviors helps businesses gain a better understanding of the needs and preferences of each customer segment. Secondly, this analysis allows businesses to save time and cost in customer management, and focus on high-potential customer segments to increase sales. In this approach, the users are then clustered into groups to their different charged usage. However, the cluster size does not correlate to the revenue contribution, which is the cause of the bias in the simple approach. The purpose of this stage is to find the segments and their spending distributions, which later will be the vital metrics for reducing imbalance.

3. Classification:

   In general, user classification will help businesses understand their customer base better and, as a result, develop appropriate business strategies, marketing tactics, and customer care approaches tailored to different user groups. Regarding our suggested process, the classification model uses the outcome of the clustering stage to train and aims to classify the users into equivalent segments based on their summary usages.

4. Regression:

   Due to the difference in size and revenue of these above clusters, we extract them into multi-independent models to avoid their interference in the process of other clusters. The classification model acts as the gateway for the regression stage, so the splitting rule should take into account the miss-classified behavior of this model. Based on the confusion matrix, clusters are grouped if they have a miss classified into one another in the group of more than 25% of the predicted number. Finally, the bias is ceased by the combination of multiple regression to increase the accuracy.

In applications, the classification module helps us to determine which revenue model should be used. In case the whole proposed pipeline has not been retrained, their historical usage is recorded but has never been clustered, this pipeline is still able to achieve the revenue prediction output.

We proposed an additional scheme to evaluate when applying the model in industrial-scale scope:

1. Preparation: similar to the preparation stage of the training process.
2. Classification: the input will be 1 feature lower than the clustering training set used to predict the segments of customers (from clustering). The prediction from this model will be enriched with other features and passed to the next stage.
3. Regression: using the label classified by the previous stage, the splitter will invoke the regression model trained for this segment to generate revenue prediction.

4   Experiment

While working on the models, the entire dataset will be used for the clustering. However, in the later stages, only 80% of the dataset is included in the training data and the other 10% will serve as testing data for both regression and classification models. Finally, we will perform the evaluation stage for the pipeline application on the remaining dataset.

4.1   User Clustering

Dataset Pre-processing and Hyperparameter Tuning. In the user clustering tasks, we used the *AVG3MONTHS* attribute as a feature for elements within the clusters. This attribute represents the average revenue from users in 3 consecutive months. For 2 types of time windows, we clustered customers by the aggregation of their expenses. Both of them will have a total of 26 KMeans models with cluster numbers ranging from 5 to 30 to cluster the data. The Inertia and Silhouette score based on the cluster numbers are as follows:

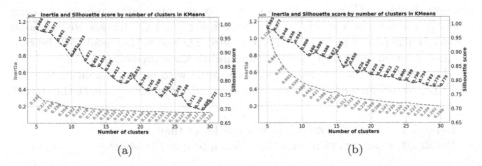

(a)   (b)

Fig. 4. Inertia and Silhouette scores by cluster numbers. (a) weekly revenue dataset. (b) monthly revenue dataset.

For each dataset, the Inertia will be measured on different cardinality of features, which lead to the dissimilar value range. The number of cluster then be selected by narrowing down the range to reach more stable changes in Inertia and Silhouette score. After selecting k for each dataset, the distribution of *AVG3MONTHS* for each cluster then be visualized in Fig. 5.

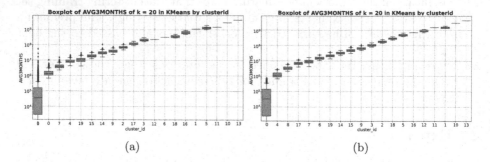

(a)   (b)

Fig. 5. Average revenue in 3 months from each user across the clusters derived from clustering model (a) trained by weekly dataset. (b) trained by monthly dataset.

Result Analysis. As in Fig. 4, the overall trend of Inertia and Silhouette score are decrease. For the weekly dataset, from $k = 12$, the improvement between two consecutive Inertia measures become less significant but Silhouette score still fluctuating, and hold the average of 0.779, which is approximate the value for 20 clusters. For the monthly dataset, the changes in both metrics marginally slower when k reach 20. Therefore, choosing $k=20$ is deemed to be an appropriate selection and these labels are added into our set of features.

Regarding clustering based on weekly revenue data, particularly with 20 clusters, based on the boxplots presented in the Fig. 5a, it can be observed that some clusters like clusters 8, 0, 7, etc., are suffering the outliers who overlap the other clusters' boxplot. Compared to using weekly revenue data from users, using monthly revenue data yields better clustering results in clusters with smaller range of outliers and higher Silhouette scores. Therefore, this result will be chosen as the baseline for user classification and revenue prediction tasks.

However, the other models require a sample with enough quantity to train and test, some clusters also contain outliers, which adversely affect the overall results.

- Remove the cluster whose cluster size is lower than 100.
- Remove the outliers of revenue by using interquartile range.

4.2 User Classification

Dataset. In the user classification problem, we will use the revenue of the two last months in 2022 to classify users into clusters. Their segment labels are inferred from the result of previous stage with $k = 20$ based on the clustering evaluation metrics.

Result Analysis. Using three algorithms: Softmax Regression, Decision Tree, and Gradient-Boosted Trees, the summarize classification results of the validation set using the key metrics of Accuracy, Precision, and Recall for each label in the table.

Table 2. Evaluation of classification algorithms

	Softmax regression		Decision Tree		Gradient-Boosted Trees	
	Accuracy: 0.9587		Accuracy: 0.9569		Accuracy: 0.9609	
Label	Precision	Recall	Precision	Recall	Precision	Recall
0	0.9882	0.9941	0.9863	0.9965	0.9862	0.9979
2	–	–	–	–	0.8788	0.9063
3	0.3279	0.3175	–	–	0.7344	0.7460
4	0.9037	0.9139	0.9119	0.9019	0.9177	0.9059
6	0.7651	0.8157	0.7244	0.8119	0.7687	0.8666
7	0.4147	0.1301	0.3355	0.1242	0.4942	0.1341
8	0.8004	0.7508	0.7758	0.7366	0.8119	0.7398
9	–	–	0.3924	0.7879	0.8188	0.7879
14	0.5995	0.5635	0.5000	0.7842	0.7825	0.8801
15	0.2678	0.1034	–	–	0.6383	0.4138
17	0.7648	0.8275	0.7545	0.7949	0.7670	0.8680
19	0.3685	0.2795	–	–	0.5990	0.3315

Overall, the accuracy of the algorithms is quite high. However, when considering the Precision and Recall metrics for the classes, it can be observed that class 0 is well-classified, while the majority of the other classes are not. This can be attributed to the imbalanced nature of the data, where class 0 dominates the dataset. In the boxplot Fig. 5b, the data in class 7 has a wide distribution and overlaps with the neighboring classes (class 17 and 6). As a result, this class also exhibits poorer classification performance compared to other classes with a high number of instances.

Based on the evaluation, it can be observed that for the classes with fewer instances (class 2, 3, 19, 15), the Softmax Regression and Decision Tree algorithms yield poor results (even failing to recognize any elements in some classes, presented by "-" in the Table 2), whereas the Gradient Boosted Tree algorithm provides better evaluation measures.

The reason why the ensembled algorithms (Gradient-Boosted Trees) perform the best is the unsuitable for handling imbalanced dataset of the others. Their poor predictive accuracy over the minority class come from the tendency to favor the samples from the majority class. On the other hand, Gradient Boosted Trees belong to the Boosting group of algorithms, which are known to handle imbalanced data well [13]. As a result, these algorithms produce more promising results compared to the others.

4.3   Regression Analysis

Baseline Result. The most elementary approach for this problem is using a single regression model for all the users. This baseline model is only one single regression without enhancement stages conducted. The result achieved by Gradient-Boosted Trees Regression recorded by Table 3.

Table 3. Performance of the simple regression approach

R^2	MAE
67.81%	500,318

The overall performance is acceptable, however, this result suffers from the imbalance ratio of income contribution and cluster size. In contrast to the biggest size of cluster 0 (81.69% customers), they only hold 8.59% of all cluster revenue. Since this characteristic can lead to a model designed for only the biggest cluster, the user segmentation result will be manipulated in the Model splitting step.

Model Splitting. Having chosen the group of 20 clusters by evaluating Silhouette and Inertia value, some of them then be removed or excluded outliers before applying the grouping logic. The final combination of these clusters will be group A includes cluster 0, group B (4, 7, 8, 17), group C (6, 19), group D (14, 15), group E (3, 9), and group F of cluster 2, ordered by cluster size (Table 4).

Table 4. Cluster size and average revenue in 3 months of each cluster group

Cluster	Total cluster size	AVG3MONTHS
0	5,159,001	79,084
4, 7, 8, 17	1,123,860	2,320,169
6, 19	24,670	16,766,213
14, 15	5,601	35,120,355
3, 9	1,933	73,260,771
2	277	172,481,584
Total	**6,315,342**	**789,318**

Data Enrichment and Preprocessing. The enriched dataset table contains 79 features about 6.9 million individual users. At the size of rows and columns, the dataset has 1 column of identify key for customers, 3 features of the revenue (with one as the dependent variable), 20 features about phone call history, 20 features of subscription, and 5 internet services columns. All data source is anonymized to ensure the customer's privacy.

Result Analysis. Altogether the regression model is the combination of 6 separated Gradient-boosted trees regression models. Cluster 0 requires the highest attributes for training and also transformation steps added. Table 5 presents the performance of each model on the test dataset (accounting for 10% total).

Table 5. Evaluation of regression models on test set

Cluster	MAE
0	70,010
4, 7, 8, 17	1,183,768
6, 19	6,515,053
14, 15	12,807,727
3, 9	23,188,716
2	34,743,690
Overall	**397,781**

According to this table, we can state that the MAE value of most models is lower than their average revenue. Even though MAE correlate to the revenue, the ratio between them is much lower for the high-spending clusters namely (3,9) or 2. In other words, this model can perform well on both the largest customer segmentation as well as the others.

The overall result is the evaluation of the predicted value of atom models with their actual value. While that may indicate a huge difference in the metrics, this is related to the varying of each model's sample size. This combined model performance also yields the MAE at **397,781**. This means the metric is reduced by about 20% compared to the base model.

Evaluation of Pipeline in Telecom's DSS. Using the trained models and 10% unused of the dataset, we test this pipeline through the classification and the invoked regression process.

The overall MAE in the application is about **422,502 VND**, approximate to the MAE of the regression evaluation result. Even though these metrics do not gain perform as well as the training stage, this prediction pipeline still achieves an improvement of **15.56%** in the MAE of the base model.

5   Conclusion

In this paper, we proposed a big data based model for analyzing telecom users' behavior. The research contribution is to enable telecom companies to understand their customers' actions, giving a closer look at what benefits they will get and may support a future recommendation system of appropriate service to each user. As the dataset is millions of people and records, the solution is implemented in a Spark cluster and takes advantage of this framework's machine learning library.

In the first analysis stage, we clustered all the people and found 20 customer segments. The clustering result was then used to build the training and testing dataset of the classification tasks to predict a cluster label for a new customer for recognizing the potential one. Acting as the classifier for the regression model, a Gradient-boosted Tree Classification is chosen to identify the segment from the customer's monthly charge.

Next, the regression analysis model set includes 6 atom models, divided by the size and their miss classified rates. They all require the enriched data of user usage summaries to pass through the preprocessing stage of feature selection and data normalization. These models will be activated by the predicted value from the classification model.

Based on the evaluation result, this research can feature in the Telecom Decision Support System with the purpose to promote the business operation. To further increase the accuracy, we recommend implementing an additional layer of churn prediction before entering the regression model. In order to improve model performance, we continue collecting the dataset over a longer period of time.

Acknowledgement. The authors acknowledge Ho Chi Minh City University of Technology (HCMUT), VNU-HCM for supporting this study.

References

1. Alapati, S.R.: Expert Hadoop administration managing, tuning, and securing spark, YARN, and HDFS. Addison-Wesley Professional (2016)
2. Chen, C.M.: Use cases and challenges in telecom big data analytics. APSIPA Trans. Signal Inf. Process. **5**, e19 (2016). https://doi.org/10.1017/ATSIP.2016.20
3. Dalvi, P.K., Khandge, S.K., Deomore, A., Bankar, A., Kanade, V.A.: Analysis of customer churn prediction in telecom industry using decision trees and logistic regression. In: 2016 Symposium on Colossal Data Analysis and Networking (CDAN), pp. 1–4 (2016). https://doi.org/10.1109/CDAN.2016.7570883
4. Felbo, B., Sundsøy, P., Pentland, A., Jørgensen, S.L., Montjoye, Y.A.: Modeling the temporal nature of human behavior for demographics prediction. Lect. Notes Comput. Sci. **10536**, 140–152 (2017). https://doi.org/10.1007/978-3-319-71273-4_12
5. Han, J., Kamber, M., Pei, J.: 10 - cluster analysis: basic concepts and methods. In: Han, J., Kamber, M., Pei, J. (eds.) Data Mining. The Morgan Kaufmann Series in Data Management Systems, 3rd edn, pp. 443–495. Morgan Kaufmann, Boston (2012). https://doi.org/10.1016/B978-0-12-381479-1.00010-1

6. Han, J., Kamber, M., Pei, J.: 6 - mining frequent patterns, associations, and correlations: basic concepts and methods. In: Han, J., Kamber, M., Pei, J. (eds.) Data Mining. The Morgan Kaufmann Series in Data Management Systems, 3rd edn, pp. 243–278. Morgan Kaufmann, Boston (2012). https://doi.org/10.1016/B978-0-12-381479-1.00006-X
7. Han, J., Kamber, M., Pei, J.: 8 - classification: basic concepts. In: Han, J., Kamber, M., Pei, J. (eds.) Data Mining. The Morgan Kaufmann Series in Data Management Systems, 3rd edn, pp. 327–391. Morgan Kaufmann, Boston (2012). https://doi.org/10.1016/B978-0-12-381479-1.00008-3
8. Meng, X., et al.: Mllib: machine learning in apache spark (2015)
9. Olle, G.: A hybrid churn prediction model in mobile telecommunication industry. Int. J. e-Educ. e-Bus. e-Manag. e-Learn. (2014). https://doi.org/10.7763/IJEEEE.2014.V4.302
10. Shaikh, E., Mohiuddin, I., Alufaisan, Y., Nahvi, I.: Apache spark: a big data processing engine. In: 2019 2nd IEEE Middle East and North Africa COMMunications Conference (MENACOMM), pp. 1–6 (2019). https://doi.org/10.1109/MENACOMM46666.2019.8988541
11. Sharma, P., Chakraborty, A., Sanyal, J.: Machine learning based prediction of customer spending score. In: 2019 Global Conference for Advancement in Technology (GCAT), pp. 1–4 (2019). https://doi.org/10.1109/GCAT47503.2019.8978374
12. Sleeman, W.C., IV., Krawczyk, B.: Multi-class imbalanced big data classification on spark. Knowl. Based Syst. **212**, 106598 (2021). https://doi.org/10.1016/j.knosys.2020.106598
13. Tanha, J., Abdi, Y., Samadi, N., Razzaghi, N., Asadpour, M.: Boosting methods for multi-class imbalanced data classification: an experimental review. J. Big Data **7**(1), 1–47 (2020). https://doi.org/10.1186/s40537-020-00349-y
14. Wang, M., Wang, Y., Wang, X., Wei, Z.: Forecast and analyze the telecom income based on arima model. Open Cybernet. Syst. J. **9** (2015)
15. Win, N.A.S., Thwin, M.M.S.: Comparative Study of Big Data Predictive Analytics Frameworks. Ph.D. thesis, MERAL Portal (2017)

An Efficient Cryptographic Accelerators for IoT System Based on Elliptic Curve Digital Signature

Huu-Thuan Huynh[1,2], Tan-Phat Dang[1,2], Trong-Thuc Hoang[3],
Cong-Kha Pham[3], and Tuan-Kiet Tran[1,2(✉)]

[1] University of Science, Ho Chi Minh City, Vietnam
{hhthuan,dtphat,trtkiet}@hcmus.edu.vn
[2] Vietnam National University, Ho Chi Minh City, Vietnam
[3] University of Electro-Communications (UEC), Tokyo, Japan
{hoangtt,phamck}@uec.ac.jp

Abstract. Given the importance of security requirements in today's Internet of Things (IoT) landscape, this study focuses on enhancing the security of IoT systems through encryption algorithms implemented on the SoC-FPGA platform. Specifically, this paper presents the development of a hardware system that generates public keys and performs digital signature generation and verification using the Elliptic Curve Digital Signature Algorithm (ECDSA) based on the SECP256K1 curve. The ECDSA encryption hardware functions as a co-processor and demonstrates a maximum operating frequency of 30 MHz. In terms of performance, the ECDSA IP achieves efficient processing speeds, taking approximately 17 ms to generate a public key and produce a digital signature and nearly 30 ms to verify the digital signature. These results showcase a well-balanced design that enables a trade-off between speed, area, and power dissipation in the proposed system.

Keywords: DSA · ECDSA · SoC-FPGA · IoT

1 Introduction

Securing the confidentiality of IoT devices with limited computing capabilities and low-power requirements is of utmost importance within the context of scientific research. These devices, including sensors, wearables, and smart home appliances, often operate in resource-constrained environments, which makes them susceptible to security risks. Integrating System-on-Chip-Field-Programmable Gate Array (SoC-FPGA) technology has emerged as a vital solution to tackle these challenges. SoC-FPGA offers a unique combination of hardware flexibility, low power consumption, and computational efficiency, making them an ideal choice for enhancing the security of IoT devices. Leveraging SoC-FPGA technology allows for implementing robust security features directly within the hardware of the IoT devices. These features include secure boot mechanisms, encryption

N. Thai-Nghe et al. (Eds.): ISDS 2023, CCIS 1950, pp. 106–118, 2024.
https://doi.org/10.1007/978-981-99-7666-9_9

algorithms, authentication protocols, and communication interfaces. The programmability of SoC-FPGAs enables the customization of security solutions to meet the specific requirements of each device, ensuring efficient resource utilization while minimizing energy consumption. Several studies, such as the works of [1,2], have highlighted various advantages of FPGA technology, including implementing custom security algorithms and protocols directly in hardware. This capability significantly reduces vulnerability to software-based attacks. IoT devices can achieve higher computational efficiency, lower power consumption, and improved resistance to side-channel attacks by offloading cryptographic operations to the FPGA. Furthermore, FPGAs provide a flexible and reconfigurable platform for implementing security features, enabling rapid updates and adaptations to evolving threats, as demonstrated in the work of Aras et al. [3].

In scientific research, extensive efforts have been devoted to exploring the implementation of Elliptic Curve Cryptography (ECC) on FPGA to enhance performance and efficiency. ECC is a public-key cryptography algorithm that relies on the computational difficulty of solving the elliptic curve discrete logarithm problem (ECDLP) to provide security. Numerous research studies have delved into various aspects of ECC implementation on FPGAs, encompassing key generation, scalar multiplication, point addition, point doubling, and other ECC operations. These investigations have proposed diverse architectures and techniques to optimize ECC performance on FPGAs. Notable approaches include parallel processing techniques [4], optimized arithmetic operations [5], efficient memory management [6], and pipelining to reduce latency and improve the overall performance of ECC operations [7]. Furthermore, researchers have also explored hardware/software co-design methodologies to strike a balance between hardware acceleration and software control, capitalizing on FPGA advantages while preserving flexibility and ease of programming, which was suggested by Tanaka et al. [8]. By optimizing the hardware implementation of ECC on FPGAs, researchers aim to enhance the efficiency, security, and performance of cryptographic applications, particularly in domains such as secure communication, digital signatures, and secure data storage. In this article, our primary objective revolves around augmenting the security of resource-constrained IoT devices characterized by limited hardware and computational capabilities. We studied and implemented a digital signature algorithm utilizing ECC on the System-on-Chip-FPGA (SoC-FPGA) platform to accomplish this goal. Specifically, we employed the SECP256K1 curve to facilitate modular arithmetic operations and point computations, thereby enabling the effective execution of the ECDSA algorithm.

The structure of this paper is as follows. Section 2 offers an overview of the DSA, providing background information on ECC and ECDSA, focusing on the SECP256K1 curve. In Sect. 3, we introduce a cryptographic system that combines FPGA's computational power and flexibility with the integration capabilities of SoC to implement digital signatures based on elliptic curves. This system incorporates the ECDSA Intellectual Property (IP), which consists of fundamental ECC point computations and modular operations. Moreover, we present the hardware design of modules such as inversion and multiplication modular and point operation modules encompassing point addition, doubling, and multiplication operations. Section 4 presents the implementation results and compares our proposed system with other research in the field. Finally, Sect. 5 concludes the paper.

2   Background

2.1   Digital Signature Algorithm (DSA)

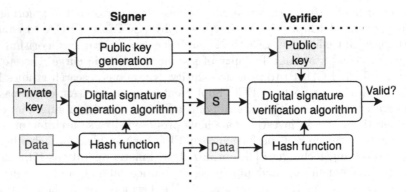

Fig. 1. DSA algorithm.

The DSA is a widely adopted cryptographic algorithm utilized to generate and verify digital signatures. Its primary purpose is to ensure digital documents' authenticity, integrity, and non-repudiation. DSA is founded upon the principles of modular exponentiation and discrete logarithms. It employs a key pair consisting of a public key, which is shared openly, and a private key, which is kept confidential. To create a digital signature using DSA, the signer first hashes the message and then combines the hashed message with their private key and a random "nonce." These components are utilized to compute the signature. Verification of the signature involves performing mathematical computations using the public key, the original message, and the signature components. The signature is considered valid if the computed values match the signature components. The security of DSA is contingent upon solving challenging mathematical problems such as the discrete logarithm problem. This algorithm finds extensive application in secure email communication, digital certificates, and authentication protocols. It provides practical and efficient means of establishing trustworthiness for digital data. Figure 1 illustrates the process of DSA between the sender and the receiver, visually representing the steps involved in generating and verifying digital signatures.

2.2   ECC and ECDSA

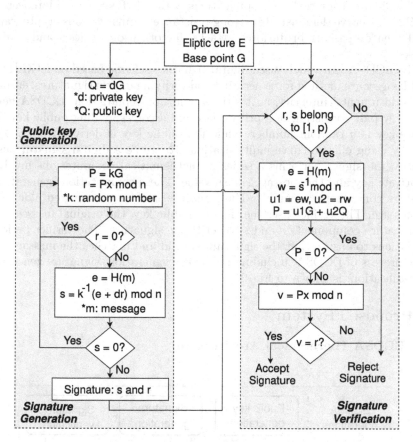

Fig. 2. ECDSA Algorithm.

The ECC algorithm, referenced in [9], represents a widely employed public-key cryptography technique in secure communication and data protection. This algorithm leverages the mathematical properties of elliptic curves over finite fields to achieve its objectives. Compared to other public-key algorithms like RSA, ECC offers robust security while employing shorter key lengths, enhancing computational efficiency, and minimizing memory requirements. Specifically, ECC operates on elliptic curves defined by an equation of $y^2 = x^3 + ax + b$ over finite fields. The selection of curve parameters and field size significantly impact ECC's security level and performance. Among the commonly utilized curves are those recommended by NIST, such as SECP256K1, described in [10]. SECP256K1 presents multiple advantages for IoT systems, including high-security levels with shorter keys, rendering it suitable for resource-constrained

IoT devices characterized by limited computational power and memory capacity. Additionally, the SECP256K1 curve exhibits fast computation speeds, making it well-suited for real-time applications within IoT systems. Moreover, the SECP256K1 curve demonstrates strong resistance against various cryptographic attacks, encompassing brute force attacks, factorization attacks, and collision attacks.

The ECDSA is a widely-used digital signature scheme based on ECC [11]. It provides a secure method for generating and verifying digital signatures, ensuring digital documents' integrity, authenticity, and non-repudiation. ECDSA generates a key pair consisting of a private key and a corresponding public key. The private key is a random number, while the public key is derived from the private key using elliptic curve multiplication. To create a digital signature using ECDSA, the signer performs a series of mathematical computations involving the private key and the message's hash value. These computations generate two signature components: r and s. The signature is then transmitted along with the message. The verifier uses the signer's public key, the original message, and the signature components r and s to verify the signature. The verifier performs calculations to ensure that the signature is valid and matches the message. The whole process of ECDSA, including public key generation, signature generation, and verification, is shown in Fig. 2.

3   Proposed System

3.1   ECDSA Hierarchical Architecture

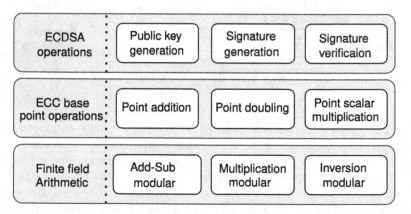

Fig. 3. ECDSA hierarchical architecture.

The ECDSA architecture, as depicted in Fig. 3, encapsulates the evolutionary journey from fundamental finite field operations to elementary ECC point operations, culminating in the ECDSA algorithm. The cornerstone of ECC computation lies in modular operations, encompassing Add-Sub modular, Multiplication modular, and Inversion modular processes. Based on these modular operations, the ECC base point operations are constructed: point addition, point doubling, and scalar point multiplication. By leveraging the ECC base point operations, the ECDSA method is formed, facilitating Public Key Generation, Signature Generation, and Signature Verification calculations.

In this paper, we put forward a novel architectural design for the Inversion modular and Multiplication modular components, integrating them to create a comprehensive point operation module that can perform point addition, point doubling, and scalar point multiplication. Additionally, we introduce the structure of the ECDSA core, comprising the point operation module and a dedicated Finite State Machine (FSM) responsible for control and coordination.

3.2 Modular Inversion

Modular inversion plays a vital role in ECC computations in our implementation. To accomplish this, we utilize the Binary Euclidean Algorithm (BEA), specifically designed to calculate the modular inversion of numbers in ECC. Modular inversion is crucial in ECC as it allows us to compute the multiplicative inverse of an element modulo a given prime. The Binary Euclidean Algorithm, an optimized version of the Euclidean Algorithm, takes advantage of binary arithmetic operations such as bit shifts and bitwise logical operations. This algorithm is highly efficient for modular inversion in ECC due to the binary nature of the underlying field operations.

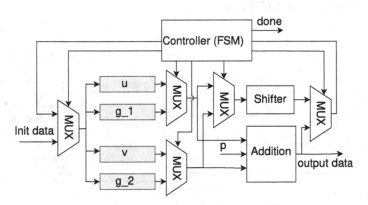

Fig. 4. Inversion modular module.

The hardware implementation of the BEA in our proposed architecture is illustrated in Fig. 4. Our BEA architecture consists of four registers that store the initial data of variables u, v, g_1, and g_2, as well as their temporary data during the calculation process. The BEA controller manages the writing of newly generated data, acquired through addition and shift operations, back into these temporary registers. Additionally, the BEA controller determines the termination condition for the inversion process. We have specifically implemented the BEA FSM based on Algorithm 1. Upon receiving the completion signal, the final result is assigned to variable y from the output of the addition submodule.

Algorithm 1. Binary Euclidean Algorithm.

Input: Prime p and $a \in [1, p-1]$;
Output: $y = a^{-1} \bmod p$;

1: $u \leftarrow a, v \leftarrow p, g_1 \leftarrow 1, g_2 \leftarrow 0$
2: **while** $u \neq 1$ and $v \neq 1$ **do**
3:     **while** $u[0] = 0$ **do**
4:         $u \leftarrow u \gg 1$
5:         **if** $g_1[0] = 0$ **then**
6:             $g_1 \leftarrow g_1 \gg 1$
7:         **else**
8:             $g_1 \leftarrow (g_1 + p) \gg 1$
9:     **while** $v[0] = 0$ **do**
10:         $v \leftarrow v \gg 1$
11:         **If** $g_2[0] = 0$ **then**
12:             $g_2 \leftarrow g_2 \gg 1$
13:         **else**
14:             $g_2 \leftarrow (g_2 + p) \gg 1$
15:     **if** $(u > v) \wedge (u \geqslant (v \ll 1) - v)$ **then**
16:         $u = u + v, g_1 = g_1 + g_2$
17:     **else**
18:         $v = u + v, g_2 = g_2 + g_1$
19: **if** u = 1 **then**
20:     **return** $y = g_1$
21: **else**
22: **return** $y = g_2$

3.3 Multiplication Modular

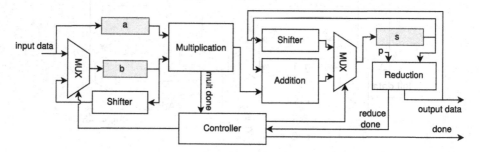

Fig. 5. Multiplication modular.

Modular multiplication is a crucial component in ECC computations, and to improve efficiency and reduce delays, we have implemented a fully pipelined approach for this operation. The structure of the modular multiplication submodule is illustrated in Fig. 5.

In our design, we divide the multiplicand b into eight segments, each consisting of 32 bits. By multiplying a 256-bit wide multiplier a with a 32-bit wide multiplicand, we obtain a product with a width of 288 bits. This product is accumulated and stored in a register s. To ensure optimal size and precision, we apply the Barrett reduction technique to reduce the accumulated value in register s to 256 bits. Once the reduction step is completed, the register b is shifted left by 32 bits to obtain a new segment for multiplication with register a. Simultaneously, the register s is also shifted left and added to the result of the modular multiplication. After the addition, the register s undergoes the reduction process again, ensuring its size remains at 256 bits. This process continues until the last segment of register b is processed.

Algorithm 2. Barrett reduction

Input: Prim p 256 bits and 512 bits s_in number
Output: $s\_out = s\_in \mod p$
1: $q = s\_in[511:256]$
2: $t0 = q \times p$
3: $t1 = s\_in[255:0] - t0[255:0]$
4: $t2 = t0[255:0] - p$
5: **if** $p \leqslant t1$ **then**
6:      $s\_out = t2$
7: **else**
8:      $s\_out = t1$
      return s_out

The Barrett reduction algorithm, which is extensively discussed in [12], is widely recognized and utilized for performing efficient modular arithmetic operations involving large numbers in the field of ECC. In our implementation, specifically when working with the SECP256K1 set in ECC, we set the values of m and k in the Reduction module as 1 and 256 respectively. Algorithm 2 provides a detailed description of the computation within the Reduction module, based on the general algorithm of the Barrett reduction algorithm. Moreover, the multiplication in the Reduce module is implemented using the pipeline method to enhance throughput and reduce delay. This particular configuration allows for a straightforward hardware implementation of the Barrett reduction algorithm, resulting in improved efficiency.

3.4 ECDSA Core

The core component of our proposed system is the ECDSA IP. Figure 6 illustrates the structure of the ECDSA core, which includes a point operation module controlled by an ECDSA FSM. This module incorporates registers to store temporary values. The point operation module consists of several modular modules, including two add-sub modules, one inversion module, and one multiplication module. The Add FSM and the Double FSM, described in Algorithm 3 and Algorithm 4 respectively, coordinate the functionality of these modules. It is important to note that Fig. 6 does not depict the temporary registers used to store computed values for submodules within the point operation module. These FSMs play a vital role in managing input data and controlling the data path for point addition and doubling operations. Moreover, the point operation module is equipped with the capability to perform scalar point multiplication using a binary method, which is outlined in Algorithm 5 [13]. The ECDSA FSM employs a multiplexer (mux) to choose the control between the Add FSM and the Double FSM. This enables the point operation module to perform diverse operations using the Double-and-Add technique for scalar point multiplication computation. Our modular point operation module has been meticulously designed to meet the requirements of compactness and low power consumption in IoT devices. It efficiently manages multiplication, inversion, and add-sub operations, allowing the point module to execute multiple operations based on the control signals received from the ECDSA FSM.

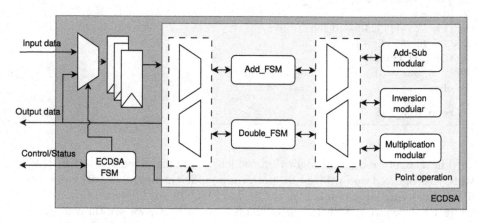

Fig. 6. ECDSA core.

Algorithm 3. ADD FSM	**Algorithm 4.** Double FSM
Input: Point $P(Px, Py)$, $Q(Qx, Qy)$	**Input**: Point $P(Px, Py)$
Output: Point $R(Rx, Ry) = P + Q$	**Output**: Point $R(Rx, Ry) = 2P$
1: $s1 \leftarrow Py - Qy$, $s2 \leftarrow Px - Qx$	1: $s1 \leftarrow Px + Px$, $s2 \leftarrow Py + Py$
2: $s \leftarrow s2^{-1} \times s1$	2: $s3 \leftarrow s1 + Px$, $s \leftarrow s3 \times s2^{-1}$
3: $p \leftarrow s \times s$	3: $p \leftarrow s \times s$
4: $Rx \leftarrow p - Px - Qx$	4: $Rx \leftarrow p - s1$
5: $s3 \leftarrow Px - Rx$	5: $s4 \leftarrow Px - Rx$
6: $p \leftarrow s \times s3$	6: $p \leftarrow s \times s4$
7: $Ry \leftarrow p - Py$	7: $Ry \leftarrow p - Py$

The ECDSA core incorporates a point operation module that handles modular arithmetic and point computations. To control input and manage output data, an FSM is employed. This enables the ECDSA core to perform various tasks, such as public key generation, signature generation, and signature verification, as outlined in the ECDSA algorithms shown in Fig. 2. By integrating the point operation module with the FSM, the hardware required for ECDSA is significantly minimized compared to using separate modules and organizing them into a scheme. Nevertheless, this approach increases processing time as it necessitates waiting to complete previous tasks before subsequent ones can be processed. To tackle this issue, we have implemented a pipeline functionality within the point operation module, enabling it to perform multiple tasks concurrently. For example, it can simultaneously calculate add-sub or inversion operations during multiplication and vice versa. This optimization effectively mitigates the augmented processing time resulting from task dependencies.

Algorithm 5. Left-to-right Point Multiplication Binary Method

---

Input: $k = \sum_{i=0}^{n-1} k_i 2^i$ and Point P
Output: $Q = kP$

1:   $Q \leftarrow O,\ G \leftarrow P$
2:   **for** $i \leftarrow n - 1$ to 0 **do**
3:     Double the point Q: $Q \leftarrow 2 \cdot Q$
4:     **if** $k_i = 1$ **then**
5:       Add the base point G to Q: $Q \leftarrow Q + G$
    return Q

3.5   Integration System

Our proposed system is illustrated in Fig. 7, where the ECDSA IP takes on a central role. This IP is responsible for executing three crucial operations of the ECDSA algorithm: public key generation, signature generation, and verification. These operations are implemented within a dedicated FPGA IP known as the ECDSA IP. To integrate this IP into the SoC-FPGA system, we employ an ARM core running the Linux operating system. This setup enables efficient control and monitoring of the ECDSA IP core's status using the Lightweight HPS-to-FPGA bridge. Additionally, the ECDSA IP core is designed to make use of direct memory access (DMA) for autonomous data retrieval from SDRAM, serving as the input and output data storage. The Read port of the ECDSA IP core is used to retrieve data, while the Write port is utilized for data storage in the SDRAM. Both ports access the SDRAM through the FPGA-to-SDRAM bridge. In order to enhance the speed of data transfer, we have opted to use a 256-bit width for the FPGA-to-SDRAM bridge, allowing the DMA to read/write each data item in a single clock cycle.

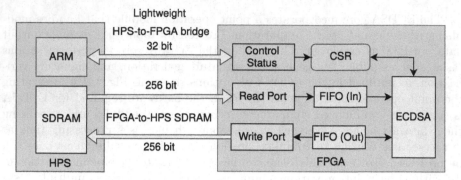

Fig. 7. Proposed system

4   Experimental Results

Regarding synthesis results, the ECDSA IP core consumed approximately 52% of the hardware resources of the Cyclone V chip (5CSXFC6D6F31C6) when used independently. Furthermore, the proposed system, depicted in Fig. 7, utilized around 64% of the Cyclone V chip's resources. This system included the ECC IP core, DMA interface, FIFOs, and bus data. As for the operating clock, our proposed system can operate at 30 MHz for the ECDSA IP. The throughput of the ECDSA IP enables computation with a maximum key width of 256 bits for public key generation and signature generation, taking approximately 17 ms. Additionally, signature verification operations require around 30 ms. Moreover, we implemented the ECDSA IP on the Stratix 10 board, where the ECC core consumed only about 2% of the hardware resources and operated at a clock frequency of approximately 62 MHz. This estimation of the ECDSA IP's processing time is based on a counter that starts counting when the ECC IP begins calculation and stops when the process is complete. Therefore, the processing time can decrease linearly with the same count value but a higher frequency.

Table 1. Implementation results and comparision.

Ref	Micro-controler		VLSI/FPGA		
	[14]	[15]	[16]	[17]	Our work
Curve	P-192	SECP256K1	SECP256R1	P-256	SECP256K1
Platform	AVR	MSP430X	Virtex-5	65-nm	Cyclone-V
F (MHz)	7.73	-	50	18.66	30
Resource	-	-	14.256 (LUT-FF)	13K (Gate)	22K (ALMs)
Sign time (ms)	-	557	7.26	563.40	17
Verify time (ms)	1,892	1,156	9.4	-	30

Table 1 presents the results achieved in our research, including the published outcomes. Compared to studies conducted using software on microcontroller platforms [14,15], our results demonstrate significantly faster processing times for signature generation and verification. However, when compared to hardware-based implementations using FPGA or ASIC technologies, our achieved processing speeds are faster than those reported in research [17] but inferior to the results obtained by [16]. Nevertheless, when considering hardware consumption on the FPGA, our results may be superior, as with 22K ALMs, using the equivalent of 5.5K LUTs, indicating a lower hardware utilization on the FPGA chip. Additionally, since our target application is IoT devices with limited hardware resources at the edge, we employ a step-wise processing mechanism controlled by an FSM. This results in lower hardware consumption but sacrifices processing time. However, the achieved results of 17 ms for public key and signature generation and 30 ms for signature verification are sufficient to meet the processing bandwidth. Moreover, the processing time is calculated by the number of clock cycles multiplied by the clock period; thus, increasing the clock frequency decreases the processing time. Therefore, with the synthesized results on the Stratix 10 board, the maximum achievable system frequency is 62 MHz, leading to an approximate processing time of 8.2 ms for public key and signature generation and 14 ms for signature verification.

5 Conclusions

This paper presents an innovative cryptosystem designed specifically for IoT systems utilizing SoC-FPGA platforms. Our main objective is to offer a streamlined and efficient cryptographic solution that addresses the hardware and computation limitations of IoT systems. To accomplish this, we have developed a tailored IP core that implements the ECDSA digital signature algorithm on SECP256K1 curves. Moreover, we have customized the Linux kernel and drivers to effectively manage and control the hardware system. Extensive evaluations of various algorithms and techniques were conducted to enable FPGA-based cryptography, prioritizing high-speed processing through custom memory modules, pipeline architecture, and DMA techniques. By adopting this approach, we guarantee swift cryptographic operations within our solution.

Acknowledgment. This research is funded by University of Science, VNU-HCM under grant number ĐTVT 2022-04.

References

1. Olazabal, O., et al.: Multimodal biometrics for enhanced IoT security. In: Proceedings of IEEE Annual Computing and Communication Workshop and Conference (CCWC), pp. 0886–0893, January 2019
2. Shrivastava, A., Haripriya, D., Borole, Y.D., Nanoty, A., Singh, C., Chauhan, D.: High performance FPGA based secured hardware model for IoT devices. Int. J. Syst. Assur. Eng. Manag. **13**, 736–741 (2022)

3. Aras, E., Delbruel, S., Yang, F., Joosen, W., Hughes, D.: Chimera: a low-power reconfigurable platform for internet of things. ACM Trans. Internet Things **2**(2), 1–25 (2021)
4. Shah, Y.A., Javeed, K., Azmat, S., Wang, X.: A high-speed RSD-based flexible ECC processor for arbitrary curves over general prime field. Int. J. Circ. Theory Appl. **46**(10), 1858–1878 (2018)
5. Järvinen, K.: Optimized FPGA-based elliptic curve cryptography processor for high-speed applications. Integration **44**(4), 270–279 (2011)
6. Sandoval, M.M., Flores, L.A.R., Cumplido, R., Hernandez, J.J.G., Feregrino, C., Algredo, I.: A compact FPGA-based accelerator for curve-based cryptography in wireless sensor networks. J. Sens. **2021**, 1–13 (2021)
7. Imran, M., Rashid, M., Jafri, A.R., Kashif, M.: Throughput/Area optimised pipelined architecture for elliptic curve crypto processor. IET Comput. Digit. Tech. **13**(5), 361–368 (2019)
8. Tanaka, K., Miyaji, A., Jin, Y.: Efficient FPGA design of exception-free generic elliptic curve cryptosystems. In: Proceedings of Applied Cryptography and Network Security (ACNS), pp. 393–414, June 2021
9. Miller, V.S.: Use of elliptic curves in cryptography. In: Williams, H.C. (ed.) CRYPTO 1985. LNCS, vol. 218, pp. 417–426. Springer, Heidelberg (1986). https://doi.org/10.1007/3-540-39799-X_31
10. Brown, D.R.L.: SEC 2: recommended elliptic curve domain parameters. Standards for Efficient Cryptography, pp. 1–37, January 2010. https://www.secg.org/sec2-v2.pdf
11. IEEE: IEEE Standard Specifications for Public-Key Cryptography. IEEE Std 1363-2000, pp. 1–228, August 2000
12. Barrett, P.: Implementing the Rivest Shamir and Adleman public key encryption algorithm on a standard digital signal processor. In: Odlyzko, A.M. (ed.) CRYPTO 1986. LNCS, vol. 263, pp. 311–323. Springer, Heidelberg (1987). https://doi.org/10.1007/3-540-47721-7_24
13. Finite field arithmetic. In: Guide to Elliptic Curve Cryptography, pp. 25–73. Springer, New York (2004). https://doi.org/10.1007/0-387-21846-7_2
14. Liu, Z., Seo, H., Großschädl, J., Kim, H.: Efficient implementation of NIST-compliant elliptic curve cryptography for 8-bit AVR-based sensor nodes. IEEE Trans. Info. Forensics Secur. **11**(7), 1385–1397 (2016)
15. Gouvêa, C.P.L., Oliveira, L.B., López, J.: Efficient software implementation of public-key cryptography on sensor networks using the MSP430X microcontroller. J. Crypto. Engi. **2**, 19–29 (2012)
16. Glas, B., Sander, O., Stuckert, V., Müller-Glaser, K.D., Becker, J.: Prime field ECDSA signature processing for reconfigurable embedded systems. Int. J. Reconfig. Comp. **2011**, 1–12 (2011)
17. Ikeda, M.: Hardware acceleration of elliptic-curve based crypto-algorithm, ECDSA and pairing engines. In: Proceedings of IEEE International Conference on ASIC (ASICON), pp. 1–4, October 2021

Improved Gene Expression Classification Through Multi-class Support Vector Machines Feature Selection

Thanh-Nghi Do[1,2]([✉]) and Minh-Thu Tran-Nguyen[1,2]

[1] College of Information Technology, Can Tho University, 90000 Cantho, Vietnam
{dtnghi,tnmthu}@ctu.edu.vn
[2] UMI UMMISCO 209 (IRD/UPMC) Sorbonne University, Pierre and Marie Curie University, Paris 6, France

Abstract. This paper proposes a new approach for gene expression classification by using a multi-class support vector machine (SVM) with feature selection. The proposed algorithm is based on the One-Versus-All (OVA) multi-class strategy, which learns binary 1-norm SVM models. As the 1-norm SVM solution is very sparse, the algorithm can automatically suppress a large number of dimensions that correspond to null weights. This feature elimination improves the classification results for high-dimensional gene expression datasets. Empirical test results on 25 gene expression datasets show that our multi-class SVM eliminates 99% of full dimensions, resulting in 7.1%, 4.03% increase in accuracy compared to training SVM, random forest models on the full dimensions of gene expression datasets, respectively.

Keywords: Gene expression classification · Feature selection · Support vector machines · Random forests

1 Introduction

Gene expression classification involves analyzing the patterns of gene activity in cells or tissues. It aims to identify the genes that are expressed or silenced in different biological conditions or diseases. Gene expression classification has important applications in many areas of biomedical research, including cancer diagnosis, drug discovery, and personalized medicine. Gene expression data can be created using various experimental techniques that measure the activity of genes in cells or tissues. The most common methods include microarray and RNA sequencing technologies, which allow for the simultaneous measurement of the expression levels of thousands of genes.

To classify gene expression patterns, various computational techniques in [5,8,10,12,19] have been developed, including machine learning algorithms, such as support vector machines (SVM [21]), random forests [2], and neural networks [14]. However, gene expression datasets have a large number of predictor variables (dimensions or features) relative to the number of observations (data-points). This can pose several issues in machine learning, including overfitting, instability of models, and difficulty in identifying important features. There are several methods for feature selection that have been developed for gene expression datasets, including the recursive feature elimination (RFE) introduced by [9,10,15,23], the SVM-1 proposed by [1,3,7,20], the approach in [4] using random forests [21], the technique [13] based on partial least squares, the software for feature selection based on machine learning approaches [15,16].

Our proposed approach involves a feature selection multi-class SVM algorithm. The multi-class SVM algorithm uses the One-Versus-All (OVA) multi-class strategy, which involves training binary 1-norm SVM models as demonstrated in prior works such as [1,3,7,20]. Since the 1-norm SVM solution is highly sparse, the algorithm can automatically eliminate a significant number of dimensions that correspond to null weights. This feature elimination enhances the classification performance for high-dimensional gene expression datasets. Experimental results on 25 gene expression datasets indicate that our multi-class SVM approach removes 99% of full dimensions, leading to 7.1%, 4.03% increase in accuracy compared to training SVM, random forest models on the full dimensions of gene expression datasets, respectively.

The remainder of this paper is organized as follows. Section 2 briefly presents the feature selection multi-class SVM algorithm. Section 3 shows the experimental results before conclusions and future works presented in Sect. 4.

2   Feature Selection Multi-class Support Vector Machines

2.1   Binary Support Vector Machines for Feature Selection

We consider a binary classification problem depicted in Fig. 1, where we have a dataset $D = [X, Y]$ containing m training datapoints $X = \{x_1, x_2, \ldots, x_m\}$ with corresponding labels $Y = \{y_1, y_2, \ldots, y_m\}$ that are either ± 1. The SVM algorithm proposed by [21], aims to find the optimal separating plane represented by a normal vector $w \in R^n$ and a bias $b \in R$, which is situated furthest from both class $+1$ and class -1. To achieve this objective, the SVM algorithm simultaneously maximizes the margin between the supporting planes for each class and minimizes errors.

An alternative approach, called SVM-1 introduced by [1,7], achieves this goal by minimizing the 1-norm of w (represented by $||w||_1$) instead of the 2-norm used by the standard SVM. Any point x_i falling on the wrong side of its supporting plane is considered to be an error, denoted by z_i ($z_i \geq 0$, when x_i falls on the correct side of its supporting plane, $z_i = 0$). The training SVM-1 algorithm yields the linear programming (1).

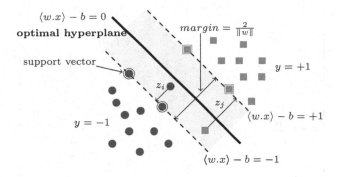

Fig. 1. Classification of the datapoints into two classes

$$min \ f(w, b, z) = ||w||_1 + \lambda \sum_{i=1}^{m} z_i$$

$$s.t. \begin{cases} y_i(\langle w \cdot x_i \rangle - b) + z_i \geq 1 \\ z_i \geq 0 \quad \forall i = 1, 2, ..., m \end{cases}$$

$$(1)$$

where the positive constant λ is used to tune errors and the margin size.

The separating plane w, b found in the linear programming (1) is used to classify a new datapoint x as follows:

$$predict(x) = sign(\langle w \cdot x \rangle - b). \tag{2}$$

As the optimal w obtained by solving the linear programming (1) for the SVM-1 is highly sparse, many dimensions will be eliminated since they have no weight in the optimal solution. Thus the SVM-1 formulation automatically performs feature selection.

2.2 Multi-class Support Vector Machines for Feature Selection

The SVM-1 model, which is initially designed for binary classification, can be extended to handle multi-class problems (c classes, where $c \geq 3$). In practice, a common approach for multi-class SVMs is to train a set of binary SVM-1 models, such as the One-Versus-All (OVA [21]) strategy. This approach, illustrated in Fig. 2, involves building c binary SVM-1 models, with each model separating one class from the others. To predict the class of a new datapoint, the model calculates the distances between the datapoint and each of the binary SVM-1 models, and selects the class with the largest distance vote.

Algorithm 1 illustrates how the multi-class SVM selects relevant features of gene expression datasets. The feature subset, represented by $Fsubset$ is initialized to the empty. The algorithm trains c binary SVM-1 models, in which the kth SVM-1 model, denoted by w_k performs the separation of class k from the remaining classes, choosing dimensions d corresponding to non-zero weights

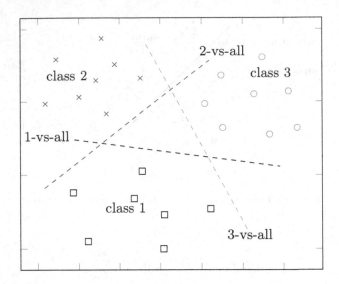

Fig. 2. Multi-class SVM (One-Versus-All)

$(w_k[d] \neq 0)$ and being not already in the feature subset $Fsubset$, to append them to the feature subset $Fsubset$.

3   Experimental Results

We are interested in experimentally evaluating the effectiveness of multi-class SVM's feature selection in the classification of gene expression datasets. Therefore, an evaluation of the performance of multi-class SVM's feature selection is necessary in terms of classification accuracy.

3.1   Experimental Setup

To implement multi-class SVM's feature selection in Python, we use NumPy [11] and SciPy [22] libraries. For training classification models on gene expression datasets, we use the state-of-the-art linear SVM algorithm LIBLINEAR [6], and random forest (RF) algorithm [2], which were implemented in Python and Scikit-Learn [18] library.

All tests are run under a machine Linux Fedora 32, Intel(R) Core i7-4790 CPU, 3.6 GHz, 4 cores and 32 GB RAM.

We conducted experiments using 25 gene expression datasets obtained from the ArrayExpress repository [17]. The datasets, as described in Table 1, exhibit characteristics such as being multi-class and having a small number of datapoints (individuals) in a high-dimensional input space (at least 11,950). With datasets having greater or equal than 300 datapoints, we used 10-fold cross-validation protocol to evaluate performance. This protocol involves random partitioning of

---

Algorithm 1: **FS-MC-SVM**(X, Y, λ) training algorithm for feature selection multi-class SVM

---

    input :

                Trainset $D = [X, Y]$ containing m datapoints $X_i \in R^n$

                with corresponding labels $y_i \in \{1, 2, \ldots, c\}$

            Positive constant $\lambda > 0$

    output:

            $F subset$

```
 1  begin
 2  │   Fsubset = {}
 3  │   for k ← 1 to c do                                  /* class k */
 4  │   │   for i ← 1 to m do
 5  │   │   │   if (y_i == k) then
 6  │   │   │   │   Y_k[i] = 1
 7  │   │   │   else
 8  │   │   │   │   Y_k[i] = −1
 9  │   │   │   end
10  │   │   end
11  │   │   w_k = SVM-1(X, Y_k, λ) ;    /* solving the linear prog. (1) */
12  │   │   for d ← 1 to n do
13  │   │   │   if (w_k[d] ≠ 0) and (d ∉ Fsubset) then
14  │   │   │   │   Fsubset.append(d)
15  │   │   │   end
16  │   │   end
17  │   end
18  │   Return Fsubset
19  end
```

---

the dataset into 10 folds, with one fold retained as the testset and the remaining nine as the trainset. The cross-validation process is repeated 10 times (corresponding to 10 folds). Training algorithms learn models from the trainset and then the resulting is used to classify the testset. The 10 test results can be then averaged to obtain the final classification results. For other datasets having less than 300 datapoints, the test protocol is leave-one-out cross-validation (loo), where the number of folds equals to the number of datapoints.

Table 1. Description of gene expression datasets

ID	Dataset	#Datapoints	#Dimensions	#Classes	Protocol
1	E-GEOD-20685	327	54627	6	10-fold
2	E-GEOD-20711	90	54675	5	loo
3	E-GEOD-21050	310	54613	4	10-fold
4	E-GEOD-21122	158	22283	7	loo
5	E-GEOD-29354	53	22215	3	loo
6	E-GEOD-30784	229	54675	3	loo
7	E-GEOD-31312	498	54630	3	10-fold
8	E-GEOD-31552	111	33297	3	loo
9	E-GEOD-32537	217	11950	7	loo
10	E-GEOD-33315	575	22283	10	10-fold
11	E-GEOD-36895	76	54675	14	loo
12	E-GEOD-37364	94	54675	4	loo
13	E-GEOD-39582	566	54675	6	10-fold
14	E-GEOD-39716	53	33297	3	loo
15	E-GEOD-44077	226	33252	4	loo
16	E-GEOD-47460	582	15261	10	10-fold
17	E-GEOD-63270	104	18989	9	loo
18	E-GEOD-63885	101	54675	4	loo
19	E-GEOD-65106	59	33297	3	loo
20	E-GEOD-6532	327	22645	3	10-fold
21	E-GEOD-66533	58	54675	3	loo
22	E-GEOD-68468	390	22283	6	10-fold
23	E-GEOD-68606	274	22283	16	loo
24	E-GEOD-7307	677	54675	12	10-fold
25	E-GEOD-73685	183	33297	8	loo

For training linear SVM models, the positive constant C is a parameter in the SVM formula for keeping a trade-off between the margin size and the errors. In all experiments, the value of C is fixed at 100,000. The RF algorithm trains 200 decision trees for classification models

Our study aims to evaluate the SVM, RF classification performance of multi-class SVM's feature selection (selected dimensions) against those without feature selection (all dimensions).

3.2   Classification Results

Table 2 shows the reported classification performance in terms of classification accuracy, with the best results highlighted in bold. Additionally, Figs. 3, 4 present a plot chart depicting the classification results obtained by SVMs, RF, comparing the use of all dimensions versus selected dimensions.

Table 2. Classification results

ID	All dimensions			Selected dimensions		
	#Dimensions	SVM (%)	RF (%)	#Dimensions	SVM (%)	RF (%)
1	54627	84.43	85.79	85	**91.07**	85.84
2	54675	67.34	74.23	43	**80.7**	76.14
3	54613	62.38	72.44	275	63.35	**72.71**
4	22283	87.34	85.44	78	**94.9**	87.18
5	22215	77.17	72.17	35	**96.1**	76.47
6	54675	92.14	88.21	42	**96.5**	92.95
7	54630	86.14	84.75	195	**98.38**	88.31
8	33297	87.39	86.49	44	**91.7**	87.16
9	11950	77.97	77.46	171	**90.2**	78.6
10	22283	**83.51**	80.33	2321	82.23	82.56
11	54675	73.15	72.84	326	73	**74.32**
12	54675	77.31	75.38	140	**89.1**	76.09
13	54675	83.1	78.52	441	**85.82**	82.3
14	33297	90.57	79.25	118	**98**	94.12
15	33252	99.56	98.32	23	**100**	**100**
16	15261	80.68	78.52	304	**85.69**	79.48
17	18989	55.12	59.71	158	**69.6**	66.67
18	54675	60.61	57.67	128	**80.8**	61.62
19	33297	74.58	61.02	45	**98.2**	92.98
20	22645	89.53	91.06	127	**92.86**	90.69
21	54675	**96.55**	89.66	10	96.4	94.64
22	22283	**97.2**	94.16	261	95.9	95.88
23	22283	**100**	**100**	263	**100**	**100**
24	54675	74.01	**82.93**	974	73.63	82.37
25	33297	80.33	77.6	95	**91.7**	85.64

As illustrated in Table 2, the use of selected dimensions by multi-class SVM's feature selection in SVMs leads to the significant removal of approximately 99% of irrelevant features from gene expression datasets. Furthermore, improvement of SVMs on selected dimensions against training SVM models on the full dimensions is 7.1% classification accuracy.

Summary of the accuracy comparison in Table 3 shows that SVMs with selected dimensions has 19 wins, 1 tie, 5 defeat (p-value = 0.0033) against SVM models on full dimensions.

RF models on selected dimensions improve 4.03% classification correctness compared to RF on full dimensions.

Summary of the accuracy comparison in Table 4 shows that RF with selected dimensions has 22 wins, 1 tie, 2 defeat (p-value = 0.0000179) versus RF models on full dimensions.

Table 3. Summary of the accuracy comparison between SVM on full dimensions and SVM on selected dimensions

	Mean (Dim.)	Mean (Acc.)	Win	Tie	Defeat	p-value
SVM (All dim.)	37916	81.52				
SVM (Selected dim.)	**268**	**88.63**				
SVM (Selected dim.) vs SVM (All dim.)			19	1	5	0.0033

Table 4. Summary of the accuracy comparison between RF on full dimensions and RF on selected dimensions

	Mean (Dim.)	Mean (Acc.)	Win	Tie	Defeat	p-value
RF (All dim.)	37916	80.16				
RF (Selected dim.)	**268**	**84.19**				
RF (Selected dim.) vs RF (All dim.)			22	1	2	0.0000179

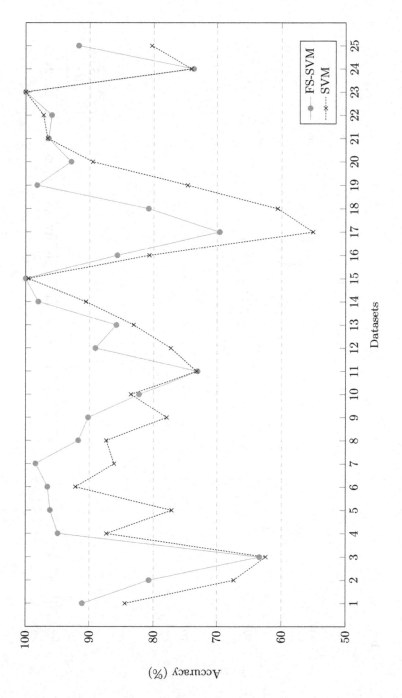

Fig. 3. Classification results of SVMs on full dimensions (denoted by SVM) and SVMs on selected dimensions (denoted by FS-SVM) for 25 gene expression datasets

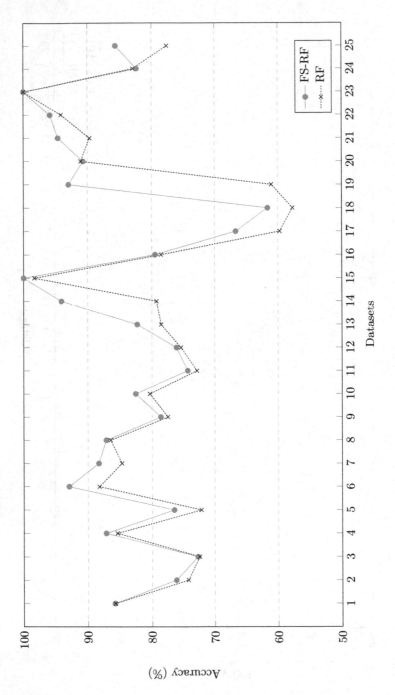

Fig. 4. Classification results of RF on full dimensions (denoted by RF) and RF on selected dimenions (denoted by FS-RF) for 25 gene expression datasets

4   Conclusion and Future Works

We have presented multi-class SVM's feature selection for efficiently classifying very-high-dimension gene expression datasets. The main idea is to use the OVA multi-class technique, which learns binary 1-norm SVM models. This multi-class SVM algorithm can significantly remove irrelevant features from large number of gene expressions since the 1-norm SVM solution being very sparse. The multi-class SVM's feature selection improves the SVM classification accuracy for very-high-dimension gene expression datasets. Empirical test results on 25 gene expression datasets show that our multi-class SVM can suppress about 99% of full dimensions while improving 7.1%, 4.03% accuracy compared to training SVM, random forest models on the full dimensions, respectively.

We plan to conduct further empirical testing on large benchmark of gene expression datasets, and make comparisons with other algorithms in the coming time.

Acknowledgments. This work has received support from the College of Information Technology, Can Tho University. We would like to thank very much the Big Data and Mobile Computing Laboratory.

References

1. Bradley, P.S., Mangasarian, O.L.: Feature selection via concave minimization and support vector machines. In: Proceedings of the Fifteenth International Conference on Machine Learning (ICML '98), pp. 82–90. Morgan Kaufmann Publishers Inc., San Francisco (1998)
2. Breiman, L.: Random forests. Mach. Learn. **45**(1), 5–32 (2001)
3. Dedieu, A.: Mit 9. 520/6. 860 project: feature selection for SVM (2016)
4. Diaz-Uriarte, R., de Andres, S.A.: Gene selection and classification of microarray data using random forest. BMC Bioinform. **7**(3) (2006)
5. Do, T.-N., Lenca, P., Lallich, S., Pham, N.-K.: Classifying very-high-dimensional data with random forests of oblique decision trees. In: Guillet, F., Ritschard, G., Zighed, D.A., Briand, H. (eds.) Advances in Knowledge Discovery and Management, pp. 39–55. Springer, Heidelberg (2010). https://doi.org/10.1007/978-3-642-00580-0_3
6. Fan, R.E., Chang, K.W., Hsieh, C.J., Wang, X.R., Lin, C.J.: LIBLINEAR: a library for large linear classification. J. Mach. Learn. Res. **9**(4), 1871–1874 (2008)
7. Fung, G., Mangasarian, O.L.: A feature selection newton method for support vector machine classification. Comput. Optim. Appl. **28**, 185–202 (2004)
8. Furey, T.S., Cristianini, N., Bednarski, D.W., Haussler, D.: Support vector machine classification and validation of cancer tissue samples using microarray expression data. Bioinformatics **16**(10) (2001)
9. Guyon, I., Nikravesh, M., Gunn, S., Zadeh, L.A. (eds.) Feature Extraction: Foundations and Applications. Springer, Heidelberg (2006). https://doi.org/10.1007/978-3-540-35488-8
10. Guyon, I., Weston, J., Barnhill, S., Vapnik, V.: Gene selection for cancer classification using support vector machines. Mach. Learn. **46**(1–3), 389–422 (2002)
11. Harris, C.R., et al.: Array programming with NumPy. Nature **585**, 357–362 (2020)

12. Huynh, P., Nguyen, V.H., Do, T.: Novel hybrid DCNN-SVM model for classifying RNA-sequencing gene expression data. J. Inf. Telecommun. **3**(4), 533–547 (2019)
13. Lê-Cao, K.A., Boitard, S., Besse, P.: Sparse pls discriminant analysis: biologically relevant feature selection and graphical displays for multiclass problems. BMC Bioinform. **12**(253) (2011)
14. LeCun, Y., Bottou, L., Bengio, Y., Haffner, P.: Gradient-based learning applied to document recognition. In: Proceedings of the IEEE, vol. 86, pp. 2278–2324 (1998)
15. Masoudi-Sobhanzadeh, Y., Motieghader, H., Masoudi-Nejad, A.: Featureselect: a software for feature selection based on machine learning approaches. BMC Bioinformatics **20**(170) (2019)
16. Mishra, S., Mishra, D., Satapathy, S.K.: Integration and visualization of gene selection and gene regulatory networks for cancer genome. Elsevier Academic Press (2018)
17. Parkinson, H., et al.: ArrayExpress - a public repository for microarray gene expression data at the EBI. Nucl. Acids Res. **33**(suppl_1), D553–D555 (2005)
18. Pedregosa, F., et al.: Scikit-learn: machine learning in python. J. Mach. Learn. Res. **12**, 2825–2830 (2011)
19. Statnikov, A., Wang, L., Aliferis, C.F.: A comprehensive comparison of random forests and support vector machines for microarray-based cancer classification. BMC Bioinformatics **9**(319) (2008)
20. Thi, H.A.L., Nguyen, M.C.: DCA based algorithms for feature selection in multiclass support vector machine. Ann. Oper. Res. **249**, 273–300 (2017)
21. Vapnik, V.: The Nature of Statistical Learning Theory. Springer, New York (1995). https://doi.org/10.1007/978-1-4757-2440-0
22. Virtanen, P., et al.: SciPy 1.0 contributors: SciPy 1.0: fundamental algorithms for scientific computing in python. Nat. Methods **17**, 261–272 (2020)
23. Zifa, L., Weibo, X., Tao, L.: Efficient feature selection and classification for microarray data. PLOS ONE **13**(8), 1–21 (2018)

Deep Learning and Natural Language Processing

Transliterating Nom Script into Vietnamese National Script Using Multilingual Neural Machine Translation

Phat Hung[1,2], Long Nguyen[1,2]([✉]), and Dien Dinh[1,2]

[1] Faculty of Information Technology, University of Science,
Ho Chi Minh City, Vietnam
nhblong@fit.hcmus.edu.vn
[2] Vietnam National University, Ho Chi Minh City, Vietnam

Abstract. Traditional methods for Nom transliteration have been relying on statistical machine translation. While there have been initial attempts to apply neural machine translation (NMT) to Nom transliteration, these approaches have not effectively utilized the shared characteristics of multilingualism. To overcome this, we propose a new method that employs a highly shared parameter multilingual NMT model, using a unified parameter set the encoder and decoder. This approach utilizes a single encoder-decoder pair for all translation directions. Experimental evaluation of our proposed method demonstrates a significant improvement of +6.3 BLEU score compared to the NMT baseline. We also conduct preliminary experiments on incorporating BERT-NMT into multilingual machine translation to enhance the performance of low-resource translation tasks in general.

Keywords: Transliteration · Nom script · Multilingual NMT

1 Introduction

Transliteration refers to the process of converting text from one writing system to another while preserving the phonetic values of the original text. As an example, "ベトナム" (meaning "Vietnam") written in Katakana can be represented in the Latin alphabet as "*betonamu*". This process can be automated via a one-to-one lookup table which maps Katakana characters to their Romaji equivalents. A similar approach can be applied to Korean Hangul or Russian Kirin scripts.

However, transliterating from the Nom script to the Vietnamese National Script is more complicated due to differences in writing systems and a many-to-many relationship. A single Nom character can have multiple National Script representations based on context, history, and region, making it more challenging. Since parallel corpus for Nom transliteration is extremely limited at the moment, findings and advancements in machine translation-based Nom transliteration can be extrapolated to benefit low-resource machine translation as well.

N. Thai-Nghe et al. (Eds.): ISDS 2023, CCIS 1950, pp. 133–147, 2024.
https://doi.org/10.1007/978-981-99-7666-9_11

Prior research on Nom transliteration has primarily focused on statistical machine translation, which has demonstrated remarkable performance compared to other rule-based methods [1]. More recently, [2] introduced a novel approach based on multilingual neural machine translation (NMT) employing a multi-encoder and multi-decoder architecture. Their method shared parameters between different translation tasks using an attention bridge, enabling the exchange of multilingual knowledge among different encoders.

However, it is important to consider that the National Script and Nom characters are just two different writing systems for the Vietnamese language. Therefore, we suspect that the knowledge used for decoding Vietnamese can also aid in encoding Nom characters. Furthermore, the transliteration model by [2] severely suffered from negative interference introduced by the noisy Nom-National Script corpus, limiting its potential scalability to other low-resource NMT tasks. Following [3], we adopt the Compact translation model which not only promotes parameter sharing among different language pairs but also across the encoding and decoding steps. Additionally, we investigate the application of multi-stage fine-tuning [4] to initialize the model using knowledge from closely related languages with richer resources. Furthermore, we also explore the potential benefits of incorporating BERT-NMT [5] in a multilingual scenario to enhance performance on low-resource tasks. In summary, our contributions are three-fold:

1. We build a new Nom-National Script transliteration system employing the compact representor and multi-stage fine-tuning method
2. We integrate BERT-NMT into a low-resource multilingual setting and carry out some exploratory experiments
3. We conduct experiments on the model and compared our results with various other approaches

2   Background and Related Works

2.1   Overview of the Nom Script

The Nom script was developed between the 10th and 12th centuries, adapting Chinese characters to represent Vietnamese sounds. This script, both ideographic and phonographic, was prominent in Vietnamese poetry and administrative writings. However, by 1945 its use had declined, and the Latin-based National Script (Chữ Quốc ngữ) became the official system for Vietnamese. Given that most modern Vietnamese can not read or write Nom, creating an automated transliteration system is crucial to preserving Vietnam's rich historical and cultural legacy over the past millennium.

Nom characters can be classified into two main categories: phonograms and ideograms. Ideographic Nom characters are borrowed, modified, or combined from Han characters, disregarding phonetic elements. For example, the Vietnamese word *trời* 坏 (sky) can be written by combining two Han characters 天 (sky) and 上 (upper). On the other hand, phonographic Nom characters are created by combining different elements representing the semantic and phonetic features of the corresponding character. For instance, the character ba 㕛 (three) is created by combining the phonetic element 巴 (pinyin: /bā/) and the semantic element 三 . Since the majority of Nom characters are phonographic, there are some Nom characters that can be mapped to two or more National Script equivalents. For the sake of convenience, from this point onwards, we will view the Nom-National Script transliteration problem as a regular translation problem, and refer to the corresponding translation direction as "Nom-Viet".

2.2   Multilingual Neural Machine Translation

Multilingual neural machine translation (multi-NMT) [6–8] can be seen as a formulation that addresses NMT from a multi-task perspective. Given K language pairs, let $\mathcal{D}_k = \{(x_k, y_k)\}$ be the training corpus for the k-th pair, where $x_k = (x_1, x_2, \dots, x_n)$ is the source sequence and $y_k = (y_1, y_2, \dots, y_m)$ is the target sequence. The encoder f_{enc} takes in the input sequences from multiple source languages and maps them into a continuous representation $h = (h_1, \dots, h_n)$. The decoder f_{dec} generates the output sequence incrementally by estimating $P(y|y_{<t}), t \in 1..m$ as follows:

$$P(y_t|y_{<t}) = \text{softmax}(f_{\text{dec}}(s, c_t))$$

where s is the hidden states of the decoder, c_t is the context vector.

$$c_t = \sum_{i=1}^{n} a_{ti} h_i$$

where a_{ti} is the attention weight, and e_{ti} is the alignment score to evaluate the correspondence of the source word at position i and target word at position j.

$$a_{ti} = \text{softmax}(e_{ti}) = \frac{\exp(e_{ti})}{\sum_{j=1}^{n} \exp(e_{tj})}$$

The encoder and decoder are jointly optimized to maximize the probability of observing the target sequence given a source sequence over K different language pairs available in the training set:

$$\mathcal{L}_{\text{trans}}(\mathcal{D}_k; \theta) = \sum_{k=1}^{K} \sum_{d=1}^{|\mathcal{D}_k|} \sum_{t=1}^{m_d} \log P(y_t|y_{<t}, x; \theta_{\text{enc}}, \theta_{\text{dec}}, \theta_{\text{attn}})$$

where m_d is the length of the d-th target sentence, and $\theta_{\text{enc}}, \theta_{\text{dec}}, \theta_{\text{attn}}$ are the parameters of the encoder, decoder and attention mechanism respectively.

The encoder and decoder can me implemented using various neural network architectures such as RNN [9], CNN [10], or Transformer [11].

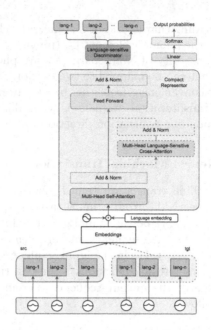

Fig. 1. The Compact translation model

2.3 Compact Translation Model

The Compact translation method [3] is an effective approach to enhance parameter sharing across translation directions. This model extensively shares weights among common sub-layers in the encoder and decoder, including self-attention, layer normalization, and feed-forward modules. This integration of shared parameters results in a *"compact representor"*. This approach also aligns with the recent advancements in language models, such as the GPT family [12], which have empirically demonstrated the capability of a standalone decoder in achieving impressive performance across a wide range of NLP tasks, including machine translation. In addition to this, the Compact model also introduces several other improvements compared to the original Transformer architecture.

Instead of using a dedicated token to indicate the source language, the model uses *"language-sensitive embedding"*, which incorporates source language information into each source token, similar to positional encoding. K_s additional learnable vectors $e_{\text{lang}}^k \in \mathbb{R}^{d_{\text{model}}}, k \in 1..K_s$ corresponding to the K_s languages present on the input side are introduced. These embeddings are then added to each input token in the corresponding source languages.

To enhance the model's language representation, a *"language discriminator"* is introduced on top of the decoder output, classifying the sentence's output language. This helps mitigate parameter interference in a multilingual setting, considering that the encoder representation in multilingual NMT isn't entirely language-agnostic [13,14]. This sub-task is jointly optimized with the translation tasks using an objective weight of λ.

We experiment with two cross-attention mechanisms proposed by the original work: *"language-sensitive attention"* (LS) and *"hybrid attention"* (HB). LS creates separate cross-attention modules for each language pair, allowing dynamic parameter selection for each translation direction. This method works better in cases when training languages possess distinct features, such as different word orders [3]. HB, on the other hand, employs a single cross-attention module for all translation directions. The mentioned modules are illustrated in Fig. 1.

2.4 BERT-Fused NMT

In low-resource translation scenarios, where parallel corpora for a specific translation direction is not widely available, knowledge augmentation on either side may also be beneficial. Integrating a pre-trained large language model (LLM) into the translation model is one approach to enhance source-side comprehension. LLMs are pre-trained on extensive unlabeled data to capture general meaningful input representations for downstream tasks, which is particularly valuable when domain-specific data is limited [15]. [5] incorporated a BERT [16] encoder into the Transformer framework through a supplementary cross-attention module. The attention score of the l-th encoder layer $\tilde{a}^{(l)}$ is calculated as follows:

$$\tilde{a}^{(l)} = \text{Attn}_S(h^{(l-1)}, h^{(l-1)}, h^{(l-1)}) + \text{Attn}_B(h^{(l-1)}, h_B, h_B) \tag{1}$$

where $h^{(l)}$ is the hidden state of the l-th encoder layer, h_B is the output of BERT, and Attn_S, Attn_B denote the self-attention and the BERT-NMT cross-attention modules respectively.

The overall data flow in BERT-NMT can be summarized as follows:

1. x is passed through BERT, yielding h_B;
2. x is processed by the encoder. h_B is also injected into the encoder via cross-attention, creating a context-rich representation of the source sentence. After L layers, we obtain the final encoder hidden state $h^{(L)}$;
3. In addition to $h^{(L)}$ as in the conventional Transformer model, h_B is also injected into the decoder via cross-attention to generate the target sequence.

Cross-attention modules for BERT incorporation can either be separated or shared between the encoder and decoder (Fig. 2).

2.5 Multi-stage Fine-Tuning

Empirical evidence has shown that fine-tuning-based transfer learning is an effective technique for transferring knowledge from language pairs with abundant resources to those with limited resources [17,18]. [4] introduced the method to first pre-train an NMT model on a task with a moderate corpus size, then use its weights to initialize the multi-NMT model. The approach can be viewed as a soft labor division: initializing the encoder or decoder with the knowledge from a medium-sized parallel corpus; and train a multi-NMT system using the weights from the pre-trained bilingual model. Experiments showed that the method yielded up to +9.67 BLEU enhancement on the IWSLT English-Vietnamese task [19], with English being the common source language across multiple stages.

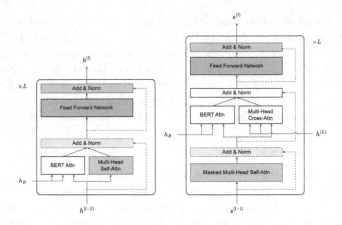

Fig. 2. The BERT-fused Multi-NMT model

3   The Proposed Model

For this particular transliteration task and also low-resource NMT in general, employing a model that both optimizes parameter usage to mitigate overfitting and effectively utilizes language commonality is crucial. While the multi-encoder, multi-decoder model by [2] offers flexibility for tuning specific architectures for each direction, it may not effectively capture common linguistic features across languages due to its distinct parameter sets for each translation direction, leading to unnecessary model complexity. Additionally, we argue that the conventional multi-NMT method may not fully recognize the equivalence between the Nom script and the National Script in representing the Vietnamese language. These insights highlight the need for utilizing the Compact translation method for our neural transliteration model and low-resource NMT in general.

While BERT-NMT has demonstrated its effectiveness in augmenting translation knowledge, prior experiments were limited to bilingual scenarios. Our model integrates the BERT-NMT algorithm with the Compact translation model, marking the first application of this approach in a multilingual context. The resulting BERT-fused Compact Translation model is illustrated in Fig. 3.

Additionally, we further generalize the method by introducing a control parameter γ into the model to control the influence of BERT in the encoding process. Equation 1 now becomes:

$$\tilde{a}^{(l)} = (1 - \gamma)\text{Attn}_S(h^{(l-1)}, h^{(l-1)}, h^{(l-1)}) + \gamma\text{Attn}_B(h^{(l-1)}, h_B, h_B) \qquad (2)$$

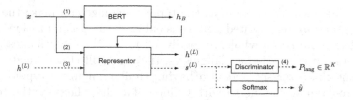

Fig. 3. The BERT-fused Compact Translation Method. Continuous lines and dashed lines represent the data flow of the encoder and decoder respectively. For simplicity, the embedding layers are omitted

The resulting model is trained using the multi-stage training strategy as illustrated in Fig. 4, which consists of two steps:

1. **Pre-training:** in this stage, the model is pre-trained on the Chinese-English task. As a significant portion of Vietnamese vocabulary consists of Sino-Vietnamese words, which share similar characters with the Chinese language when written in Han-Nom scripts, utilizing Chinese as a helper language can potentially enhance the model's performance in the Nom transliteration task;
2. **Mixed fine-tuning:** during this stage, a variety of bilingual corpora are utilized, with Vietnamese serving as the target language. The source languages include Chinese, English, and Nom script; each serves a distinct purpose. The Chinese-Vietnamese corpus could aid the model in mapping the Sino-Vietnamese vocabulary into their respective target words, while the English-Vietnamese could expose the model to a wider range of natural and authentic Vietnamese sentences.

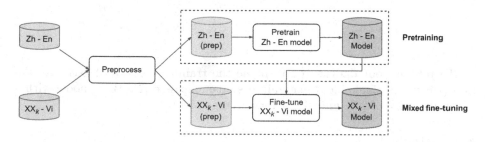

Fig. 4. The multi-stage fine-tuning method. The XX_k set consists of the following languages: Chinese, English, and Nom

Our approach differs from [4] in that the target optimization pair Nom-Viet has no common language with the pre-training pair Chinese-English. While the auxiliary datasets may not closely align with the literary style of classical Nom poetry, their inclusion in the model offers several benefits. Firstly, considering the data scarcity of Nom-Vi, the model may encounter challenges in approximating the target distribution of Vietnamese sentences. Incorporating additional language pairs with Vietnamese as the target language can make the distribution more generally represented, and thus enhance the generalization capability across different pairs as a whole. Secondly, although Nom/Vietnamese and Chinese corpora represent distinct languages, they are still closely related, sharing a significant character set and also having similar linguistic typology. Consequently, pre-training the model with Chinese data has the potential to capture common linguistic properties, ultimately enhancing the model's performance.

4   Experiments

4.1   Datasets

We use the following language pairs and corpora in our experiments

- For the Chinese-English (Zh-En) pair, we employ the IWSLT17 [20] dataset
- For the English-Vietnamese (En-Vi) pair, we use the IWSLT15 [19] dataset
- For the Chinese-Vietnamese (Zh-Vi) pair, we use the dataset provided by the CLC*. This dataset consists mostly of sentence-aligned news articles and Chinese literature translated into Vietnamese
- Lastly, for the Nom-Vietnamese (No-Vi) pair, we use the dataset provided by the CLC, consisting mostly of medieval Vietnamese poetry written in Nom and their respective transliteration in the modern National Script (Table 1)

Table 1. Dataset statistics

Language pair	Zh-En	Zh-Vi	En-Vi	No-Vi
Dataset	IWSLT 2017	CLC	IWSLT 2015	CLC
Training set	229K	48K	133K	6.5K
Validation set	1K	2.7K	1.5K	700
Test set	1K	2.7K	1.2K	400

To tokenize our dataset, we combine the training corpora of each translation pair into a single file, then train a `sentencepiece` [21] BPE model with a vocabulary size of 20K and character coverage of 1.0.

4.2   Setup

We conduct four experiment groups to demonstrate the effectiveness of our method by incrementally introducing the aforementioned enhancements into the multilingual translation model. First, four bilingual (labeled BL) experiments are carried out, corresponding to the language pairs presented in Sect. 4. Second, we train multilingual models employing the Compact architecture (labeled C) with three language pairs: Zh-Vi, En-Vi, and No-Vi. Third, we conduct experiments with the multi-stage fine-tuning method on the Compact model (labeled CM). Finally, we incorporate BERT into our multi-NMT system (labeled CMB).

---

* Computational Linguistics Center. University of Science, Ho Chi Minh City.

Table 2. Configuration for bilingual NMT models (BL).

Language pair	Zh-En	Zh-Vi	En-Vi	No-Vi
Batch size	4096	2048	2048	1024
Attention heads	8	4	8	2
Training epochs	30	20	30	20
Warm up steps	4000	4000	4000	1000

Common Configuration. We use our own implementation of the Compact translation model in `fairseq`. All of our models are trained with the Adam optimizer with `betas = (0.9, 0.98)` and `lr = 3e-4`. For multilingual experiments, we use inverse square root learning rate scheduling with `warmup_steps = 4000`, `init_lr = 1e-7` and `min_lr = 1e-9`. The language judgment layer is implemented with a simple feed-forward network with 1024 hidden neurons. Inferencing is carried out with `beam_size = 4` and `length_penalty = 0.6`.

BL. In this experiment, we train four NMT models corresponding to the four aforementioned language pairs. Most of our parameters follow the fairseq configuration `transformer_iwsl_de_en`, with the exception of those explicitly mentioned in Table 2.

C. In this experiment, we use the Compact translation model without applying the multi-stage fine-tuning strategy. We first train the model on all four language pairs, and then only on three pairs listed in the *"mixed fine-tuning"* stage to investigate the impact of using a consistent target language. The language judgment ratio is set to $\lambda = 0.1$. The models are trained for 30 epochs for language-sensitive attention and 15 epochs for hybrid attention.

CM. In this experiment, we use the Compact translation model with the multistage fine-tuning strategy. The first stage is trained on Zh-En, while the second stage is trained on {Zh, En, No}-Vi. We use a drop-out ratio of 0.2 for the first stage and 0.3 for the second stage. For each stage, the models are trained for 30 epochs for language-sensitive attention or 15 epochs for hybrid attention.

CMB. This experiment follows a similar setup to Experiment 3, with the difference being the utilization of the BERT-fused Compact model instead of the original Compact model. We set the drop-net ratio [5] $p_{net} = 0.3$ for all of our experiments. We also experiment with two different values of the control ratio γ, which are 0.35 and 0.5. The BERT cross-attention modules are separated between the encoder and decoder. In this experiment, we only employ the hybrid attention mechanism, which has already yielded more promising results in CM.

4.3 Results

We measure the performance of our experiments in BLEU score [22] and summarized them in Table 3. The experiment that obtained the highest BLEU score within each group is selected for inference on the test set, with results presented in Table 4. For the multilingual experiments (C, CM and CMB), the hybrid attention mechanism is selected. For CMB, the chosen BERT weight γ is 0.35 for an optimal score on the Nom-Viet pair.

Additionally, we include the results of the ATB model, a multi-encoder and multi-decoder architecture proposed by [2], which achieved marginally higher Nom transliteration performance compared to our method. However, it is important to note that the parallel Nom-Viet corpus used in training is exceptionally limited, and NMT models trained on such data are highly susceptible to noise. As the precise train-test split utilized in their experiments is unknown, direct comparisons of the results may not be feasible. On the other hand, their model suffered from significant negative interference in other language pairs, posing challenges when adapting it to other low-resource NMT tasks.

Table 3. Validation BLEU scores for the conducted experiments. "LS" represents language-sensitive cross-attention, while "HB" represents hybrid cross-attention. The dash symbol indicates that the language pair was not included in the training process of the corresponding experiment/stage.

Exp	Cross-attn	γ	Stage	Zh-En	Zh-Vi	En-Vi	No-Vi
BL				**16.4**	26.6	26.2	69.0
C	LS			–	28.1	25.1	75.9
	HB			–	26.3	24.6	78.1
				12.7	27.0	24.1	74.2
CM	LS		1	16.1	–	–	–
			2	–	28.2	26.5	77.1
	HB		1	15.7	–	–	–
			2	–	27.4	25.0	**78.5**
CMB	HB	0.5	1	16.2	–	–	–
			2	–	**30.1**	**26.6**	77.1
		0.35	1	15.4	–	–	–
			2	–	28.2	25.6	78.3

4.4 Discussion

We briefly analyze the outcome along the following axes: model, choice of attention mechanism and training strategy.

Table 4. Test BLEU score for the conducted experiments

Exp	Zh-Vi	En-Vi	No-Vi
BL	23.9	**29.6**	68.7
C	26.0	26.9	72.8
CM	27.2	27.3	73.4
CMB	**28.0**	28.0	**75**
ATB	9.69	2.23	76.1

Model. In our experiments, the enhancement in performance of multilingual models compared to bilingual NMT models is not consistent. Notably, the No-Vi pair demonstrated the most significant improvement, achieving a +9.5 BLEU score gain over the NMT baseline. On the other hand, scores for En-Vi constantly fluctuate and often fall below the NMT baseline. When incorporating BERT-NMT into our models, we observe a consistent increment in BLEU scores across all language pairs. Notably, Zh-Vi showed the most significant improvement, benefiting from both BERT's additional language knowledge and the model's initialization from the Zh-En NMT task. Adjusting the BERT-NMT ratio had varying effects: lowering the ratio decreased scores for Zh-Vi and En-Vi but increased scores for No-Vi. This difference can be due to the fact that BERT has never been trained on Nom data, so it may encounter challenges in distinguishing between Nom and Chinese words that share similar characters but have different meanings. It is noteworthy that CMB still yields better results than CM on Zh-Vi despite using a low BERT ratio, proving the effectiveness of leveraging BERT in a low-resource setting. However, BERT-NMT nearly doubles the training duration, posing challenges when experimenting various approaches in our research process.

Training Strategy. This section focuses on experiments C and CM as the BERT-NMT model was not trained without multi-stage fine-tuning. For the same cross-attention mechanism, multi-stage fine-tuning yields better results for all language pairs compared to the C model with 3 language pairs. While the model tends to forget English decoding knowledge learned in the Zh-En task during stage 2, the Chinese encoding knowledge carries over to subsequent stages, as evidenced by the increase in BLEU scores for Zh-Vi and No-Vi. Training C with 4 pairs improves performance for Zh-Vi but negatively impacts No-Vi. This could be due to the model paying more attention to encoding Chinese than Nom, as Chinese and Nom share a significant character set overlap.

Attention Mechanism. In this section, we analyze the performance of two distinct cross-attention mechanisms utilized in our experiments. LS constructs separate cross-attention modules for each language pair, allowing the model to dynamically adapt its parameters for source and target languages with distinct features. This approach can also mitigate confusion when dealing with destination languages of different word orders.

On the other hand, HB uses a universal cross-attention module shared across all translation pairs. Given our experimental setup with a single target language (Vietnamese), the issue of varying word orders, often encountered in many-to-many setups, becomes irrelevant. Additionally, since our main focus is on Nom-Vi, LS can introduce additional complexity and lead to overfitting. Moreover, parameter sharing also allows the transliteration task to benefit more from the Vietnamese decoding knowledge gained from other translation tasks. These factors likely contributed to HB's superior performance on the No-Vi pair.

4.5   Error Analysis

This section highlights the errors observed in the proposed Nom transliteration model. By examining the generated prediction sentences from our experiments on the test set, we have identified the five most prevalent errors made by our transliteration models. The illustrative examples are given in Table 5.

- **Skipping words:** word omission in source sentence due to unseen or infrequent words in test set (as with the word *"dù"* in examples 3). The model also sometimes skips one syllable of a reduplicative word (as with the word *"lẫn lẫn"* in example 9)
- **Repeating words:** words are repeated multiple times (as with the word *"xa"* in example 5)
- **Wrong transliteration:** preferring popular transliteration candidates over less frequent ones in the training set (as with the word *"dìu giặt"* in example 3 and *"dường"* in example 4)
- **Translation instead of transliteration:** words which are not available in the No-Vi training set but occur frequently in the Zh-Vi training set are "translated" as if those words are Chinese (as with the word *"bỉ thử"* in examples 2 and *"bóng"* in example 7)
- **Unknown translation:** non-sensical prediction for words which are not available in the No-Vi training set and occur infrequently in the Zh-Vi training set (as with the word *"tù"* in example 10)

5   Conclusion and Future Work

In this work, we propose a new approach that leverages the Compact translation model to build a new multilingual NMT system to address the parameter sharing problem in Nom-National Script transliteration. We also conduct a wide range of experiments and give an exploratory insight into leveraging the multi-stage fine-tuning method, as well as incorporating BERT into a multilingual setting. Our finding is that multi-stage multilingual translation does significantly improve the BLEU score for most of the tasks, especially when trained with the guidance of an external pre-trained BERT model.

Table 5. Several incorrectly transliterated examples

src		㛪伶伶唪戈時時催		蟻戈蚁吏㛪乘醜車	
truth		đã tu tu trót qua thì thì thôi		tu kiến khiêu ngưu chức nữ tinh	
bl	1	đã tu tu tuân qua trót thì thì thôi	5	chuả kén tựa ngưu nữa tinh	
c		đã tuất tuót chót qua thì thôi		thẹn kẽn vưống ngưu huy nữa tinh	
cm		đã tuôn tuôn trót qua thì thôi		thẹn kiêu ngưu mệnh nữ tinh	
cmb		đã tuộc tuôn chót qua thì thì thôi		thẹn kén mác ngưu tinh nữa tinh	
src		翁浪彼此一時		羞見牽牛織女星	
truth		ông rằng bỉ thử nhất thì		lậu thủy đinh đinh chúc ảnh hồng	
bl	2	ông rằng dữ nhất thì	6	là thuỷ đinh đinhm hồng hồng	
c		ông rằng lẫn nhất thì		lậu thuỷ đinh đinh ngòi hồng	
cm		ông rằng lẫn nhất thì		lậu thuỷ đinh đinh cúi hồng	
cmb		ông rằng lẫn nhất thì		lầu thuỷ đinh đinh chau bóng hồng	
src		泛弹迢迭瓶仙		漏水丁丁燭影紅	
truth		phím đàn dìu dặt tay tiên		quả mai ba bảy đương vừa	
bl	3	phím đàn dù giật tay tiên	7	quả mai ba bảy bảy đang bề	
c		phím đàn dù giật tay tiên		quả mai ba bảy đương bề	
cm		phím đàn giậm tay tiên		quả mai ba bảy đương bề	
cmb		phím đàn dìu giật tay tiên		quả mai ba bảy đang vừa	
src		春准瑶池兼溦襀		果梅呸毕當皮	
truth		xuân chốn dao trì nhường bể rộng		nghìn năm dẳng dặc quan giai lần lần	
bl	4	xuân chốn dao trì dường bể rộng	8	nghìn năm dẳng dẳng quan giai lần	
c		xuân chốn dao trì dường bể rộng		nghìn năm dẳng dẳng dẳng giai lần	
cm		xuân chốn dao dường bể rộng		nghìn năm dẳng dẳng quan giai lần	
cmb		xuân chốn dao trì dường bể rộng		nghìn năm dẳng dặc quan giai lần lần	

For future research, three primary avenues emerge: First, to maintain suitable sentence lengths, constraints should be introduced in the transliteration process from the Nom script to the National Script. Second, there's a need to refine transliteration models to effectively group Nom characters that have analogous pronunciations and to discern similarities among homophones. Finally, building on the initial positive outcomes from integrating BERT-NMT into multilingual translation, a deeper evaluation of its efficacy on standardized benchmarks is essential to fully understand the benefits of employing large language models in multilingual machine translation.

References

1. Dinh, D., Nguyen, P., Nguyen, L.H.B.: Transliterating nôm scripts into Vietnamese national scripts using statistical machine translation. Int. J. Adv. Comput. Sci. Appl. **12**(2) (2021). https://doi.org/10.14569/IJACSA.2021.0120205
2. Nguyen, H.B.L., Trang, M.C., Nguyen, T.H., Dinh, D.: Nôm-scripts transliteration using multilingual neural machine translation approach. In: Proceedings of the 14th National Conference on Fundamental and Applied Information Technology Research, Ho Chi Minh City, Vietnam: Nhà xuất bản Khoa học tự nhiên và Công nghệ,, December 2021 (2021). https://doi.org/10.15625/vap.2021.0091

3. Wang, Y., Zhou, L., Zhang, J., Zhai, F., Xu, J., Zong, C.: A compact and language-sensitive multilingual translation method. In: Proceedings of the 57th Annual Meeting of the Association for Computational Linguistics, pp. 1213–1223 (2019)

4. Dabre, R., Fujita, A., Chu, C.: Exploiting multilingualism through multistage fine-tuning for low-resource neural machine translation. In: Proceedings of the 2019 Conference on Empirical Methods in Natural Language Processing and the 9th International Joint Conference on Natural Language Processing (EMNLP-IJCNLP), pp. 1410–1416 (2019)

5. Zhu, J., Xia, Y., Wu, L., et al.: Incorporating BERT into neural machine translation. arXiv preprint arXiv:2002.06823 (2020)

6. Dong, D., Wu, H., He, W., Yu, D., Wang, H.: Multi-task learning for multiple language translation. In: Proceedings of the 53rd Annual Meeting of the Association for Computational Linguistics and the 7th International Joint Conference on Natural Language Processing (Volume 1: Long Papers), Beijing, China: Association for Computational Linguistics, July 2015, pp. 1723–1732 (2015). https://doi.org/10.3115/v1/P15-1166

7. Luong, M.-T., Le, Q.V., Sutskever, I., Vinyals, O., Kaiser, L.: Multi-task sequence to sequence learning (2016). arXiv: 1511.06114 [cs.LG]

8. Johnson, M., Schuster, M., Le, Q.V., et al.: Google's multilingual neural machine translation system: enabling zero-shot translation. CoRR, abs/1611.04558 (2016). arXiv: 1611.04558

9. Cho, K., van Merriënboer, B., Bahdanau, D., Bengio, Y.: On the properties of neural machine translation: encoder-decoder approaches. In: Proceedings of SSST-8, Eighth Workshop on Syntax, Semantics and Structure in Statistical Translation, Doha, Qatar: Association for Computational Linguistics, October 2014, pp. 103–111 (2014). https://doi.org/10.3115/v1/W14-4012

10. Gehring, J., Auli, M., Grangier, D., Yarats, D., Dauphin, Y.N.: Convolutional sequence to sequence learning (2017). arXiv: 1705.03122 [cs.CL]

11. Vaswani, A., Shazeer, N., Parmar, N., et al.: Attention is all you need (2017). https://doi.org/10.48550/ARXIV.1706.03762

12. Brown, T.B., Mann, B., Ryder, N., et al.: Language models are few-shot learners. CoRR, vol. abs/2005.14165 (2020). arXiv: 2005.14165

13. Clinchant, S., Jung, K.W., Nikoulina, V.: On the use of BERT for neural machine translation. In: Proceedings of the 3rd Workshop on Neural Generation and Translation, Hong Kong: Association for Computational Linguistics, November 2019, pp. 108–117 (2019). https://doi.org/10.18653/v1/D19-5611

14. Chiang, T.-R., Chen, Y.-P., Yeh, Y.-T., Neubig, G.: Breaking down multilingual machine translation. In: Findings of the Association for Computational Linguistics: ACL 2022, Dublin, Ireland: Association for Computational Linguistics, May 2022, pp. 2766–2780 (2022). findings-acl.218. https://doi.org/10.18653/v1/2022

15. Peters, M.E., Neumann, M., Iyyer, M., et al.: Deep contextualized word representations. In: Proceedings of the 2018 Conference of the North American Chapter of the Association for Computational Linguistics: Human Language Technologies, Volume 1 (Long Papers), New Orleans, Louisiana: Association for Computational Linguistics, June 2018, pp. 2227–2237 (2018). https://doi.org/10.18653/v1/N18-1202

16. Devlin, J., Chang, M.-W., Lee, K., Toutanova, K.: BERT: pre-training of deep bidirectional transformers for language understanding. arXiv preprint arXiv:1810.04805 (2018)

17. Nguyen, T.Q., Chiang, D.: Transfer learning across low-resource, related languages for neural machine translation. In: Proceedings of the Eighth International Joint Conference on Natural Language Processing (Volume 2: Short Papers), Taipei, Taiwan: Asian Federation of Natural Language Processing, November 2017, pp. 296–301 (2017)
18. Zoph, B., Knight, K.: Multi-source neural translation. In: Proceedings of the 2016 Conference of the North American Chapter of the Association for Computational Linguistics: Human Language Technologies, San Diego, California: Association for Computational Linguistics, June 2016, pp. 30–34 (2016). https://doi.org/10.18653/v1/N16-1004
19. Luong, M.-T., Manning, C.D.: Stanford neural machine translation systems for spoken language domain. In: International Workshop on Spoken Language Translation, Da Nang, Vietnam (2015)
20. Cettolo, M., Federico, M., Bentivogli, L., et al.: Overview of the IWSLT 2017 evaluation campaign. In: Proceedings of the 14th International Conference on Spoken Language Translation, Tokyo, Japan: International Workshop on Spoken Language Translation, December 2017, pp. 2–14 (2017)
21. Kudo, T., Richardson, J.: SentencePiece: a simple and language independent subword tokenizer and detokenizer for neural text processing. In: Proceedings of the 2018 Conference on Empirical Methods in Natural Language Processing: System Demonstrations, Brussels, Belgium: Association for Computational Linguistics, November 2018, pp. 66–71 (2012). https://doi.org/10.18653/v1/D18-2012
22. Papineni, K., Roukos, S., Ward, T., Zhu, W.-J.: BleU: a method for automatic evaluation of machine translation. In: Proceedings of the 40th Annual Meeting of the Association for Computational Linguistics, Philadelphia, Pennsylvania, USA: Association for Computational Linguistics, July 2002, pp. 311–318 (2002). https://doi.org/10.3115/1073083.1073135

FDPS: A YOLO-Based Framework for Fire Detection and Prevention

Tan Duy Le[1,3], Huynh Phuong Thanh Nguyen[2(✉)], Duc Tri Tran[1,3],
An Mai[1,3], Kha Tu Huynh[1,3], and Sinh Van Nguyen[1,3]

[1] School of Computer Science and Engineering, International University, Ho Chi
Minh City, Ho Chi Minh, Vietnam
[2] Japan Advanced Institute of Science and Technology, Nomi, Ishikawa, Japan
s2210406@jaist.ac.jp
[3] Vietnam National University, Ho Chi Minh City, Ho Chi Minh, Vietnam

Abstract. Accidental fires or explosions pose a significantly threat to human life and social safety, making them a major concern for humanity. In this study, we propose FDPS, an inexpensive and efficient machine learning-based framework for the early detection and extinguishing of small fires using a minimal amount of Internet of Things equipment, such as a NodeMCU, camera, and pump. The system functions through two primary phases: fire detection and fire extinguishing. During the initial phase, the camera rotates horizontally while continually analyzing images that were collected and processed by a pre-trained YOLO model to detect the presence of fire. The second phase activates when the fire is detected, and the system transmits signals to adjust the camera's angle and activates the water pump, which then extinguishes the fire. Our experiment results demonstrate that the proposed framework can accurately detect fires, achieving an average precision of nearly 90% and running inference in about five milliseconds. By detecting fires quickly, emergency responders can be alerted to respond promptly, potentially reducing property damage and saving lives.

Keywords: Fire Detection · Internet of Things (IoT) · Image
Processing · Fire Extinguishing · YOLO

1 Introduction

Nowadays, along with social and economic development, there are more and more industrial structures, export processing zones, and high-tech zones. Process lines, goods, and supplies have become modern and expensive. Those new process lines, materials, and equipments are highly flammable and explosive. Besides, the world has been urbanizing rapidly, especially in Vietnam. Many residential areas, high-rise, and complex buildings have been constructed using combustible materials, increasing the fire risk. Fires and explosions have always been severe problems faced by countries worldwide. In Vietnam, they happen every year and cause tremendous damage to lives and property. Moreover, fires and explosions in

N. Thai-Nghe et al. (Eds.): ISDS 2023, CCIS 1950, pp. 148–160, 2024.
https://doi.org/10.1007/978-981-99-7666-9_12

industrial structures and plants can even hamper the functioning of the country's economy.

According to statistics from the Vietnam Fire and Rescue Police Department, in 2021, there were 2,245 fires reported across the country, 85 people died, and direct property losses totaled 374.42 billion VND [1]. Most fires are caused by electrical system failures, irresponsible use of fire and heat, and other factors. The pressure and difficulty of fire prevention and control will continue to rise, necessitating predicting dangers and studying preventive measures in advance. Detecting and extinguishing a fire soon can reduce the loss of lives and property.

Machine learning is a branch of artificial intelligence where machines learn from data without explicit programming. It automates the development of analytical models and utilizes deep learning for various applications such as fraud detection, object recognition, and speech-to-text. The Internet of Things (IoT) connects billions of physical devices to the internet, allowing them to collect and exchange real-time data without human involvement. This connects the digital and physical worlds, creating a more intelligent and responsive environment.

In recognition of stated pressing issues, as well as the introduction and development of the mentioned advanced technologies, this paper developed FDPS, a low-cost, easy-to-use fire detection and prevention framework that meets the fundamental functionality criteria. The system is built based on Internet of Things (IoT), image processing, and object detection with the mission of applying science and technology to real life and contributing to controlling the risk of fires and explosions. The system combines state-of-the-art technological concepts such as the IoT and Convolutional Neural Network, which helps improve the system's performance and increase accuracy in fire detection. Including low-cost IoT devices lowers the cost of developing and operating the system and makes it more portable. When a fire gets detected, the system automatically adjusts the water pipe to point towards the fire, thus focusing on the amount of water required to extinguish it. Furthermore, the system's model can detect fire in real time at a reasonable frame rate and requires less computational power for inference.

The compact design of this system allows for installation in high-risk areas such as offices, hallways, lobbies, and homes, where fire and explosion risks are prevalent. The system is easy to install, maintain, and repair due to its simple design. It can be widely used in production or serve as a foundation for further research and technology innovation in fire detection and prevention at universities, research centers, and application centers.

2   Literature Review

Fire detection is a critical issue. Therefore, a system that can limit the harm caused by the vast number of fire accidents occurring every day is urgently needed. Initially, the researchers focused on the color and motion features of flame detection to build customized algorithms for fire detection. Thou-Ho et al. [2] used both dynamic and chromatic attributes of smoke and fire for accurate flame detection. Two separate color schemes are used by Celik et al. [3] to distinguish fire from smoke. Zaidi et al. [4] employed the YCbCr color space with some adjustments to solve the shortcomings of the prior technique by developing more generic rules to detect fire. Still, the associated drawbacks were a high false detection rate and the ability to detect the fire only at a feeble distance. In certain publications, the motion property has been used as a criterion for catching fire apart from the color attribute. In certain publications, apart from the color attribute, the motion property has been used as a criterion for detecting fire. Rafiee et al. employed the dynamic and static qualities of fire, and smoke [5]. However, the false-negative rate remains a concern due to other objects with identical color attributes to fire pixels.

For flame detection, Qiu et al. [6] suggested an auto-adaptive edge detection approach. Rinsurongkawong et al. [7] employed the dynamic features of fire to detect flames, but this method did not work with photos that contain pseudo-fire-like items in the background. The authors of [8] proposed two optical flow estimators to distinguish fire from non-fire objects, which solved the misidentification problem. Safe from fire, a fire detection system developed by Mobin et al. Author in [9] used multiple sensors to detect and distinguish between fire and smoke. However, the system became more expensive due to the utilization of several sensors.

Rishika Yadav and Poonam Rani [10] suggested an IoT-based fire alarm system that detects fire early on, created an automated alarm and alerted the remote user or fire control station. The system also attempted to put out the fire. With the assistance of a fire and a smoke sensor, using Arduino was proposed to detect the surroundings for fire occurrence. The system succeeded in detecting fire and allowing users to control the system remotely. However, there was a difference in the time needed to receive signals from both sensors because the temperature increase takes a significant time while the natural circulation of smoke takes less time. The signal from the smoke detector was obtained within 45 s, while it took around 4 min and 40 s to get the signal from the temperature sensor. It is hazardous as the fire can spread rapidly during that time. On the other hand, those sensors can only give the best output when the environment temperature is high, and smoke density is dense.

These fire detection techniques, such as the Raspberry Pi, are inexpensive and can be implemented on low-cost hardware with a reasonable frame rate. Still, they have the drawback of requiring human feature extraction from raw pictures. Hand-engineering becomes exceedingly time-consuming and inefficient due to this problem, mainly when the dataset comprises many photos. Automatic feature extraction is an advantage of DL-based systems, making the procedure significantly more efficient and dependable than traditional handmade image processing techniques. However, these deep learning algorithms, on the other hand, need a lot of computing resources during training and when the learned model is deployed to hardware to perform a specific task. Because the device must be physically and economically equal to a traditional fire detector, the algorithm's ability to run on any heavy hardware, such as a personal computer, is worthless in fire detection.

Nowadays, wet fire sprinkler systems [11] are installed and used widely to prevent fire. When the fire is detected, it ejects water when the temperature exceeds the preset value. Sprinkler systems provide early fire control or extinguishment. The sprinkler systems are designed based on the ceiling height as these systems are installed on the rooftops. The different types of sprinkler systems are the dry pipe, wet pipe, pre-action, and deluge. The wet-pipe sprinkler system includes an alarm check valve, valve tamper switches, water flow switches, and other values, and the basic requirement for this system to work efficiently is the continuous flow of water supply and sufficient pressure to manage the sprinkler. The system, however, sprays water over the entire space it occupies while there are spaces not on fire. Wasting water may lead to the need for more water for big fires. Moreover, water spraying manner can cause damage to documents and facilities in offices and houses.

3   Methodology

3.1   Framework Architecture

The overview of the FDPS framework can be seen in Fig. 1. The FDPS - fire detection and prevention system framework consists of three sequential operations: Receiver, External Computing, and Output. In the Receiving process, an embedded system is responsible for controlling the camera device to rotate and scan the surrounding areas. This embedded system plays a pivotal role in creating the video output by capturing raw video data from the sensor and processing it into a video format. This system is responsible for receiving raw video data from the camera sensor, performing necessary processing like color correction, compression, and outputting the formatted video. Additionally, this system controls many aspects of the camera operation, such as rotation, scan, focus, or zoom. It may also incorporate features such as image stabilization or noise reduction. As a result of this process, the video is produced as the first process's output.

In the External Computing process, the video output is used as input data. Image frames from the video of the space are taken, and the Image Processing stage analyzes them. In this stage, the video frame will be transformed into training data for the YOLO model. A large number of fire-related images are used to be the training dataset. Therefore, the image processing stage will capture the images that contain the fire object from the video frame set to be the output of the image processing stage. These images constitute the Fire Detection Model input to determine the fire position. This process is crucial for accurate fire detection, and its significance cannot be overstated.

Consequently, after calculating the fire position, the information is sent to the Output process. In this process, the results are displayed on the screen, and the pump system is activated to extinguish the fire. When the system is turned on, the stepper motor rotates, causing the camera to scan the space around it.

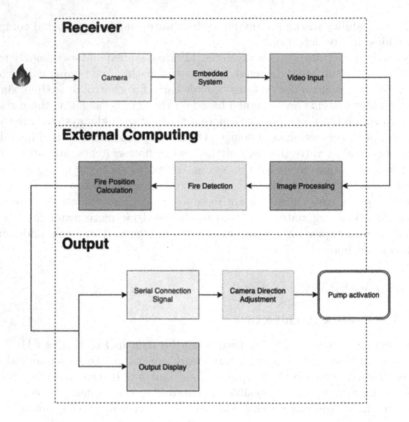

Fig. 1. Overview of FDPS - fire detection and prevention system framework.

When a fire emerges in the camera frame, the image recognition system placed on the computer recognizes the location of the fire. It then sends a signal to the processor based on the position of the fire. The processor receives a signal that adjusts the camera-oriented motor, so the fire is located in the center of the camera's observable frame. Once the camera is properly positioned to see the fire in the center, the CPU transmits a control signal that activates the pump via the Relay. The pump will be turned on until the fire goes out and then stop. The camera then returns to the scanning location, continuing to scan until there is a fire, then re-performing the cycle above. This system is an essential tool for fire protection and can save many lives and billions of dollars in damages.

3.2 Embedded System

Figure 2 shows how the hard wares are connected, and the water pump is connected to any water source with enough water to put out a small fire through water pipes. The hardware's program is written using the Arduino integrated development environment software using the processing language. The smaller the step angle of the stepper motor, the more steps it takes on each revolution,

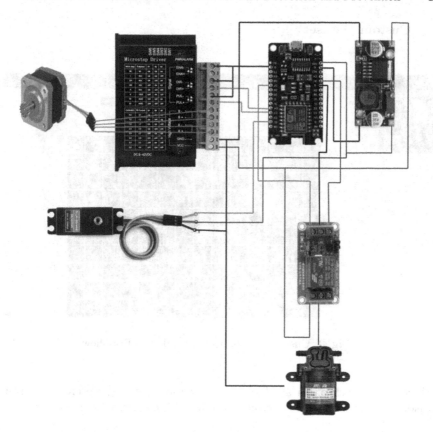

Fig. 2. Circuit diagram.

and the higher the precision of the resulting location. The stepper motor of the system is adjusted to achieve the 1:16 micro-step resolution (SW1 ON, SW2 ON, SW3 OFF). This means the stepper motor will rotate 0.1° for each pulse and need 3600 pulses to rotate 360°. This makes video processing easy, and the fire detection model operates efficiently but still ensures the speed and time required to adjust the camera direction when there is a fire. Because the rotation angle allows the system to be put in any room area and is not damaged due to its present design, it is designed to spin 90° with 900 pulses. For every 30 ms, a pulse signal is transmitted, and as a result, the camera rotates 1.8°/16. The whole process takes about 27 s for the camera to rotate 90° forwards and backward. During the procedure, if the system receives a signal transmitted through the serial connection on the computer, meaning that a fire is detected, the system will adjust its motion accordingly to the defined pattern:

– Signal 1 or −1: change the camera's horizontal direction
– Signal 0: 1 or 0: −1: change the camera's vertical direction
– Signal 0: 0: activate the water pump to extinguish the fire

3.3   Custom Dataset

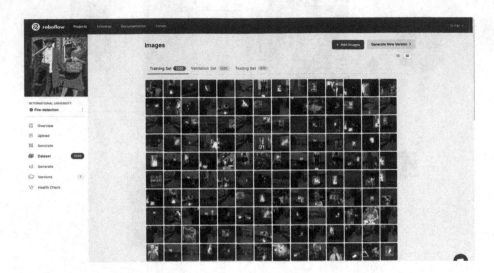

Fig. 3. Custom fire dataset uploaded to Roboflow

Due to the lack of an available fire detection dataset, we create a training dataset by collecting many custom fire-related images from the internet.

Prepare Data: The dataset's images are collected from available datasets such as COCO, taken by the authors, and taken from sources on the Internet. Then, the images are saved at different sizes and applied with augmentation methods such as crop, rotation, and cutout to increase the size of the dataset. The images are then labeled along with the coordinates of the object (with or without the presence of fire). Finally, the data set consists of approximately five thousand photographs, suitable for use in the train model in this project. Besides, after the dataset is uploaded to the Roboflow platform, it gets resized to 416×416 pixels to be consistent.

Label Data: LabelImg is used to label data collected. LabelImg is a visual image annotation tool with boxes covering objects that can be labeled in an image. It is written in Python and has a graphical user interface built with Qt. The annotations are recorded as XML files in the PASCAL VOC format, which ImageNet uses. In addition, it supports the YOLO format.

Upload Data to Roboflow: Since Roboflow supports converting datasets to YOLO Darknet format, which YOLOv4 tiny model uses, the fire detection dataset is uploaded to Roboflow as shown in Fig. 3. Besides, Roboflow also provides an annotation tool and other data preprocessing tools, making it more convenient to process data for future use. In this project, the images uploaded to Roboflow are resized to 416 × 416.

3.4 Fire Detection Model

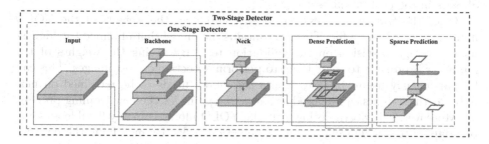

Fig. 4. One-Stage Detector, Two-stage detector [12]

This research uses YOLOv4 as the core model for fire detection, as shown in Fig. 4. YOLO [13] stands for You Only Look Once. It is an object detection model that uses features learned by a deep neural network to detect objects. YOLO is a CNN-based network model that detects, identifies, and classifies objects. It consists of a mix of linked layers and convolutional layers. The fully connected layers estimate the subject's probability and coordinates, while the convolutional layers extract the image's features. The YOLO model architecture follows a similar pattern of components:

- **Head**: Includes neck characteristics and bounding box predictions. The detection operation is completed by classifying and regressing the characteristics and bounding box coordinates. Generated values include x, y, width, and height.
- **Neck**: A group of layers integrate and blend attributes before transmitting them to the prediction layer. Feature pyramid networks, such as FPN, PAN, and Bi-FPN, are included.
- **Backbone**: A convolutional neural network that gathers and creates different sizes and forms of visual features. Feature extractors like Visual Geometry Group, EfficientNet, and ResNet are employed in classification models.

The following steps are taken to implement YOLOv4 on the custom dataset: 1) Set up a GPU environment in Google Colab. 2) Install the YOLOv4 Darknet training environment. 3) Clone the custom dataset for the YOLOv4 model prepared by Roboflow and set up necessary directories. 4) Create a specific YOLOv4 training configuration file for Darknet. 5) Train the YOLOv4 object detector with the custom dataset. 6) Reload the YOLOv4 trained weights and perform inference from test images.

Based on the number of classes in the custom dataset, YOLO models slightly modify the architecture design, and the training time should be adjusted accordingly. This means that depending on the dataset, a set of custom variables must be created in order to ensure optimal performance. These variables include the number of classes, maximum batches, layer filters, and iteration steps. By tuning these variables, the YOLO model can be trained to better recognize and classify objects in the custom dataset.

Once the training process has been completed, the weights of the YOLO model are saved into a file. This file contains the model that obtained the highest mean average precision on the validation test. By saving the weights of the model, it is possible to reuse them to perform detection on test images. This can be particularly useful in cases where the model needs to be retrained on new data, as the weights can be used as a starting point for the new training process.

Moreover, it is worth noting that the YOLO model is highly flexible and can be adapted to a wide range of image detection tasks. By modifying the network design and adjusting the training time, the model can be fine-tuned to better suit specific needs. This means that with the right adjustments, the YOLO model can be trained to effectively detect and classify objects in a variety of contexts, from simple object detection to complex scene understanding.

3.5   Video Processing and Inference

The process of fire detection and control involves several steps. First, the system camera captures video and reads each frame. Then, the frames are analyzed to detect the presence of fire. If a fire is detected, the program calculates its position and continuously transmits signals to a NodeMCU through a serial connection. The NodeMCU then adjusts the camera direction to ensure that the fire is in the right position in the frame.

To reduce image noise, the program applies a blurring technique to the frames. Additionally, the program uses the YOLOv4 tiny model, which has been previously trained, to perform inference during the fire detection process.

Once the fire is in the correct position in the frame, the pump is turned on and water is sprayed toward the fire until it is extinguished. The program continuously checks the fire and sends a signal to the system to confirm that the fire has been controlled once it has been completely extinguished.

A detailed illustration of the process of image capturing and fire detection can be seen in Fig. 5.

4   Result and Evaluation

4.1   Result

After completing the configuration of the environment, data, and training settings, the custom YOLOv4 small model can begin the training process. During

(a)

(b)

Fig. 5. The fire detection result a) before and b) after detection.

this process, the model achieves a mean average accuracy of 89.49% due to the training dataset. Once the training procedure is completed, the performance of the model is evaluated using the test data. The results show that the model can detect fires with over 80% confidence. The inference time for a batch size of 1 and GPU Tesla T4 is also relatively fast, taking only about five milliseconds to run.

To manage the entire system process over the internet, an application is developed for smartphones using Blynk. Blynk [14] is an IoT platform that allows for remote hardware control, displaying sensor data, storing and visualizing data, and performing other exciting functions. To set up our project, we created an app on the blynk server using port 9443. Whenever a fire is detected, notifications will be sent to our phones to keep us informed.

4.2 Evaluation

We have successfully developed a functional system that adequately fulfills its intended purposes. As this research focuses on constructing a fire detection and prevention system, it is essential to adhere to such a system's relevant standards and requirements. The following features were successfully incorporated into the final working product:

- Detect fire precisely and its position by the camera from early stages
- Consume very few memory resources
- Localize the fire and then spray water effectively to extinguish the fire
- Low-cost, robust, portable
- Announce users when there is a fire through applications
- Can be applied to the real world and contribute to minimizing the risk of fires and explosions

After conducting many experiments, the system takes about 30 s to detect a fire, adjust the direction and extinguish the fire. Depending on the circumstances, the camera can put out the fire. For example, if the source of the fire comes from a cigarette or candle, the time required for the fire to spread out will be slow, then the camera will have enough time to identify and extinguish it immediately. On the contrary, if the source of the fire is from the gas explosion and the fire spreads quickly, the camera may need to be fixed. This situation can be improved by further research and implementation in the future.

Fire detection system based on the YOLO model is a relatively new approach to fire detection that uses deep learning algorithms to identify flames in images and videos. Table 1 shows the comparison between our fire detection system and other standard fire detection systems. Overall, fire detection based on the YOLO model has several advantages over other fire detection systems, including higher accuracy, real-time detection, and the ability to distinguish between different types of fires. However, YOLO-based systems may be more expensive than Conventional Fire Detection and require more computing resources than other systems. They may not be suitable for all environments or applications. Ultimately, the choice of fire detection system depends on the specific needs and requirements of the building or facility being protected.

5   Discussion and Conclusion

The community is approaching social and economical resolution rapidly, increasing industrial buildings and the number of zones. All of them have become more highly combustible and explosive. Building a reliable fire detection system for preventing a flammable accident is a trivial activity since it invokes detecting fire penitential soon and resulting extinguishing at the first stage. This paper presents FDPS, a cost-effective and efficient machine learning-based framework for early fire detection and suppression with a minimal set of Internet of Things (IoT) devices. The system comprises two primary phases, namely fire detection and extinguishing. In the first phase, the camera rotates horizontally while continuously analyzing images captured by a pre-trained YOLO model to detect any signs of fire. The second phase is triggered when a fire is detected, and the system sends signals to adjust the camera's angle and activates the water pump to extinguish the fire.

Table 1. Comparison between YOLO Fire Detection System and other standard fire detection systems.

	Functionality	Accuracy	Time	Cost
FDPS (our proposed framework)	Use deep learning algorithms to detect flames in images which allows for accurate detection of different types of fires, real-time data, and fast response times	High	Fast	Cheap
Conventional Fire Detection [15]	Basic and can only detect the presence of smoke or heat, but cannot distinguish between different types of fires or provide real-time data	Low	Slow	Cheap
Addressable Fire Detection [16]	Allow for more control and flexibility: the ability to locate the source of the fire	Low	Slow	Expensive
Analog Fire Detection [17]	Detect changes in the environment that may indicate the presence of a fire	Low	Slow	Expensive
Wireless Fire Detection [18]	Offers flexibility but may not be as accurate or reliable as YOLO-based systems	Low	Slow	Expensive

The results of our experiments demonstrate that our model can detect fires in images with high accuracy, achieving a mean average precision of nearly 90%, and the inference takes about five milliseconds to run. This fire detection model has potential applications in various settings, including public safety, industrial facilities, and homes. With the ability to detect fires quickly, emergency responders can be alerted to respond promptly, potentially minimizing property damage and saving lives. In future work, our model could be optimized and evaluated for its performance in real-world settings. A plan is designed to integrate more state-of-the-art object detection models into the system so that it can be used to serve as a multi-function surveillance system. It will also be scaled by adding more sensors and data sources to enhance the accuracy and efficiency of our fire detection system.

Acknowledgment. We gratefully acknowledge AIoT Lab Vietnam for their invaluable support in this research.

References

1. V. News: Summer rise in fatal fires leads to calls for caution. https://www.viet namnews.vn/society/1250808/summer-rise-in-fatal-fires-leads-to-calls-for-caution. html. Accessed 27 Feb 2023
2. Chen, T.-H., Wu, P.-H., Chiou, Y.-C.: An early fire-detection method based on image processing. In: 2004 International Conference on Image Processing, ICIP 2004, vol. 3, pp. 1707–1710 (2004)
3. Çelik, T., Özkaramanli, H., Demirel, H.: Fire and smoke detection without sensors: image processing based approach. In: 2007 15th European Signal Processing Conference, pp. 1794–1798 (2007)
4. Zaidi, N., Lokman, N., Daud, M., Achmad, M., Khor, A.: Fire recognition using RGB and YCBCR color space, vol. 10, pp. 9786–9790, January 2015
5. Rafiee, A., Dianat, R., Jamshidi, M., Tavakoli, R., Abbaspour, S.: Fire and smoke detection using wavelet analysis and disorder characteristics. In: 2011 3rd International Conference on Computer Research and Development, vol. 3, pp. 262–265 (2011)
6. Qiu, T., Yan, Y., Lu, G.: An autoadaptive edge-detection algorithm for flame and fire image processing. IEEE Trans. Instrum. Meas. **61**(5), 1486–1493 (2012)
7. Rinsurongkawong, S., Ekpanyapong, M., Dailey, M.N.: Fire detection for early fire alarm based on optical flow video processing. In: 2012 9th International Conference on Electrical Engineering/Electronics, Computer, Telecommunications and Information Technology, pp. 1–4 (2012)
8. Mueller, M., Karasev, P., Kolesov, I., Tannenbaum, A.: Optical flow estimation for flame detection in videos. IEEE Trans. Image Process. Publ. IEEE Signal Process. Soc. **22**, 04 (2013)
9. Iftekharul, M., Ar Rafi, A., Neamul, M., Rifat, M.: An intelligent fire detection and mitigation system safe from fire (SFF). Int. J. Comput. Appl. **133**, 1–7 (2016)
10. Yadav, R., Rani, P.: Sensor based smart fire detection and fire alarm system. In: Proceedings of the International Conference on Advances in Chemical Engineering (AdChE) (2020)
11. W. The free encyclopedia: Fire sprinkler system. https://www.wikiwand.com/en/ Fire_sprinkler_system
12. Bochkovskiy, A., Wang, C.-Y., Liao, H.-Y.M.: YOLOv4: optimal speed and accuracy of object detection. arXiv preprint arXiv:2004.10934 (2020)
13. Redmon, J., Divvala, S., Girshick, R., Farhadi, A.: You only look once: unified, real-time object detection. In: Proceedings of the IEEE Conference on Computer Vision and Pattern Recognition, pp. 779–788 (2016)
14. Blynk intro. https://docs.blynk.cc/#intro
15. F. System: What is the difference between addressable and conventional fire alarm panels? https://firesystems.net/2020/11/17/what-is-the-difference-between-addressable-and-conventional-fire-alarm-panels/
16. Protec: Addressable or conventional fire alarm system? https://www.protec.co.uk/ latest-news/addressable-or-conventional-fire-alarm-system/
17. F. Security: What is an analogue-addressable fire alarm system? https://www.ae-fire.co.uk/analogueaddressablefirealarm/
18. S. F. S. Ltd.: Wireless fire alarm systems. https://surreyfire.co.uk/wireless-fire-alarm-systems/

Application of Swin Transformer Model to Retrieve and Classify Endoscopic Images

Ngo Duc Luu[1]([✉]) and Vo Thai Anh[2]

[1] Bac Lieu University, Bac Lieu, Vietnam
ndluu@blu.edu.vn
[2] Can Tho University, Can Tho, Vietnam

Abstract. The machine learning community is very interested in image classification and retrieval, especially in the area of computer vision and with an emphasis on medical image retrieval. Numerous machine learning approaches have been used for image retrieval problems and have made as a result of the ongoing developments in techniques like Convolutional Neral Networks (CNN) and Vision Transformers with quite good performances. The Swin Transformer model is used to create a specialized medical image retrieval system in this paper that is well suited to gastric endoscopic pictures. The suggested technique takes advantage of the Swin Transformer model's classification process to create feature vectors by combining fragmented image segments collected from local windows, making it easier to calculate similarity on the Kvasir dataset that we have added some additional images. Empirical results show that the Swin Transformer model retrieves endoscopic images with a remarkable classification accuracy of 90.5% and an 85% mean average precision at top 20 (mAP@20).

Keywords: Image Retrieval · Classification · Vision Transformer · Swin Transformer · Endoscopic Images

1 Introduction

A big database of images can be retrieved using the concept of image retrieval. The increased use and storage of digital photographs in the medical industry has made it challenging to search these sizable databases. Because of this, the adoption of a content-based picture retrieval system is becoming more and more popular. A system for searching and retrieving images from a sizable database of digital photographs is known as an image query system. Adding metadata to the photos in the form of subtitles, keywords, or descriptions in order to enable retrieval appears in the footnotes is the most conventional and traditional technique of image retrieval. It takes much time, money, and labor to manually annotate photographs. To solve this problem, there has been a lot of research to implement automatic image annotation.

In recent years, the number of people suffering from ColoRectal Cancer (CRC) is increasing, accounting for one-third of all cancers in the world for many consecutive years [1]. However, according to medical organizations, the key issue to overcoming this

N. Thai-Nghe et al. (Eds.): ISDS 2023, CCIS 1950, pp. 161–173, 2024.
https://doi.org/10.1007/978-981-99-7666-9_13

situation is that we need to diagnose and prevent this disease from the very beginning. Several studies have demonstrated that nearly 95% of CRCs are from glandular polyps [2]. Removal of polyps can reduce the risk of developing CRC. Even so, the best way to deal with CRCis to diagnose and treat it as soon as possible. Nowadays, with the development of an increasing number of CRC patients, digital image storage is applied to store endoscopic images [3]. However, the doctors found it very difficult to query the database because the number of images in the database was too large.

With the strong development of CNN [4, 10, 16–18], there have been many architectural models applied in the feature vector generation process such as ResNet, DenseNet, and EfficientNet,... Together With many new models such as Vision Transformer [21–25], Mixer-MLP [6] has made the image retrieval model more and more diversified. In early March 2021, Microsoft's AI research team in Asia introduced a new version of Vision Transformer. It is the Swin Transformer [7] for image classification and has been awarded the best paper at the International Conference on Computer Vision (ICCV) 2021.

In this paper, we build an endoscopic image retrieval system with image data training based on Swin Transformer architecture and through this method generate feature vectors representing images. Calculate the similarity between the query image features and the features of the images in the dataset. The architecture of this Swin Transformer differs from the original paper in that it has removed the classification layer and added an embedding layer to generate feature vectors.

This article consists of 6 parts. The first part gives a general introduction to the article. The second section presents relevant studies. The third part presents the algorithm used for the search system. The fourth section presents data and methods. The fifth presents the experimental results. Finally, the comment section and give the conclusion of the article.

2 Related Research

The Swin Transformer [7] model is an improvement on the Vision Transformer (ViT)[5] and adds some important new methods, the Shift Window technique (there are 2 forms of padding and cyclic). This improvement has resulted in a better model that uses the Attention mechanism for image data.

The ViT model provided the ability to use the Transformer as a backbone for visual tasks. However, because a model such as a transformer conducts the Global Attention mechanism, its complexity increases exponentially with image resolution. This makes ViT ineffective for image segmentation tasks other complex tasks.

Because Swin transformer is an improved form of Vision transformer, based on that calculation platform, change is further developed. Parts not mentioned in Swin Transformer will remain as in the ViT model.

There are 3 main changes:

- Network Architecture
- Self -Attention in Non-overlapped windows
- Shifted Windows

These components have differentiated and mastered the creation of maps characterized by a single low resolution and high computational complexity due to Global Attention computation.

Network Architecture

There are 4 main components (See Fig. 1) in the Swin Transformer network: Patch Partition; Linear Embedding; Swin Transformer Block; Patch Merging. The special thing is that there are two important techniques that create efficiency in creating feature vectors of Swin Transformer: Self-attention in Non-overlapped Windows and Shifted Windows.

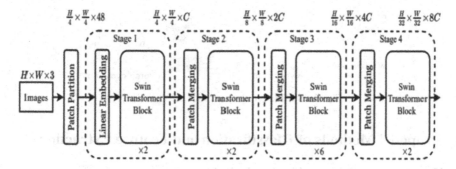

Fig. 1. Swin Transformer network architecture [7]

Self-attention in Non-overlapped Windows

One of the major contributions of the Swin Transformer is that it proposes to implement Self-Attention in a local window instead of globally (each bordered in red as shown below). The windows are arranged to evenly partition the image in a non-overlapping way, and each contains arrays $M \times M$ patch ($M = 7$ in the original paper). Swin Transformer is a better model than ViT because it does not use fixed patches to pass through Transformer Encoder classes. Which will create merged hierarchical featured maps.

Shifted Windows

The window-based on Self-Attention module lacks connectivity between windows, which limits its modeling capabilities. To internally connect multiple windows while maintaining efficient calculations of non-overlapping windows, the method of shifted window partitioning(See Fig. 2) is proposed. This method alternates between two partition configurations in successive Swin Transformer blocks.

To process the window at the boundary of the image, cyclic shift (See Fig. 3) is used. With cyclic variation, the batch window count will revert to the same number as the number of regular window partition windows and is therefore also more effi-cient. It is said to be more efficient and accurate than the sliding window method or the padding window method.

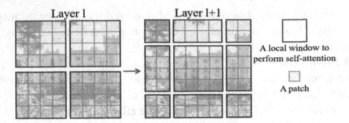

Fig. 2. Shifted window (padding) technique [7]

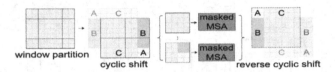

Fig. 3. Shifted window (cyclic) technique [7]

3 Proposed Methods of Classification and Feature Extraction

For Transformer models (See Fig. 4), when extracting features in the final decoder, the image patches after passing through the model will combine into features based on the original sequence numbers. From there, it will remain with the Transfer learning model that will use the MLP (MultiLayer Perceptron) layer to classify. In this MLP class will include components such as the GELU class, the Dense class, and the Dropout class. This MLP model is very familiar with the network model as Resnet50, because it is highly layered and ensures the characteristic value of the image to ensure an accurate reading of the model.

With Transformer models, the Decoder will be cut out, leaving only the feature at the heads where the image patches have been calculated after passing the model. These patches will be rejoined to return to their original position, thereby creating full characteristics at each end of the image. Feature words at the ends of the images will be used to create vectors characteristic of the Transformer model.

For the models in the experimental part of this query, there will be a Vision Transformer but extracted specifically by the DeiT [9] model. Because of the layering model of the Vision Transformer of the Hugging Face library, the library does not yet have an image-specific extraction feature for querying by content. This was replaced by the optimal model for extraction characterized by DeiT (Distilled Data-efficient Image Transformer) and obtained in the form of the Small model. This is also another form of Vision Transformer.

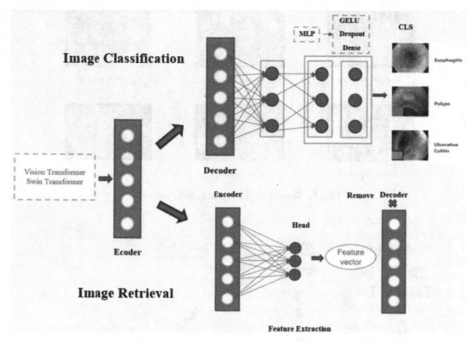

Fig. 4. Image Retrieval and Classification with Swin Transformer

4 Experimental Data and Method

4.1 Experimental Data

The dataset used to evaluate this query system is the Kvasir version 1 dataset [8, 15, 19]. The Kvasir dataset was collected using an endoscope at the Vestre Viken Health Trust (VV for short) in Norway. In which VV includes 4 hospitals and healcare for 470000 people. One of these hospitals has Baerum Hospital which operates a gastroenterology department where data is collected and provided. Furthermore the images were carefully annotated by one or more VV medical professionals and The Cancer Registry of Norway (CRN). Our data has been supplemented with 1000 endoscopic images from Soc Trang General Hospital.

The experimental dataset consists of 8000 images in 10 folds for cross validation during training and evaluation. 8000 images are divided into eight layers with 1000 images for each. Figure 5 illustrates some example layers: dyed-lifted-polyps, dyed-resection-margins (esophagitis), esophagitis (normal jejunum), normal-cecum (normal pylorus), normal-pylorus (normal glands), normalz-line, polyps (mucosa lesions) and ulcerative-colitis (ulcerative colitis).

4.2 Experimental Method

The experimental method referenced [6] (See Fig. 6) consists of 3 main parts:

– Extract Feature Retrieval Dataset.

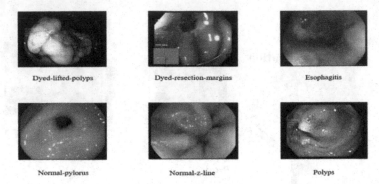

Dyed-lifted-polyps   Dyed-resection-margins   Esophagitis

Normal-pylorus   Normal-z-line   Polyps

Fig. 5. Some images in the Kvasir dataset

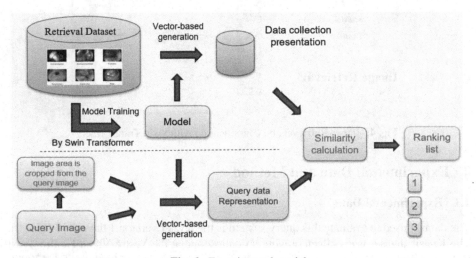

Fig. 6. Experimental model

- Extract Feature Query Image.
- Similarity Calculation and Rank List.

Extract Query Data Feature

With the dataset, the image query of gastroscopy will be fed into the Swin Transformer model and generate baseline feature vectors. Then reduce the image features to a fixed form for presentation and similarity calculation.

Extract Query Image Data

Since this query image is unique compared to previous system articles and the image we wish to query, we will immediately crop the part of the image that contains the most important elements. The result that follows will be the one that is most similar to the query image and in particular that location.

Similarity Calculation and Rank List

After obtaining the feature from the query image and retrieval dataset, we start performing cosine similarity measurements. After calculating the similarity, we will output images from dataset, ranking images with similarity from high to low.

4.3 Image Retrieval System Evaluation Measurement

The measurements evaluate the image search system based on the matrix of confusion based on content and similarity in Table 1. Here, specify that the similarity query element (similarity factor $>=$ 0.5) relative to the input image is positive (Positive), the non-homomorphic query element (similarity factor $<$ 0.5) compared to the input image is negative (Negative).

Table 1. Confusion matrix

	Predict Positive	Predict Negative
The fact is Positive	TP	FN
The fact is Negative	FP	TN

Where:

TN: The number of query elements is not similar and the content is incorrect.

FN: The number of query elements is not similar but has the correct content.

TP: The number of query elements is similar and has the correct content.

FP: The number of query elements is similar but has incorrect content.

- **Accuracy** - The accuracy of the image retrieval system by content is calculated by the number of query elements whose content is correct with the queried image, the formula is as follows:

$$Accuracy = \frac{TP + TN}{TP + FP + TN + FN} \quad (1)$$

For content-based search systems, determining accuracy must be based on the label of the data or the user's senses. Therefore, calculating Accuracy will not be able to evaluate the prediction of the system.

Therefore, to evaluate the effectiveness of the image retrieval system, different measurements are also used as follows:

- **Precision -** This measure calculates the proportion of query elements of the positive class that are correctly classified out of the total number of elements predicted to be positive classes, the formula is as follows:

$$Precision = \frac{TP}{TP + FP} \qquad (2)$$

- **Recall or SE -** This measure the proportion of query elements of the positive class is determined to have the correct content out of the total number of elements of the positive class, the formula is as follows:

$$Recall = \frac{TP}{TP + FN} \qquad (3)$$

- **F-Measure or F1-Score** - The measurement is calculated based on 2 measurement degrees, *precision* and *recall*, *F-Measure* is calculated according to the following formula:

$$F_1 = 2\frac{Precision * Recall}{Precision + Recall} \qquad (4)$$

- **mAP (Mean Average Precision) -** This is a measure of the aggregate results of multiple queries applied to the search system. To calculate, we must have *AP (Average Precision)* as the average of the precisions at the threshold points returned by each correct result, written with the following formula:

$$AP = \sum_{k=0}^{k=n-1} [Rs(k) - Rs(k + 1)] * Ps(k) \qquad (5)$$

recalls(n) = Rs(n) = 0, precisions(n) = Ps(n) = 1, n = threshold coefficient.
Once AP is available, the formula for *mAP* is written as follows:

$$mAP = \frac{1}{n} \sum_{k=1}^{k=n} AP_k \qquad (6)$$

AP_k = value AP của lớp k, n = number of class.
- **AUC:** ROC is a curve representing probability, and AUC represents the grading level of the model. AUC-ROC is also known as AUROC (**Area Under The Receiver Operating Characteristics**). The meaning of AUROC can be interpreted as follows: It is the probability that a randomly taken positive sample will be ranked higher than a randomly taken negative sample. Represented by the formula, we have **AUC = P(score(x +) > score(x-))**. The higher the AUC index, the more accurate the model is in classifying classes.

In this image search system, we use algorithms to compare the similarities of vectorized images [20] by calculating similarities between characteristic vectors. Then give suggestions to the system for images with high similarity. In practice there are many different degrees of homologation. However, within the scope of this study, we use the

Cosine similarity measure because it is very commonly used to measure the degree of similarity between two vectors, A is the feature vector of the image to be quered, B is the feature vector of the image in the experimental data, the specific formula is as follows:

$$CosineSimilarity(A, B) = \frac{A.B}{||A||||B||} = \frac{\sum_{i=1}^{n} A_i B_i}{\sqrt{\sum_{i=1}^{n} A_i^2} \sqrt{\sum_{i=1}^{n} B_i^2}} \tag{7}$$

The algorithm in the retrieval system is detailed as below.

- **Input**: *Query image*
- **Output**: *Top 20 recommended images*
- **Process**:

{

    prepare image database;

    data feature = model(SWIN).predict(image database);

    //fine-tuning

    query feature = model(SWIN).predict(query image)

    cosine(query_feature,data_feature)

    show recommend image retrieval

}

We recommend using one input image as the output of 20 suggested images. With this number of output images, it will be easier to evaluate and more applicable to real-world search systems.

5 Results

Table 2 shows the results of models in the experimental dataset. Experimentally, focusing on comparisons between convolutional models such as VGG16 [13], VGG19 [12], ResNet50[], InceptionV3 [14], Xception with Transformer models we have Vision Transformer and Swin Transformer. During the experiment, we adjusted the layering of the networks so that the results were most suitable for each structure. Optimize accuracy when starting to evaluate the model after being trained. Evaluating models based on Accuracy(ACC), Precision, Recall, and F1-Score (F1) indicators, models with competitive results give comments on data and models.

Comparing many classification results of models, the Model has the highest classification result of **Swin_Base** with AUC (98.6%) Accuracy (90.5%) Precision (91.4%) Recall (89.3%) F1 (90.2%) that highest. is a **Swin_Tiny** model with AUC (98%) Accuracy (88.7%) Precision (88.5%) Recall (88.2%) F1 (87.5%) this is the accuracy of Transformer models. The model with the highest accuracy among CNN models is **Xception** with AUC (99.2%) Accuracy (87%) Precision (87.2%) Recall (87.3%) F1 (87.2%).

Table 2. Results of evaluation of classification models.

Model	ACC (%)	PRECISION (%)	RECALL (%)	F1 (%)	AUC (%)
VGG16	85.9	85.8	85.9	85.9	98.7
VGG19	84.6	85.1	84.7	84.6	98.6
ResNet50	86.5	87.5	86.7	86.5	99.1
InceptionV3	82.6	83.5	82.6	82.6	98.4
Xception	**87.0**	**87.2**	**87.3**	**87.2**	**99.2**
ViT_Tiny	84.9	84.1	84	84.8	91.1
ViT_Base	86.8	86.2	86.7	86.4	91.8
Swin_Tiny	**88.7**	**88.5**	**88.2**	**87.5**	**98**
Swin_Base	**90.5**	**91.4**	**89.3**	**90.2**	**98.6**

Showing the difference in classification results of the best models, we can feel that the difference is not too outstanding, but we can still see the difference between the CNN model and the Transformer model in the classification problem. Experimental data set.

Table 3. Average accuracy of retrieval image experiments

Model	mAP@20 (%)	mAP@30 (%)	mAP@40 (%)	mAP@50 (%)
VGG16	71.5	70.3	66.5	63.4
VGG19	72.0	71.7	67.4	65.6
ResNet50	75.0	74.3	71.3	68.7
InceptionV	73.0	70.7	62/0	59.5
Xception	**78.0**	**75.3**	**67.6**	**66.5**
DeiT_Small	71.0	68.5	63.0	59.5
Swin_Tiny	**80.0**	**78.6**	**75.2**	**72.4**
Swin_Base	**85.0**	**79.5**	**74.8**	**72.2**

The results (See Table 3) of the internal retrieval of the models will be calculated by the average accuracy of mAP and calculated by calculating the AP accuracy of each search right with the class of the query image. The AP calculation of the experiment is used from the Scikit-learn library. With the change in numbers 20, 30, 40 and 50 here is about the number of result images returned for each query. But often the mAP@20 return is more noticeable. With this evaluation index, we will get the results as shown in Table 4 showing that the **Swin_Base** model has the highest mAP@20 (85%) compared to other models. Next is **Swin_Tiny** with the next highest mAP@20 (80%), finally **Xception** with mAP@20 (78%). With this index, the more results returned, the smaller the accuracy is because the occurrence of the returned results is not of the same type as the query image.

In addition to evaluating accuracy in the query domain, there are other evaluation metrics such as Feature Dimensions, Memory Cost, Extraction and Query Time. From the above criteria will help the experiment to easily identify the advantages and disadvantages of each model. From the results of (see Table 4) we can see that the best memory cost is for the **Swin Transformer** model (Tiny 30MB) (Base 40MB) and the CNN models with at least 125MB capacity of **VGG16**, **VGG19** and **InceptionV3**. The **DeiT_Small** model has the lowest characteristic dimension of 386. The Swin Transformer model has a relatively low number of characteristic dimensions with (Tiny 768) and (Base 1024) while the **CNN** models have two forms, 512 and 2048.

Table 4. The results of the models on the retrieval system

Model	Dimensions	Extraction Time	Query time	Memory cost	mAP @20
VGG16	512x512	16.5 m	1.84 s	125MB	72%
VGG19	512x512	18.3 m	2.16 s	125MB	73%
ResNet50	2048x2048	11.9 m	2.32 s	720MB	75%
InceptionV3	2048x2048	4.7 m	1.52 s	125MB	72%
Xception	2048x2048	4.2 m	1.4 s	720MB	78%
DeiT_Small	386x386	10.2 m	2.2 s	83MB	71%
Swin_Tiny	768x768	19.3 m	1.32 s	30MB	80%
Swin_Base	1024x1024	20.8 m	1.22 s	40MB	85%

Therefore, it cannot be concluded that the Transformer model has superior search time than the CNN model. The above results show 2 outstanding advantages of Swin Transformer: low memory cost and fast and accurate query time. The downside is that the extraction time is longer compared to CNN models. The extraction time of CNN models is much better than that of Swin Transformer. With the fastest time is 4.2 m and the slowest is 16.5 m and with the lowest Swin Transformer time is 18.3 m for Tiny. With query time, Swin Transformer has (Tiny 1.32 s) (Base 1.22 s) faster than Xception (1.4 s) and InceptionV3 (1.52 s). In fact, this query time difference is not significant compared to the search problem.

6 Conclusion

In general, our research proposes a method of using the Swin Transformer model to extract features of endoscopic images for an image retrieval system. Our method achieves competitive results in terms of content-based retrieval. By using the Swin Transformer model and slicing the classifier, we can create a vector to represent the data feature. That will allow a new approach to content retrieval. Furthermore, our method will serve as a good reference for many new models that will later evolve from the Vision Transformer model.

From this research in the coming time, the research team will develop applications to support the treatment of stomach diseases, and further develop identification of diseased areas based on endoscopic images.

References

1. Tang, J., Qu, M., Wang, M., Zhang, M., Yan, J., Mei, Q.: Line: large-scale information network embedding. In: Proceedings of the 24th International Conference on World Wide Web, pp. 1067–1077. ACM (2015)
2. Rao, N., Jiang, H., Luo, C.: Review on the applications of deep learning in the analysis of gastrointestinal endoscopy images. Article in IEEE Access, September 2019
3. Sommen, F., Zinger, S., Schoon, E.J. (eds.) Computer-aided detection of early Cancer in the Esophagus Using HD endoscopy images. In: Medical Imaging 2013: Computer-Aided Diagnosis, vol. 8670. International Society for Optics and Photonics, Florida (2013)
4. Hu, H., et al.: Content-based gastric image retrieval using convolutional neural networks. Accepted 20 July 2020
5. Dosovitskiy, A., et al.: An image is worth 16×16 words: transformers for image recognition at scale. Submitted on 22 Oct 2020 (v1)
6. Trinh, Q.-H., Nguyen, M.-V.: Endoscopy image retrieval by mixer multi-layer perceptron. Computer Science and Information Systems, pp. 223±226. ACSIS. ISSN 2300-5963
7. Liu, Z., et al.: Swin transformer: hierarchical vision transformer using shifted windows. Submitted on 25 Mar 2021 (v1)
8. Pogorelov, K., Randel, K.R., Griwodz, C., Eskeland, S.L., de Lange, T., Johansen, D., et al. (eds.) Kvasir: A multi-class image dataset for computer aided gastrointestinal disease detection. Paper presented at: Proceedings of the 8th ACM on Multimedia Systems Conference. ACM (2017)
9. Touvron, H., Cord, M., Douze, M., Massa, F., Sablayrolles, A., Jégou, H.: Training data-efficient image transformers & distillation through attention (2020)
10. Zeiler, M.: ADADELTA: An adaptive learning rate method. Endoscopic Image Classification and Retrieval use of the Clustered Convolutedonal Features, p. 1212 (2012)
11. Dubey, S.R., Singh, S.K., Chu, W.-T.: Vision transformer hashing for image retrieval, 26 September 2021
12. Xia, X., Xu, C., Nan, B.: Inception-v3 for flower classification, pp. 783–787 (2017). https://doi.org/10.1109/ICIVC.2017.7984661
13. Chebbi, I.: VGG16: VGQR (2021)
14. Chollet, F.: Xception: deep learning with depthwise separable convolutions, pp. 1800–1807 (2017). https://doi.org/10.1109/CVPR.2017.195
15. Pogorelov, K., et al.: KVASIR: a multi-class image dataset for computer aided gastrointestinal disease detection (2017). https://doi.org/10.1145/3083187.3083212
16. Maruyama, T., et al.: Comparison of medical image classification accuracy on the machine learning methods. J. X-ray Sci. Technol. **266**, 885, 893 (2018)
17. Yadav, S.S., Jadhav, S.M.: Deep convolutional neural network based medical image classification for disease diagnosis. J. Big Data **6**, 1–18 (2019)
18. Ahmad, J., Muhammad, K., Baik, S.: Medical image retrieval with compact binary codes generated in frequency domain using highly reactive convolutional features. J. Med. Syst. **42**, 119 (2017). https://doi.org/10.1007/s10916-017-0875-4
19. Shamna, P., Govindan, V.K., Nazeer, K.A.: Content-based medical image retrieval by spatial matching of visual words. J. King Saud Univ. Comp. Inf. Sci. **34** (2018). https://doi.org/10.1016/j.jksuci.2018.10.002

20. Image content based retrieval system using cosine similarity for skin disease images. ACSIJ Adv. Comput. Sci. Int. J. **2** (2013)
21. Song, C., Yoon, J., Choi, S., Avrithis, Y.: Boosting vision transformers for image retrieval (2022)
22. El-Nouby, A., Neverova, N., Laptev, I., Jégou, H.: Training vision transformers for image retrieval (2021)
23. Thakrar, A., et al.: Semantic retrieval of similar radiological images using vision transformers (2023). https://doi.org/10.1101/2023.02.16.23286056
24. Feng, Q., et al.: EViT: Privacy-preserving image retrieval via encrypted vision transformer in cloud computing (2023)
25. Tang, T., et al.: Learning self-regularized adversarial views for self-supervised vision transformers (2022). https://doi.org/10.48550/arXiv.2210.08458

Face Emotion Using Deep Learning

Chien D. C. Ta[⊠] and Nguyen Thanh Thai

Faculty of Information Technology, Industrial University of Ho Chi Minh City, Ho Chi Minh City, Vietnam
{taduycongchien,ngththai}@iuh.edu.vn

Abstract. After the COVID-19 pandemic, online helps people get used to shopping online, doing online work, giving online classes. Dedication of pregnancy due to the student's performance during and after a session is an important factor in online learning. The effect of helping learners to learn is expected to be two-way, as a result of the learning curve of the learners, due to the different approaches to learning, through better learning and better understanding. After learning about his performance, he is the next student in his face, the analysis is positive through the image in his face, and because of his high self-esteem. Ever since then, a method of analyzing students' commitment through the process of improving the model after VGG-16 has shown a high level of understanding in the student's commitment. The result is understood to be higher than that of a normal VGG-16. Furthermore, we evaluate the ontological structure and the relations of some terms of the ontology.

Keywords: Face emotion · Deep learning · Image processing

1 Introduction

The COVID-19 pandemic that has been ongoing for the past few years has made online learning a popular method. Thanks to online learning, students were able to attend classes during the pandemic, and classes were still conducted normally, which has made students increasingly adapt to online learning. Nowadays, even though the pandemic has passed, some schools still maintain online learning methods. Convenient softwares such as Zoom meeting, Google Meet, and Microsoft Meeting have made online learning easy and convenient. However, one of the difficulties of online learning is evaluating the level of concentration and satisfaction of learners during and after each class, as lecturers and students do not have direct face-to-face interactions, which can result in lower learning effectiveness.

Measuring and evaluating the emotions and learning attitudes of students is important in the teaching process, as it helps educators understand the psychology of students and adjust their teaching methods to suit the learners.

To improve the effectiveness of online teaching, we propose a facial emotion detection model to assess the level of satisfaction of students based on the VGG-16 deep learning model with some improvements over traditional models, which has achieved

N. Thai-Nghe et al. (Eds.): ISDS 2023, CCIS 1950, pp. 174–184, 2024.
https://doi.org/10.1007/978-981-99-7666-9_14

promising results. We collected over 20,000 images of students through Zoom meeting software to serve as training, validation, and testing data.

Our key contributions are as follows: (i) improving VGG-16 model with high accuracy; (ii) creating a dataset of over 20,000 images with six different states to serve the purpose of training, validation, and testing.; (iii) we also evaluate our methodology by comparing VGG-16 model with AlexNet.

The rest of this paper is organized as follows: Sect. 2 examines related work and overviews a sample of approaches; Sect. 3 introduces the proposed methodology; Sect. 4 illustrates the experimental results; Sect. 5 discusses the conclusions and future works.

2 Related Works

The detection of facial emotions is a further development after facial recognition, however, there are many perspectives in defining the concept of emotions, which is often unclear. According to Matsumoto [1], he divides facial emotions into 7 key categories such as happiness, surprise, satisfaction, sadness, anger, disgust, and fear. However, Mase et al. [2] believe that only 4 types of emotions are clearly expressed, which are happiness, surprise, anger, and disgust. Other types of emotions are often unclear and depend largely on the observers. S. Ren et al. [3] proposal algorithms to hypothesize object locations. They introduce a Region Proposal Network (RPN) that shares full-image convolutional features with the detection network, thus enabling nearly cost-free region proposals. An RPN is a fully convolutional network that simultaneously predicts object bounds and abjectness scores at each position. The RPN is trained end-to-end to generate high-quality region proposals, which are used by Fast R-CNN for detection. Furthermore, according to Unit of the European Data Protection Supervisor (EDPS) [4], they define face emissions recognition as facial emotion recognition is a technology used for analyzing sentiments by different sources, such as pictures. It belongs to the family of technologies often referred to as 'affective computing', a multidisciplinary field of research on computer's capabilities to recognize and interpret human emotions and affective states and it is often built on Artificial Intelligence technologies. L.S. Parrett et al. [5] survey examples of this widespread assumption, which we refer to as the common view, and we then examine the scientific evidence that tests this view, focusing on the six most popular emotion categories used by consumers of emotion research: anger, disgust, fear, happiness, sadness, and surprise. S. Du et al. [6] propose an important group of expressions, which they call compound emotion categories. Compound emotions are those that can be constructed by combining basic component categories to create new ones. For instance, happily surprised and angrily surprised are two distinct compound emotion categories. The present work defines 21 distinct emotion categories. Sample images of their facial expressions were collected from 230 human subjects. Tianrong Rao et al. [7] propose a new deep network that learns multi-level deep representations for image emotion classification (MldrNet). Image emotion can be recognized through image semantics, image aesthetics and low-level visual features. H.R. Kim et al. [8, 9] combined the different levels of features, we build an emotion-based feed forward deep neural network which produces the emotion values of a given image. The output emotion values in our framework are continuous values in the 2-dimensional space which are more effective than using a few number of emotion categories in describing emotions from both global and local views. Hao Zhang et al. [9] proposed the model which provided a description from the deep semantic representation to shallow visual representation. Additionally, several feature fusion approaches are analyzed and discussed to optimize the deep model. Extensive experiments on several image emotion recognition datasets show that our model outperforms various existing methods.

In general, the above researchers have investigated in computing the image emotions by using various features extracted from images and they used the different methodologies to recognize face emotion states. In this paper, we experimented with a Deep Learning architecture based on multiple convolutional layers (ConvNet) to detect facial expressions. The data was obtained from Zoom Meeting software's camera, capturing students' faces in six different states, including neutral, disgust, angry, happy, sad, and surprised. Then, the data was processed with SoftMax output, returning the probability of three emotions computed by the system.

3   Preliminaries

3.1   Overview of VGG-16 Model

With the development of deep learning models, it is easier to analyze emotions based on facial images. However, the evaluation of emotions is highly dependent on the training and testing data sets. In this paper, we use the VGG-16 model, one of the deep learning models, which is an effective model in analyzing learners' emotions through images of students' faces by the ZOOM meeting software.

VGG-16 [10] is a convolutional neural network (CNN) architecture proposed by the Visual Geometry Group (VGG) at the University of Oxford in 2014. It consists of 16 layers, including 13 convolutional layers and 3 fully connected layers, and is mainly used for image recognition tasks. VGG-16 is characterized by the use of small 3x3 convolutional kernels and a depth of 16 layers. The architecture of VGG-16 is also used as a feature extraction module for other deep learning models. VGG-16 has achieved excellent results in famous image recognition competitions such as the ImageNet Large Scale Visual Recognition Challenge (ILSVRC) in 2014. In the following section, we use VGG-16 with some changes to improve the process of face emotion recognition.

3.2   Methodologies and Algorithms for Facial Emotion Recognition

In order to increasing the precision of face emotion regconition, besides convolutional layers and fully connected layers, we added some of layers such as pooling layers and using dropout to solve overfitting.

The pooling layer is used to reduce the size of the input image while still retaining important information and increasing computational speed. There are three commonly used types of pooling layers including Max pooling, Average pooling and sum pooling. With Max pooling, it returns the maximum value from the portion of the image covered by the filter. On the other hand, Average pooling returns the average value of all the values from the portion of the image covered by the filter and Sum pooling returns the total value of all the values from the portion of the image covered by the filter. in this paper, we use Max pooling. The max-pooling is performed over a (2 × 2) pixel window, with stride size set to 2.

To solve the overfitting problem in neutral network, dropout layers are used. At each step of the training process, when performing forward propagation to a layer that uses dropout, instead of computing all the units on that layer, we randomly select a subset of units to compute instead of all the units in that layer. The way dropout works is to achieve the average result when training many sub-networks in the network by assuming the hidden of a certain percentage of units instead of relying on the result of training a single network.

The proposed model of VGG-16 is shown in Fig. 1

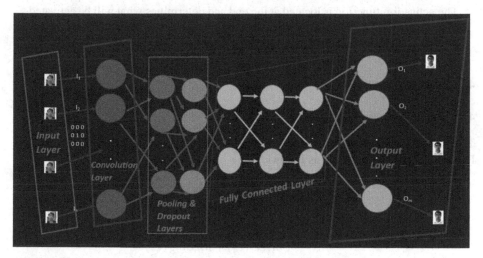

Fig. 1. The proposed model of VGG-16.

Where,

- The input layer includes the images of the training set. The set of images are divided into three subsets including training set, validation set, and test set. The images in the training set will get preprocessing, including image classification (7 categories: 0.angry, 1.disgust, 2.fear, 3.happy, 4.sad, 5.surprise and 6.neutral), conversion of colored images to grayscale, and resizing to a uniform size of 48x48.
- Convolution layer. Instead of 13 convolution layers on VGG-16, we use only 4 convolution layers which are the first layers of the model used to extract various features of the input image. In those layers, a convolution operation is performed between the input image and a filter of size MxM by sliding the filter along the input image. The output is called a feature map, providing us with information about the image such as edges and corners. This feature map will be passed to other layers to learn more features of the input image. In this case, we use the filter window with 3x3 size to reduce the parameters number of this model instead of 5x5 size. Furthermore, we use

Max pooling to reduce the size of images after each convolution layer and apply the Rectified Linear Units (ReLU) activation function for convolution layers.
- Pooling and Dropout layers are mentioned above.
- Fully connected layer. These layers contain weights and biases, as well as neurons, which are used to connect two different layers of the network. These layers are usually placed before the output layers and form the final layers of the CNN architecture. After being flattened into a vector, these layers are inputted into a fully connected layer to connect with one another and create a completed model. Finally, we use a softmax function to classify the output.

Additionally, we use activation functions to optimize our proposed model. Activation functions are added to the model to enable them to learn complex data. The activation function is located at the end and determines what will be activated to the next neuron layer. Some commonly used activation functions are ReLU, Softmax, sigmoid, and Tanh. We use ReLU in this case.

4 Experiment

4.1 Evaluating Based on Testing Data Set.

To evaluate the proposed model, we use three measures: Precision (P), Recall (R) and F-score. They are calculated as follows.

$$P(C_i) = \frac{Correct(C_i)}{Correct(C_i) + Wrong(C_i)} \tag{1}$$

$$R(C_i) = \frac{Correct(C_i)}{Correct(C_i) + Missing(C_i)} \tag{2}$$

$$F - score(C_i) = 2\frac{P(C_i) \times R(C_i)}{P(C_i) + R(C_i)} \tag{3}$$

where C_i represents an emotion category such as happy, angry or neutral; wrong, missing represent the number of terms, which are correct, wrong, missing, respectively.

After applying the formulars (1), (2), (3) to our data set, the results show as follows (Fig. 2).

	precision	recall	f1-score	support
0	0.437	0.499	0.466	491
1	0.441	0.545	0.488	55
2	0.463	0.424	0.443	528
3	0.754	0.803	0.778	879
4	0.463	0.389	0.423	594
5	0.724	0.745	0.735	416
6	0.522	0.511	0.517	626
accuracy			0.576	3589
macro avg	0.543	0.560	0.550	3589
weighted avg	0.571	0.576	0.572	3589

Fig. 2. The experiment results on our data set

According to Fig. 2, the precision and recall measures of happy emotion are highest and the next is fear emotion because these images of two categories are clear and sharp.

Additionally, we selected randomly from the testing data set, the results show as Fig. 3.

Fig. 3. The experiment results on images which are selected randomly.

As the results of the experiment show as Fig. 3, the precision on the testing data set is not high because face emotions of images which were collected from students are not clear and sharp. The red highlight indicated the wrong prediction of face emotions.

However, the results are lower when we apply with the same model on the testing data set of FER2013 [11] shown as Fig. 4.

Fig. 4. The experiment results on FER2013's images which are selected randomly.

The results in Figs. 3 and 4 show that our proposed method achieves high precision and the student images which we collect are correct.

4.2 Comparative Evaluation Method.

To compare our proposed model with another one, we use the AlexNet [12] model for result comparison.

The AlexNet model participated in the ImageNet Large Scale Visual Recognition Challenge 2012 (ILSVRC 2012) and was commonly known as AlexNet that is the name of the first author, Alex Krizhevsky. The input of AlexNet is an RGB image with a size of $224 \times 224 \times 3$, but this one is replaced with a size of $227 \times 227 \times 3$ for both the training and testing datasets. AlexNet has about 56 million parameters and 650000 neurons, including 5 convolutional layers, one of which is followed by a max-pooling layer, and 3 fully connected layers. Finally, a SoftMax function outputs the probability of the image belonging to 7 classes (AKIEC, BCC, BKL, DF, MEL, NV, VASC) of the data. To reduce overfitting, the authors added a regularization method called "dropout" to the Convolutional layer. However, we use grayscale images replacing the RGB images for AlexNet model.

The result experiments on testing data set by picking the random images, the results shown as Fig. 5.

Fig. 5. The AlexNet's experiment results which are selected images randomly.

In Fig. 5, there are 6 images in 24 images (25%) which are picked randomly are predicted correct where there are 11 images in 24 images (46%) in Fig. 3 are predicted correct. Therefore, the experiments show that our proposed model outperforms AlexNet model, respectably.

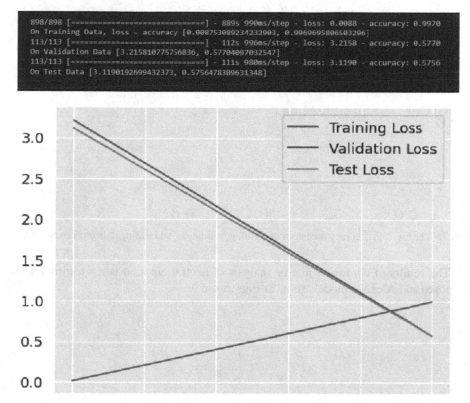

Fig. 6. The loss and accuracy measures of training, validation, and testing data on our proposed model

The loss and accuracy measures of training, validation, and testing data set show as Figs. 6 and 7.

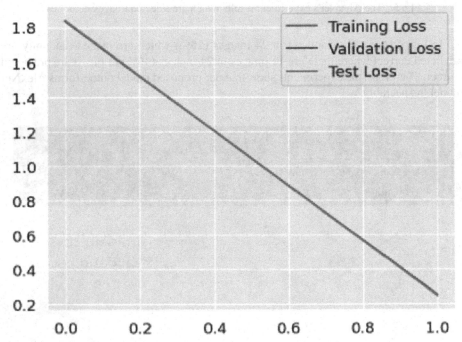

Fig. 7. The loss and accuracy measures of training, validation, and testing data on AlexNet model

The results of confusion matrix analysis of neutral emotion when testing on our proposed and AlexNet models show as Figs. 8 and 9.

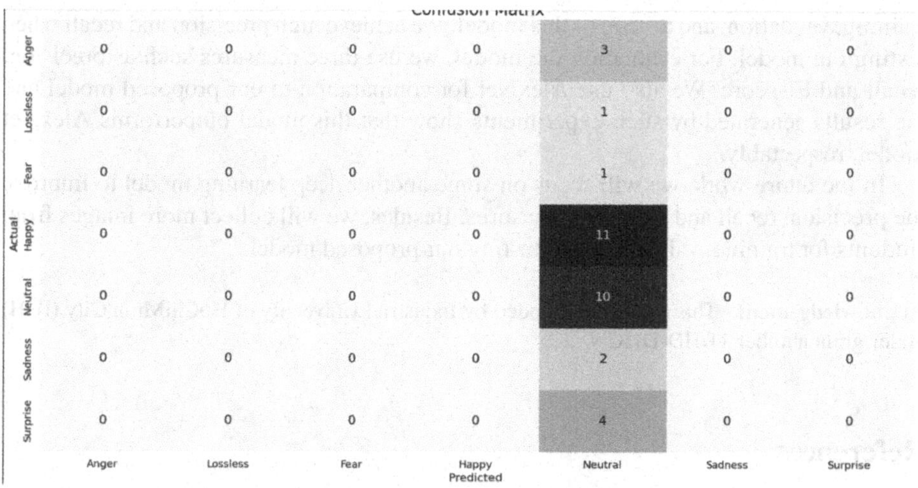

Fig. 8. The Confusion matrix analysis of neutral emotion when testing on our proposed model.

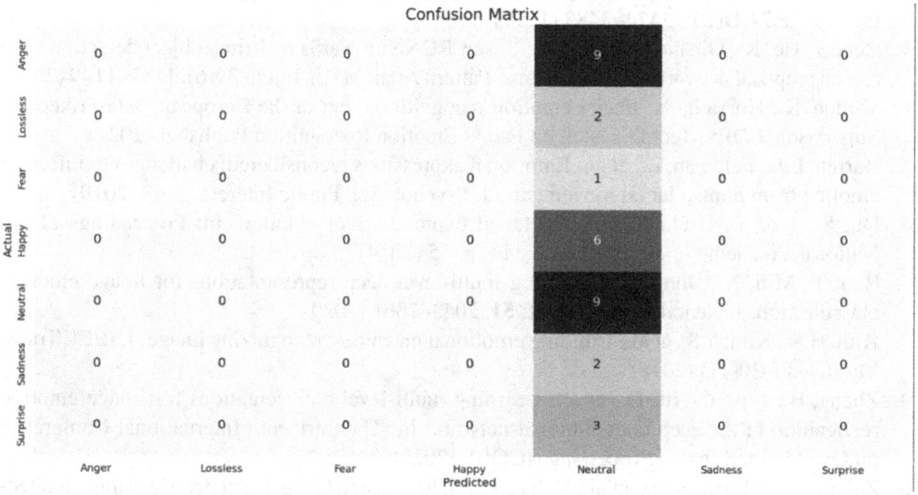

Fig. 9. The Confusion matrix analysis of neutral emotion when testing on AlexNet model.

The scores reported in Figs. 8 and 9 reveal that our proposed model outperforms the AlexNet model, respectably.

5 Conclusions

In this paper, we dealt with the problem of face emotion detection. There are many algorithms to solve this problems, especially using deep learning models. In this paper, we propose a deep learning model which is improved from VGG-16 model. We collect more than 20000 images from students by Zoom meeting application and Google meeting for training, validation, and testing of this model. We achieve high precision and recall when testing our model. For evaluation the model, we use three measures such as precision, recall and F1-score. We also use AlexNet for comparation to our proposed model and the results generated by such experiments show that this model outperforms AlexNet model, respectably.

In the future work, we will focus on some another deep learning model to improve the precision, recall and F1-score measures. Besides, we will collect more images from students for training, validation and testing our proposed model.

Acknowledgement. This research is funded by Industrial University of HoChiMinh City (IUH) under grant number 11/HĐ-ĐHCN.

References

1. Matsumoto, D., Hwang, H.S.: Reading facial expressions of emotion. J. Psychol. Sci. **25**(5), 10–18 (2011)
2. Mase, K., Pentland, A.: Recognition of facial expression from optical flow. J. IEEE Trans. Inf. Syst. **E74-D**(10), 3474–3483 (1991)
3. Ren, S., He, K., Girshick, R., Sun, J.: Faster RCNN: towards real-time object detection with region proposal networks. J. IEEE Trans. Pattern Anal. Mach. Intell. **39**(6), 1137–1149 (2017)
4. Vemou, K., Horvath, A.: Facial emotion recognition. Unit of the European Data Protection Supervisor (EDPS) TechDispatch on Facial Emotion Recognition Publisher (2021)
5. Barrett, L.F., Feldman, L., et al.: Emotional expressions reconsidered: challenges to inferring emotion from human facial movements. J. Psychol. Sci. Public Interest, 1–68 (2019)
6. Du, S., Tao, Y., et al.: Compound facial expressions of emotion. In: Proceedings of the National, Academy of Sciences, vol. 111, p. 15 (2014)
7. Rao, T., Min, X., Dong, X.: Learning multi-level deep representations for image emotion classification. J. Neural Process. Lett. **51**, 2043–2061 (2020)
8. Kim, H.R., Kim, Y.S., et al.: Building emotional machines: recognizing image. J. IEEE Trans. Multimedia **20**(11) (2018)
9. Zhang, H., Liu, Y., Xu, D., et al.: Learning multi-level representations for image emotion recognition in the deep convolutional network. In: The thirteenth International Conference on Graphics and Image Processing (ICGIP) (2022)
10. Zakaria, N., Mohamed, F., et al.: VGG16, ResNet-50, and GoogLeNet deep learning architecture for breathing sound classification: a comparative study. In: The International Conference on Artificial Intelligence for Cyber Security Systems and Privacy (AI-CSP) (2021)
11. Kaggle. https://www.kaggle.com/c/challenges-in-representation-learning-facial-expression-recognition-challenge/data. Accessed 2023
12. Krizhevsky, A., Sutskever, I., et al.: ImageNet classification with deep convolutional neural networks. J. Commun. ACM **60**, 84–90 (2012)

The Efficiency of Japanese Character Classification by Attention Mechanism

Phuc Van Nguyen[1]($\boxtimes$), Minh Hoang Le[1], Bao Khanh Duong[1],
Thanh Hien Le Nguyen[1], and Duc Vinh Vo[2]

[1] FPT University, Ho Chi Minh City, Vietnam
vannpse172344@fpt.edu.vn
[2] Ho Chi Minh City University of Banking, Ho Chi Minh City, Vietnam

Abstract. To improve model performance in the Japanese handwritten character classification task. We designed the most miniaturized but also high-performance CNN-based model. In our initial efforts, we designed a Vanilla CNN architecture and used an attention-based classification head (ML Decoder) to improve accuracy. With 3.6 million parameters, this model achieved an accuracy rate of 99.02% and 97.93% on the test set of the 10-class and 49-class, respectively. After that, we create a network named Residual CNN using the Vanilla CNN model combined with the *residual connect* of the ResNet network to retain information from the previous layers of the model. After changing the Vanilla CNN backbone to Residual CNN, the model included 3.8M parameters and achieved an accuracy rate of 99.08% and 98.11% on the test set of the 10-class and 49-class, respectively. In addition, the Grad-CAM method is used to improve the explanatory power of how the model makes decisions.

Keywords: Kuzushiji-MNIST · ML-Decoder · Model explanation

1 Introduction

Cursive Japanese character is getting more and more attention. This paper will explore and train the Kuzushiji-MIST dataset [1] consisting of three parts: Kuzushiji-MNIST, Kuzushiji-49 and Kuzushiji-Kanji. This is a single-label classification task, and we will use CNN models to solve it. CNN-based is a simple model but achieves high performance and accuracy in real-world Japanese character image classification tasks. However, these CNN models often require models with hundreds of millions or even billions of parameters, demanding significant computational resources and memory. To address this memory-intensive nature of large CNN models, we propose a solution that leverages a smaller CNN model with only a few million parameters combined with the SOTA model ML-Decoder [8] for the classification head. We enhance the interpretability of the model by employing the Gradient-weighted Class Activation Mapping (Grad-CAM) method [9].

N. Thai-Nghe et al. (Eds.): ISDS 2023, CCIS 1950, pp. 185–193, 2024.
https://doi.org/10.1007/978-981-99-7666-9_15

For an overview, we organize the paper into the following main ideas: Sect. 2, reviews the dataset, SOTA result, and some explanation methods of the Kuzushiji-MNIST dataset. Section 3, defines the task and explains the technique in detail. Section 4, implements the method, chooses the hyper-parameters, and compares the results. Section 5, discusses to identify the pros and cons of this method. Section 6, summarizes the methodology and discusses prospects.

2   Related Work

2.1   SOTA on the Kuzushiji-MNIST Dataset

The VGG-5 (Spinal FC) model was introduced in [3] has 3.630B parameters and achieved an impressive accuracy of 99.1% ($\pm 0.05\%$) on the test set. Many parameters combined with the SOTA methods gave high results, but it also has some computational cost and memory limitations. In addition, using pre-trained models is a popular approach but highly effective. According to [11], the RESNET-152 model leveraging pre-trained weight, attained an accuracy of 98.69% ($\pm 0.10\%$), while INCEPTION V3 achieved an accuracy of 98.51% ($\pm 0.11\%$) on Kuzushiji 10-class dataset. The pre-trained models arc often trained on large and diverse datasets but are not always suitable for a particular task, so using a pre-trained model may not be sensitive to cursive Japanese characters and can lead to performance degradation in this dataset.

2.2   CAM to Grad-CAM

Besides using big models or SOTA methods, we must also explain how the model makes decisions. The basic method uses Class Activation Map (CAM) [13]. CAM requires the network to apply the Global Average Pooling (GAP) [4] layer on the feature map and then forward it to the fully connected layer. From that, CAM can only be used on models that use the GAP layer before forwarding to the classification head. Therefore, we propose a more advanced solution, a Gradient-weighted Class Activation Map (Grad-CAM) [9] to enhance the explainability of the model. Grad-CAM has only one weakness: the layer behind the Convolution layer must be differentiable to get the gradient value of the feature map.

3   Method

3.1   Baseline Single-Label Classification

Classification models often have two main components: the backbone and the classification head. The backbone is a CNN-based model that processes an input image into a feature map $A \in R^{H \times W \times C}$. In this experiment, an attention-based classification head is applied to transform the feature map A into N logits $\{\hat{y}_i\}_{i=1}^{N}$, where N is the number of classes and $\hat{y}_i$ is the probability that the input image falls into the ith class. We utilize the Cross-Entropy Loss [6] function, and the objective of the problem is to optimize the parameters in the model to minimize the loss function below:

$$L(\vec{\hat{y}}, y) = -log\left(\frac{exp(\vec{\hat{y}}_y)}{\sum_{i=0}^{N} exp(\hat{y}_i)}\right) \tag{1}$$

3.2 Modifying the Architecture of CNN Models

In the Kuzushiji-MNIST dataset, the input image, $x \in R^{28 \times 28 \times 1}$, is unsuitable for some CNN models with many Convolution layers in the network. After forwarding some small input images, the Convolution or Pooling layer in the model will reduce the image size by each layer, and the shape of the features map A_k can become zero. Some experiment has demonstrated if using the layer number of Convolution layer in the network can make a *Vanishing Gradient* or if using too few Convolution layers can make *Exploding Gradients*. This paper will propose 2 types of CNN-base models with structures like ResNet [2] and Vgg [10].

We propose a Vanilla CNN model with three Convolution blocks to reduce the number of Convolution layers in the model like Vgg. Each Convolution block contains a Convolution layer, ReLU activation, and Batch Normalize. This simple model makes observing and explaining how the model makes decisions more accessible.

In some methods applied for the Convolution blocks in Vanilla CNN, we leverage the *residual connect* of the ResNet to design our model (called Residual CNN) without using a down-sample layer in ResNet. Using a *residual connect* in ResNet, we can reuse previous information, which will help the model avoid lost information over time. From that, we create a model that has 3 main Residual Blocks. Each Residual block contains 2 Convolution blocks like our Vanilla model Structure. See the pseudo-code of Residual blocks forwarding in our network below:

Algorithm 1 Residual Blocks Forwarding

Require: $x \in R^{N \times C_1 \times W \times H}$ ▷ N is batch_size
Ensure: $C_1 = C3$
1: $residual \leftarrow x$
2: $out1 \leftarrow Convolution\_layer1(x)$
3: $out1 \leftarrow ReLU(BatchNorm\_layer1(out1))$ ▷ $out1 \in R^{N \times C_2 \times W \times H}$
4: $out2 \leftarrow Convolution\_layer2(out1)$
5: $out2 \leftarrow BatchNorm\_layer2(out2)$ ▷ $out2 \in R^{N \times C_3 \times W \times H}$
6: $out \leftarrow ReLU(out2 + residual)$
7: **Return** out

This pseudo-code shows that the output is made by the sum of the input and output in Residual Blocks, from that when we do back forwarding, the grad-value will contain the information of previous blocks. To make the sum of the input and output in Residual Blocks, we ensure that the values of W and H in the shape of each out_i are the same. To ensure that, we will use *padding* in the input to maintain the output shape like the input shape when applying the Convolution filter on the input. Read the implementation for details.

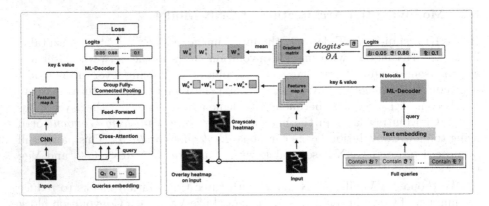

Fig. 1. Workflow overview. The left side is the workflow of the training phase. The right side is the workflow of the evaluation phase.

3.3 Attention-Base Classification Head

In the classification head, we utilize the attention-based classification head (ML-Decoder [8]) to improve the model's performance instead of using a classification head with a fully connected layer. This network contains N ML-Decoder blocks ($N > 0$). See the ML-Decoder on the left side of Fig. 1. When using the ML-Decoder for the classification head, with the applying self-attention removal for all ML-Decoder blocks, the time complexity in this ML-Decoder decreases from $\mathcal{O}(\mathbf{N^2})$ to $\mathcal{O}(\mathbf{N})$ [8].

For the queries, the results from [8] and [5] show that using either learnable or fixed queries gives similar results. To minimize the memory while training, we decided to use fixed queries. A pre-trained CLIP [7] model is used for embedding the query from the text to get the vector embedding.

3.4 Grad-CAM for Model Explainability

With the attention-based classification head, it is unsuitable to use Class Activation Map (CAM) [13] to explain the model because CAM is only applied in the model that must use GAP. Gradient-weighted Activation Map (Grad-CAM) [9] is an update of CAM, and it can be applied to models without necessarily using (GAP) [4]. To use Grad-CAM, we had to ensure that all layers after the

Convolution layer in the network should be differentiable to get the gradient. See the right side of Fig. 1. After taking the derivative, we will archive the gradient matrix and calculate the α_i value by getting the average value in each gradient matrix. Finally, we generate the grayscaled heatmap by multiplying the $\alpha_{i\,i=1}^{D}$ values with the features map $A_{i\,i=1}^{D}$, where D is the number of output channels of the Convolution layer we want to observe. See the formula below:

Get the derivative of *logits* as $c = name\_of\_class$ respective to the feature map A_i. After that, we can achieve the gradient matrix $G_{i\,i=1}^{C}$:

$$G_i = \frac{\partial logits^{c=Class}}{A_i} \qquad (2)$$

The α value at index k^{th} is equal to the average value of the matrix G_k:

$$\alpha_k = \sum_{i=1}^{n}\sum_{j=1}^{m} G_{i,j}^{k} \qquad (3)$$

After having the α_i and the feature maps A_i. We calculate the grayscaled_heatmap by taking the multiply of α_i and A_i and then adding them all together:

$$grayscaled\_heatmap = \sum(\alpha_i \times A_i) \qquad (4)$$

4 Experiment

To evaluate our method on the Kuzushiji-MNIST dataset [1], we will run and report on two types of datasets for 10 and 49 classes, respectively. The Kuzushiji-MNIST 10 classes have 60,000 for the train set and 10,000 images for the test set. The Kuzushij-MNIST 49 classes have 232,365 and 38,547 images for train and test sets, respectively. All images are in format (28×28 grayscale).

4.1 Implementation

Vanilla CNN: In the Vanilla CNN model, we design a model that has three blocks and accepts 1-channel color images; each block contains the Convolution layer, ReLU activation, and Batch normalize layer, with the output channel of each block equal to 64, 128, and 128, respectively. For the filter in Convolution of each block, we set it to 5, 5, and 3 for particular blocks, respectively.

Residual CNN: For a model like ResNet [2], we set up the first Convolution layer of the model to accept 1-channel color images and change the filter size to 3. We use the structure of the BasicBlock without a down-sample layer in the ResNet implementation from the GitHub of torch-vision to design our Residual CNN with three blocks. And then, we set up the output channel of each block in our network as 64, 64, and 128, respectively. To maintain the shape of the output is always equal to the shape of the input, we must add padding. We can easily present the padding used for the input by the equation: **padding = $\lfloor$ kernel_size/2 $\rfloor$**.

ML-Decoder Classification Head: We use only 1 decoder layer for this method, in that we set up the $num\_head = 4$ for Residual CNN, $num\_head = 8$ for Vanilla CNN, and we use the $d\_model = 512$, $dim\_feedforward = 2048$ with $drop\_out = 0.1$ for both model.

Training: We set up the image input size equal to the original data as 28×28, and the training process contains 50 epochs on both methods (Vanilla CNN and Residual CNN), with the total update step being 15,000 and 58,100 steps for the Kuzushiji dataset having 10 and 49 classes respectively. Adam optimizer with a learning rate of 0.0001 was used to optimize the model parameters during the training. We applied the OneCycleLR with the $max\_lr = 0.0001$ and $pct\_start = 0.2$ for learning rate adjusting. GPU T4 of Google Colab was chosen for running this experiment.

Table 1. The table shows the accuracy of two types of Kuzushiji datasets and the number of parameters (P). The first 9 models were reported in the original paper, in which Shake-Shake is the most prominent method on the set of 49 classes. Besides that, VGG-5 (Spinal FC) also has achieved the highest accuracy on set 10 classes. Our model also performs well when the accuracy is less than 1% of SOTA methods.

Method	10 class	49 class	P
4-Nearest Neighbour Baseline [1]	92.10%	83.65%	
PCA + 4-kNN [1]	93.98%	86.80%	
Tuned SVM (RBF kernel) [1]	92.82%	85.61%	
Keras Simple CNN Benchmark [1]	94.63%	89.36%	
PreActResNet-18 + Manifold Mixup [12]	98.83%	97.33%	11,2M
ResNet18 + VGG Ensemble [1]	98.90%		
DenseNet-100 (k = 12) [1]		97.32%	
Shake-Shake-26 2x96d (cutout 14) [1]		**98.29%**	26,2M
Shake-shake-26 2x96d (S-S-I, Cutout 14) [1]	99.34%		26,2M
VGG-5 [3]	98.94%		3,646B
VGG-5 (Spinal FC) [3]	**99.68%**		3,630B
Vanilla CNN + ML-Decoder [our]	99.02%	97.93%	**3,5M**
Residual CNN + ML-Decoder [our]	99.08%	98.11%	3,8M

4.2 Compared with SOTA on Kuzushiji-MNIST

Our method uses a small CNN-base model that has 4M or lower parameters combined with the ML-Decoder and gives the extreme results with the best accuracy in the test set of Kuzushiji 10 classes is 99.08%, 49 classes is 98.11% (see the Table 1). This training consumes about 3 GB GPU and takes about

5 h for 50 epochs on the Kuzushiji 49 classes. It achieves high accuracy on both types of datasets. The accuracy of the model is only lower than the other 2 methods: VGG-5 (Spinal FC), the model has 3,630B parameters and training in 200 epochs, and the method from the original paper of the dataset is "Shake-Shake-26 2x96d Cutout 14" has 26,2M the number of parameters in the network. Both two method is set up on a big model that has a large of parameters, and the time of the training process is long. The original GitHub shows that the Shake-Shake method was trained in 7 h on GeForce GTX 1080 Ti.

4.3   Model Evaluation

The model achieved SOTA results using the small CNN-base model, with the highest accuracy on a test set (99.08% for 10 classes and 98.11% for 49 classes). We want to explain how the model makes a decision. See the model visualization in Fig. 2. We can see that the visualized grayscale heatmap is focused precisely on the feature of the character in Japanese for making the decisions in the class with the highest probability. The performance when applied on both Vanilla CNN and Residual CNN models is the same. From that, our method uses a small model that has optimized the time and memory when the model parameters are approximately 4 Million. And this visualization shows that the model is very confident in its decision.

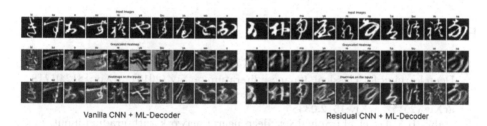

Vanilla CNN + ML-Decoder   Residual CNN + ML-Decoder

Fig. 2. Grad-CAM on Vanilla CNN and Residual combine with ML-Decoder.

5   Discussion

As a result of the Table 1, our model is 1000 times more minor, but given the accuracy is comparable with the Sate-Of-The-Art model on the Kuzushiji-MNIST dataset and the time training was shorter than the SOTA method. This research study has provided a general understanding of the classification task while using the model no more than 4M parameters. The simple classification head was replaced with the attention-based classification head to improve the model's performance. Grad-CAM also demonstrated the model's explainability very well without using the GAP layer in the network. With a few parameters, we can easily implement, and this model is suitable for Fine-grained classification tasks on Japanese poetry or manga images. Additionally, we believe using the data augmentation method in this task will enhance accuracy and enable the model to make more generalized predictions.

6   Conclusions and Future Work

Our method had success in terms of using a small CNN-based model to improve performance, and it has given SOTA results on the Kuzushiji-MNIST dataset except for the two more optimal but more expensive: (VGG-5 + Spinal) [3] and (Shake-Shake-26 2x96d) [1].

In the future, we will focus on augmenting the data to make the diverse Japanese character dataset and train the model for multi-task classification on the Hiragana, Katakana, and Kanji characters. In addition, we will also make time improvements to make the model more suitable for some mobile applications.

Acknowledgment. Thank all members for the positive findings and contributions to building this project: N.P Van for the main idea and code of this paper, L.M Hoang for the visualization and the method for explaining the model, D.B Khanh & L.N.T Hien for the running code and making content style for writing the paper. See all implementation scripts at https://github.com/sunny442k3/KMNIST.

References

1. Clanuwat, T., Bober-Irizar, M., Kitamoto, A., Lamb, A., Yamamoto, K., Ha, D.: Deep learning for classical Japanese literature. arXiv (2018). arXiv:1812.01718
2. He, K., Zhang, X., Ren, S., Sun, J.: Deep residual learning for image recognition. arXiv (2015). arXiv:1512.03385
3. Kabir, H.M.D., et al.: SpinalNet: deep neural network with gradual input. arXiv (2022). arXiv:2007.03347
4. Lin, M., Chen, Q., Yan, S.: Network in network. arXiv (2014). arXiv:1312.4400
5. Liu, S., Zhang, L., Yang, X., Su, H., Zhu, J.: Query2Label: a simple transformer way to multi-label classification. arXiv (2021). arXiv:2107.10834
6. Mao, A., Mohri, M., Zhong, Y.: Cross-entropy loss functions: theoretical analysis and applications. arXiv (2023). arXiv:2304.07288
7. Radford, A., et al.: Learning transferable visual models from natural language supervision. arXiv (2021). arXiv:2103.00020
8. Ridnik, T., Sharir, G., Ben-Cohen, A., Ben-Baruch, E., Noy, A.: ML-Decoder: scalable and versatile classification head. arXiv (2021). arXiv:2111.12933
9. Selvaraju, R.R., Cogswell, M., Das, A., Vedantam, R., Parikh, D., Batra, D.: Grad-CAM: visual explanations from deep networks via gradient-based localization. arXiv (2019). arXiv:1610.02391
10. Simonyan, K., Zisserman, A.: Very deep convolutional networks for large-scale image recognition. arXiv (2015). arXiv:1409.1556

11. Studer, L., et al.: A comprehensive study of ImageNet pre-training for historical document image analysis. arXiv (2019). arXiv:1905.09113
12. Verma, V., et al.: Manifold mixup: better representations by interpolating hidden states. arXiv (2019). arXiv:1806.05236
13. Zhou, B., Khosla, A., Lapedriza, A., Oliva, A., Torralba, A.: Learning deep features for discriminative localization. arXiv (2015). arXiv:1512.04150

Using Deep Learning to Build a Chatbot Supporting the Promotion of Speciality Dishes in Mekong Delta

Quoc-Khang Tran and Nguyen-Khang Pham[✉]

College of Information and Communication Technology, Can Tho University, Cantho 92000, Vietnam
pnkhang@ctu.edu.vn

Abstract. Based on the images of dishes in the Mekong Delta along with questions about the dishes such as: What is the name of this dish? Where is it famous? What are the main ingredients? How is it made? An application chatbot will be built to promote the speciality dishes of the Mekong Delta. This report outlines a method for training a Visual Question Answering (VQA) model for classification tasks using Transformer-based models, such as ViT for image data, BERT/PhoBERT for text data, or ViLT for simultaneous processing of image and text data. After that, a Visual Encoder-Decoder model for the task of generating sentences will be built using the VQA model as a Visual Encoder and a GPT-2 as a Decoder. The experimental dataset, which includes 7,694 photos of dishes from the Mekong Delta, is a subset of the datasets 30VNFoods and VinaFood21. The accuracy metric was used to evaluate the VQA models, and the results were relatively good. For Model 1: ViT and BERT, the accuracy scores for English and Vietnamese are 94% and 95%, respectively, while the accuracy score for Model 2: ViLT is over 92% on English-only. According to the ROUGE evaluation method, Model 3's answer sentence generation model, on English only, which used ViLT along with GPT-2, yielded results of 49.92, 39.26, and 47.53 for the ROUGE-1, ROUGE-2, and ROUGE-L, respectively. Finally, the trained models were applied to build a chatbot.

Keywords: Transformer-based · Mekong Delta · Chatbot · GPT · ViT · BERT · PhoBERT · 30VNFoods · VinaFood21

1 Introduction

The Mekong Delta, a renowned and captivating tourist destination in Vietnam, is known for its diverse natural landscapes, rich culture, fascinating history, and delightful cuisine. Among its attractions, the speciality cuisine holds a significant allure.

The Mekong Delta boasts a wide array of unique dishes and culinary treasures, including renowned specialties like banh-xeo, banh-tet, com-tam, and

N. Thai-Nghe et al. (Eds.): ISDS 2023, CCIS 1950, pp. 194–203, 2024.
https://doi.org/10.1007/978-981-99-7666-9_16

more. To introduce these delicacies to visitors from around the world, we developed a chatbot that supports the promotion of speciality dishes in the region. This chatbot offers tourists the opportunity to explore and savor the distinctive flavors of this land.

The development of this chatbot not only contributes to the growth of the Mekong Delta's tourism industry but also plays a role in preserving and promoting its cultural values, history, and authentic cuisine.

The primary aim of this report is to explore a method for constructing a chatbot dedicated to promoting speciality dishes in the Mekong Delta. This chatbot is based on the Transformer's Encoder [11] model, utilizing BERT [3] or PhoBERT [4] for text processing, and ViT [5] for image processing, along with the ViLT [7] model for handling both text and images. For answer generation, the chatbot utilizes the Transformer's Decoder model, specifically GPT-2 [6], which extracts features from food images and processes food-related questions to either classify or generate suitable answers.

2   Mekong Delta Specialties Images Classification

2.1   The Mekong Delta Specialties Images Dataset

Collecting Dataset. The image dataset of Mekong Delta specialties is part of the VinaFood21 [1] and 30VNFoods [2] datasets collected by the authors [1,2] from various sources on the internet.

The data is classified into 9 classes: banh-canh, banh-khot, banh-mi, banh-pia, banh-tet, banh-xeo, com-tam, ga-nuong, pho. The total number of images on the train dataset is 5655, the test dataset is 2039 (Fig. 1)

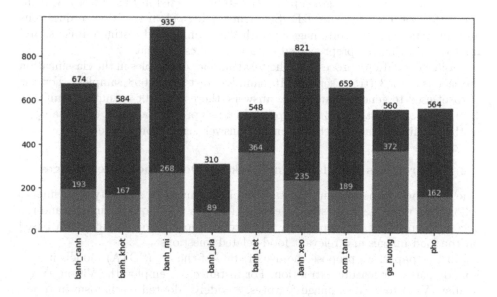

Fig. 1. Mekong Delta specialties dataset.

Pre-processing the Dataset. Data preprocessing is a crucial step in any problem-solving task before feeding the data into the model training process, as it directly impacts the results of the trained model.

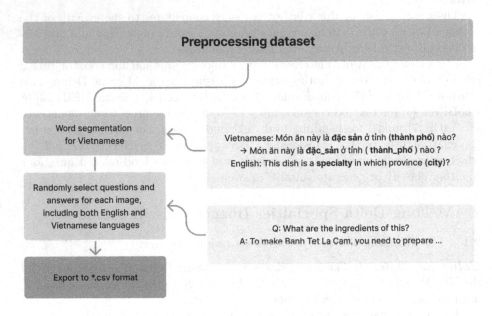

Fig. 2. The data preprocessing process to create a *.csv file.

The data preprocessing comprises the steps depicted in Fig. 2: word segmentation for Vietnamese using RDRSegmenter [8], random selection of questions and answers for each food image in both the training and testing datasets, and finally, exporting the preprocessed data to the *.csv format.

Following data preprocessing, the total number of samples in the classification dataset is 23,013 (train set: 16,915 samples, test set: 6,098 samples). For the dataset used to generate automatic answers, there are 23,064 samples (train set: 16,949 samples, test set: 6,115 samples). Figure 3 presents a subset of the samples in the training dataset of the automatic answer generation dataset.

2.2   Approaches for Classifying Mekong Delta Specialties Images

The approach to performing the task of classifying local specialty food images follows two main steps. Firstly, we extract features from both the images and the accompanying text. Subsequently, the second step involves classification based on the food images and relevant food-related questions.

In this paper, we propose the use of state-of-the-art (SOTA) models in various domains for feature extraction. For instance, we employ the Vision Transformer (ViT) to extract image features, a widely adopted mechanism in recent years. Additionally, we utilize BERT and PhoBERT for extracting features from

image_id	question_vi	label_vi	question_en	label_en	class_name	question_type
0	train/banh_canh/823.jpg	Nguyên_liệu của món ăn này là gì ?	Nguyên_liệu_chính để làm bánh_canh gồm bột gạo...	What is this food made of?	The main ingredients to make Banh Canh include...	banh_canh	ingredient
1	train/banh_canh/189.jpg	Cách chế_biến món ăn này ?	Để làm món bánh_canh , đầu_tiên người ta phải ...	How do you cook this?	To make Banh Canh, first, people have to prepa...	banh_canh	recipe
2	train/banh_canh/77.jpg	Món ăn này làm từ những gì ?	Nguyên_liệu_chính để làm bánh_canh gồm bột gạo...	What are the ingredients of this meal?	The main ingredients to make Banh Canh include...	banh_canh	ingredient
3	train/banh_canh/837.jpg	Món ăn này là đặc_sản ở đâu ?	Món ăn này nổi_tiếng ở miền Tây , đặc_biệt là ...	Where can I get specialty meal?	It is popular in the Mekong Delta, especially ...	banh_canh	place
4	train/banh_canh/638.jpg	Món ăn này nổi_tiếng ở đâu ?	Bánh_canh cũng là món ăn phổ_biến ở miền Tây ,...	What is this food's origin?	This is popular in the Mekong Delta, especiall...	banh_canh	place
5	train/banh_canh/162.jpg	Cách chế_biến món ăn này ?	Bánh_canh là món ăn phổ_biến và rất được ưa_ch...	How do you make this?	Banh Canh is a popular and very popular dish i...	banh_canh	recipe
6	train/banh_canh/178.jpg	Món ăn này là từ tỉnh (thành_phố) nào ?	Bánh_canh cũng là món ăn phổ_biến ở miền Tây ,...	What province (city) is this food from?	This is popular in the Mekong Delta, especiall...	banh_canh	place
7	train/banh_canh/88.jpg	Đây là món ăn gì ?	Món ăn này được gọi là bánh_canh	What do you call this dish?	Banh Canh	banh_canh	name
8	train/banh_canh/348.jpg	Đây là món ăn gì ?	Món ăn này được gọi là bánh_canh	What do you call this meal?	This food is called Banh Canh	banh_canh	name
9	train/banh_canh/360.jpg	Tên của _ăn này là gì ?	Món ăn này được gọi là bánh_canh	What is this?	This food is called Banh Canh	banh_canh	name

Fig. 3. A subset of the samples in the training dataset for the automatic answer generation dataset

questions. For classification, we employ fully connected layers and activation functions such as ReLU [12] and GELU [13], complemented by techniques like Dropout [14] and LayerNorm [15].

For the task of generating answer sentences, we employ a Visual Encoder-Decoder model. In this context, we utilize the GPT-2 language model as a decoder, combined with a Visual Encoder, specifically Model 2 - ViLT.

For Models 1 and 2, we use the Cross Entropy loss function. We evaluate our approach using the ROUGE metric for Model 3, which is used for answer sentence generation.

Model 1. In Model 1, as illustrated in Fig. 4, we developed a Visual Question Answering (VQA) model by combining ViT and BERT (or PhoBERT for Vietnamese). In this setup, BERT (or PhoBERT) was used to extract features from the question, while ViT was employed to extract features from the food images.

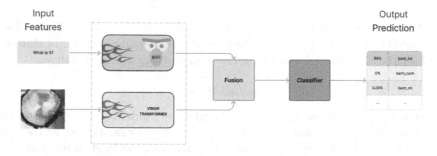

Fig. 4. Model 1: ViT and BERT (PhoBERT for Vietnamese).

Since both BERT (PhoBERT) and ViT are based on the Transformer's Encoder architecture, the output of the last Encoder layer comprises a vector representing the entire input data. Thus, we created a Single Layer Perceptron (SLP) with a ReLU activation function, referred to as the fusion layer, to combine these two feature vectors. The output of the fusion layer was then fed into the classifier layer for the classification process.

In mathematical terms, we can describe the process using the formulas below.

$$F(x, y) = classifier(fusion(B(x), V(y)))$$ (1)

In Eq. 1, let F(x, y) represent the result generated by Model 1. Where x and y represent the image of the food and the question related to the food, respectively. B(x) and V(y) are the feature vectors of the question and image, respectively, extracted using BERT and ViT. In fact, they represent the output vectors at the final layer of the Encoder.

$$fusion(a, b) = Dropout(ReLU(SLP(Concat(a, b))))$$ (2)

The fusion (Eq. 2) takes two input vectors a and b are combined through a series of operations. Initially, these vectors are concatenated to form a unified representation. This concatenated vector then undergoes a linear transformation using a Single-Layer Perceptron (SLP) before applying the Rectified Linear Unit (ReLU) activation function. Following that, dropout is applied for regularization to enhance the stability and generalization of the fused output.

$$classifier(c) = SLP(c)$$ (3)

The classifier (Eq. 3) takes the fused vector c and further processes it using a Single-Layer Perceptron (SLP). This SLP serves as a classifier to make predictions or decisions based on the fused features.

To train the VQA model (Fig. 4) on the English classification dataset, we utilized the pre-trained google/vit-base-patch16-224-in21k[1] for the ViT model and the pre-trained bert-base-uncased[2] for the BERT model. For the Vietnamese classification dataset, we replaced BERT with the pre-trained vinai /phobert-base-v2[3] combined with ViT, following the same architecture as Fig. 4.

Model 2. For Model 2, we adopted the architecture of the ViLT model and fine-tuned the classifier layer to accommodate the required number of classes.

Similarly, the Model 2 can be represented mathematically as follows:

$$G(x, y) = classifier(H(x, y))$$ (4)

Let G(x, y) represent the result generated by Model 2. H(x, y) represents the output of the ViLT model for inputs x and y and classifier further processes this output. In which, x denotes the image of the food, while y signifies the question pertaining to the food image (Eq. 4)

$$classifier(c) = SLP(GELU(LayerNorm(SLP(c))))$$ (5)

classifier(c) in Eq. 5 is a series of operations applied to the input c to make predictions. These operations include Single-Layer Perceptron (SLP),

---

[1] https://huggingface.co/google/vit-base-patch16-224-in21k.
[2] https://huggingface.co/bert-base-uncased.
[3] https://huggingface.co/vinai/phobert-base-v2.

Gaussian Error Linear Unit (GELU) activation, and Layer Normalization (LayerNorm) applied successively.

Throughout the training process, we leveraged the pre-trained `dandelin/viltb32-finetuned-vqa`[4] model and further trained it using our own dataset.

Model 3. Model 3 is built on the architecture of the Vision Encoder-Decoder[5] model integrated into the Huggingface Transformers library, which permits the utilization of any Transformer-based Vision model as the Encoder. Likewise, the Decoder is designed as a language model.

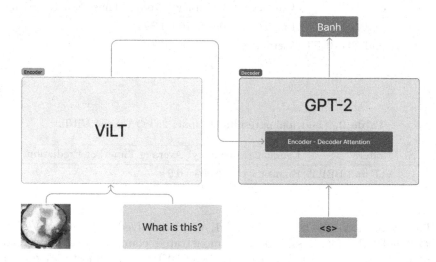

Fig. 5. Model 3: ViLT and GPT-2.

However, during the training of Model 3, we adapted the Vision Encoder-Decoder Model in the Huggingface Transformers library to accommodate input data comprising both images and text, as opposed to the default input of only images for the Vision Encoder-Decoder model (Fig. 5).

The Vision Encoder-Decoder model follows a seq2seq [10] approach, and we employed the pre-trained `dandelin/vilt-b32-finetuned-vqa` from the ViLT model as the Encoder, and the pre-trained `gpt2`[6] from the GPT-2 model as the Decoder.

3 Experimental Results

In this subsection, we will outline the main components of the proposed framework, which encompass the classification of specialties and the development of a basic chatbot.

---

[4] https://huggingface.co/dandelin/vilt-b32-finetuned-vqa.

[5] https://huggingface.co/docs/transformers/main/model_doc/vision-encoder-decoder.

[6] https://huggingface.co/gpt2.

Specialties Classification. During the model training phase, we employed PyTorch in combination with the Huggingface Transformers library to utilize pre-trained models accessible on this platform. Additionally, Google Colab PRO version was utilized to train all three models. Models 1 and 2 underwent training on GPU V100, while model 3 was trained on GPU A100.

Table 1. The training results of Model 1: VQA using ViT and BERT/PhoBERT.

No	Model	Language	Accuracy	Average Time Per Prediction
1	ViT and BERT	English	0.942276	0.2 s
2	ViT and PhoBERT	Vietnamese	0.948180	0.2 s

Table 2. The training results of Model 2: VQA using ViLT.

No	Model	Language	Accuracy	Average Time Per Prediction
1	ViT and BERT	English	0.928009	0.2 s

The evaluation outcomes of Model 1, which involved VQA using ViT and BERT/PhoBERT for classification, demonstrated commendable performance with 94% Accuracy for English using ViT and BERT, and almost 95% for Vietnamese using ViT and PhoBERT (Table 1). Furthermore, Model 2, which was based on VQA using ViLT for English, also achieved an Accuracy of approximately 93% (Table 2).

Table 3. The training results of Model 3: Generate answers with ViLT and GPT-2.

No	Model	Language	ROUGE-1	ROUGE-2	ROUGE-L	Average Time Per Prediction
1	ViLT and GPT-2	English	49.921200	39.263800	47.534200	0.8 s

Regarding Model 3, the answer generation model using ViLT and GPT-2 yielded results based on the ROUGE [9] evaluation method, with scores of 49.92, 39.26, and 47.53 for ROUGE-1, ROUGE-2, and ROUGE-L respectively (Table 3).

Build a Simple Chatbot. To build a simple chatbot application, we utilized the following frameworks and libraries:

Server-Side: To establish the server-side functionality, we employed Flask[7] to create an API that accepts a pair of data: a food image and a question related to the food. The API responds with an answer relevant to the food question. Figure 6 illustrates the detailed operational process.

1. The server receives input from the client via the POST method, consisting of three parts: the food-related question (e.g., "What is this dish?"), the language (en or vi), and the food image.
2. The preprocess function is responsible for converting the input data from step 1 into the standardized input format required by the model.
3. The predict function:
   - For the classification model: It takes input from step 2 and selects the class with the highest probability. Finally, it randomly chooses an answer from the database to provide as the response.
   - For the answer generation model: It takes input from step 2 and generates the answer using either the Greedy Search or Beam Search algorithm. We employed the .generate()[8] method to generate the answer.

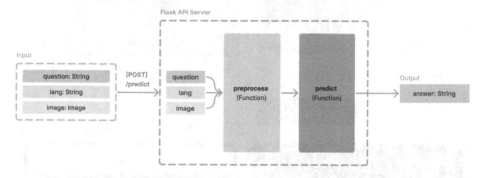

Fig. 6. The data reception and processing process on the server

Client-Side: To develop our mobile application with a chat interface, we utilized React Native[9] in combination with the Flyer Chat[10] library. This allowed us to establish seamless communication with the server-side through the API provided by the server.

- For the chatbot application, we employed the ViLT model with GPT-2 (Model 3) to handle English queries. While the application also supports Vietnamese, the model used for the Vietnamese language is a classification model comprising ViT and PhoBERT (Model 1), distinct from the generative model used for English. (Refer to Fig. 7).

---

[7] https://flask.palletsprojects.com/en/2.3.x/quickstart.
[8] https://huggingface.co/docs/transformers/main_classes/text_generation.
[9] https://reactnative.dev.
[10] https://docs.flyer.chat/react-native/chat-ui.

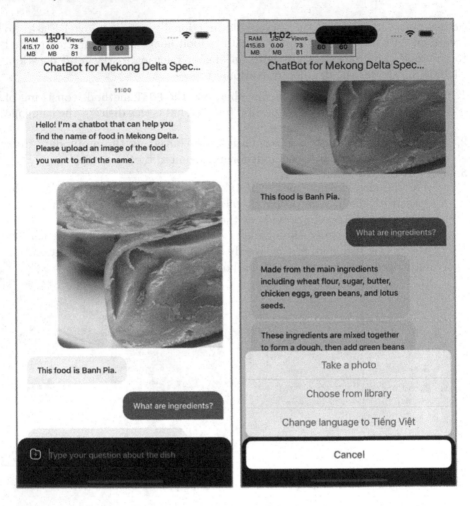

Fig. 7. The chatbot application supports both Vietnamese and English languages

4   Conclusion

In this paper, our primary contribution lies in presenting a framework for preserving local specialties in the Mekong Delta region, with a specific emphasis on supporting travelers in their tourism experiences.

Moreover, we have developed a chatbot to provide assistance in answering tourists' inquiries. The implementation involved Python for the server-side and React Native for the mobile platform.

Moving forward, we plan to broaden the range of food items to encompass all provinces and regions of Vietnam. Our aim is to enhance the entire system with more comprehensive information. Additionally, we have intentions to implement a multi-language system for greater accessibility.

References

1. Nguyen, T.T., et al.: VinaFood21: a novel dataset for evaluating Vietnamese food recognition. In: Proceedings of the 15th IEEE-RIVF International Conference on Computing and Communication Technologies (RIVF) (2021)
2. Do, T.-H., Nguyen, D.-D.-A., Dang, H.-Q., Nguyen, H.-N., Pham, P.-P., Nguyen, D.-T.: 30VNFoods: a dataset for Vietnamese foods recognition. In: 2021 IEEE International Conference on Communication, Networks and Satellite (COMNET-SAT) (2021)
3. Devlin, J., Chang, M.-W., Lee, K., Toutanova, K.: BERT: pre-training of deep bidirectional transformers for language understanding (2019)
4. Nguyen, D.Q., Nguyen, A.T.: PhoBERT: pre-trained language models for Vietnamese (2020)
5. Dosovitskiy, A., et al.: An image is worth 16×16 words: transformers for image recognition at scale (2021)
6. Radford, A., Wu, J., Child, R., Luan, D., Amodei, D., Sutskever, I.: Language models are unsupervised multitask learners. OpenAI Blog **1**(8), 9 (2019)
7. Kim, W., Son, B., Kim, I.: ViLT: vision-and-language transformer without convolution or region supervision (2021)
8. Nguyen, D.Q., Nguyen, D.Q., Vu, T., Dras, M., Johnson, M.: A fast and accurate Vietnamese word segmenter. In: Proceedings of the 11th International Conference on Language Resources and Evaluation (LREC 2018) (2018)
9. Lin, C.-Y.: ROUGE: a package for automatic evaluation of summaries. In: Text Summarization Branches Out, Barcelona, Spain, pp. 74–81. Association for Computational Linguistics (2004)
10. Sutskever, I., Vinyals, O., Le, Q.V.: Sequence to sequence learning with neural networks (2014)
11. Vaswani, A., et al.: Attention is all you need (2017)
12. Agarap, A.F.: Deep learning using Rectified Linear Units (ReLU) (2018)
13. Hendrycks, D., Gimpel, K.: Gaussian Error Linear Units (GELUs) (2016)
14. Srivastava, N., Hinton, G., Krizhevsky, A., Sutskever, I., Salakhutdinov, R.: Dropout: a simple way to prevent neural networks from overfitting. J. Mach. Learn. Res. **15**(2014), 1929–1958 (2014)
15. Ba, J.L., Kiros, J.R., Hinton, G.E.: Layer normalization (2016)

An Approach for Traffic Sign Recognition with Versions of YOLO

Phuong Ha Dang Bui, Truong Thanh Nguyen, Thang Minh Nguyen, and Hai Thanh Nguyen[(✉)]

Can Tho University, Can Tho, Vietnam
nthai.cit@ctu.edu.vn

Abstract. Recognizing traffic signs is essential in guaranteeing traffic safety and reducing the risk of traffic accidents. This study proposes a deep learning-based approach that attempts various YOLO architecture versions to perform common traffic sign recognition in Vietnam. First, data collection is conducted by collecting images taken on roads in Can Tho City and Vinh Long province and then combining them with ZaloAI dataset of Vietnamese traffic signs in 2020. Next, a data augmentation process is deployed to form an enhancement dataset. Then, two versions of YOLO, the YOLOv5 model and the YOLOv8 model, are applied to the enhancement dataset for recognizing traffic signs and comparing the effectiveness of the two approaches. The experimental results show that although the YOLOv5 model takes more training time and has fewer parameters than the YOLOv8 model, the former can perform better in traffic sign recognition tasks.

Keywords: Traffic signs · Recognition · YOLO architecture

1 Introduction

Traffic signs include signs placed at road sections, junctions, or locations that must be notified to guide and warn traffic vehicles about the conditions, regulations, and specific directions. Traffic signs may contain pictures, writing, or special symbols to help drivers and motorists understand easily and comply with traffic regulations. Traffic sign recognition is the ability to recognize and comprehend the meaning of traffic signs on the road for adhering to rules to improve traffic safety. Road users need to know the traffic sign types, their purposes, and regulations to recognize traffic signs. Traffic signs are divided into five categories, encompassing prohibitory signs, warning signs, mandatory signs, indication signs, and additional panels.

Traffic sign recognition is crucial in guaranteeing traffic safety and avoiding the risk of traffic accidents. Deep learning has recently exhibited extraordinary potential in the research of traffic sign recognition. In this study, we propose a deep learning-based approach that explores various versions of YOLO architecture to perform Vietnamese common traffic sign recognition. First, we collect data by collecting images taken on roads in Can Tho City and Vinh Long

N. Thai-Nghe et al. (Eds.): ISDS 2023, CCIS 1950, pp. 204–211, 2024.
https://doi.org/10.1007/978-981-99-7666-9_17

province and then combining them with the ZaloAI dataset of Vietnamese traffic signs in 2020. Next, we deploy a data augmentation process to form an enhancement dataset. Then, we apply two versions of YOLO, the YOLOv5 model and the YOLOv8 model, to the enhancement dataset for recognizing traffic signs and comparing the effectiveness of the two approaches. We can see from the experimental results that the YOLOv5 model can achieve better accuracy than the YOLOv8 model in traffic sign recognition tasks. However, the former has fewer parameters and a longer training time.

The remainder of this paper is organized as follows. The related work which applied deep learning for traffic sign recognition is briefly presented in Sect. 2. Section 3 proposes a deep learning approach based on two versions of YOLO, the YOLOv5 model and the YOLOv8 model, for recognizing traffic signs. Section 4 presents the results obtained by applying the YOLOv5 model and the YOLOv8 model on the enhancement dataset and compares the performance of the two models. Finally, the conclusion is discussed in Sect. 5.

2   Related Work

Deep learning approaches have recently been applied in numerous traffic sign recognition studies. In a study [1], Abedin et al. developed a traffic sign detection and recognition system using a fuzzy rules-based color segmentation method and artificial neural network. Kamal et al. [2] proposed a traffic sign detection and recognition approach by applying SegU-Net, a new network that was formed by merging SegNet and U-Net architectures, to detect traffic signs and then modifying the Tversky loss function with L1-constraint to train the network. The work [3] presented by Khan et al. proposed a lightweight CNN model, trained on the German Traffic Sign Recognition Benchmark and Belgium Traffic Sign datasets, to recognize traffic signs in urban road networks. Another work [4] applied a deep learning model using a refined Mask Region-based Convolutional Neural Network (R-CNN) to detect and recognize Indian road traffic signs. In a study [5], Boujemaa et al. presented two deep learning models to see and recognize traffic signs. The first model used the color segmentation technique and CNN, and the second model was based on the Fast R-CNN. A real-time traffic sign recognition model [6] developed by Li and Wang used a detector that combined Faster R-CNN with MobileNet and a classifier based on an efficient CNN with asymmetric kernels. Zhang et al. [7] proposed a storage-efficient SNN-CNN hybrid network, in which one spiking neural network (SNN) executed coarse classification in the first stage and six CNNs conducted fine variety in the second stage, combining with RRAM-implemented weights for traffic signs recognition.

YOLO (You Only Look Once), a series of end-to-end deep learning frameworks designed to detect objects, was applied in many research on traffic sign recognition. Such an exciting work in [8] proposed a traffic sign recognition and classification method using Single Shot Detector (SSD), Faster R-CNN, and YOLOv2 architectures and then compared the performance of the three architectures. Shahud et al. [9] presented a CNN-based model using YOLOv3 to detect and recognize Thai traffic signs. In a study [10], the authors developed a real-time detection and classification model, which trained on the German Traffic Sign Detection Benchmark dataset applied the YOLOv3 framework to detect and classify traffic signs. The work [11] presented by Tai et al. proposed a deep learning model using the YOLOv3 SPP model, which implemented the spatial pyramid pooling (SPP) principle to boost YOLOv3's backbone network. Zhu and Yan [12] presented a traffic sign recognition method that applied the YOLOv5 and SSD models on their dataset and then compared the effectiveness of the two models.

Although numerous studies based on YOLO algorithms have been proposed in the traffic sign recognition domain, examining the effects of different YOLO framework versions still needs to be thoroughly investigated. Therefore, our study has attempted the various versions of YOLO architecture and evaluated their impact on accuracy and training time performance in traffic sign recognition tasks.

3 Methods

3.1 Data Collection

This research collects data by collecting images taken with mobile phones on roads in Can Tho City and Vinh Long province and then combining them with the ZaloAI dataset[1] of Vietnamese traffic signs in 2020, which includes 4500 images.

Furthermore, we use the Roboflow [13] tool, which allows us to create and store datasets efficiently to label data. The label bounding box hugs the sign, avoids the label being more comprehensive and does not label the entire part of the sign. The resolution distribution and position of the bounding box are illustrated in Fig. 1. Marking the signs from 0 to 20 m away from the vehicle location is executed. The reason is that when identifying road users, we need to locate and notify them before passing the sign so that pedestrians can adjust their speed and pay more attention. This will be ignored for the signs with a loss of more than 40% of the area. The image can lose some important features, making the model learn incorrectly. For each image after labeling, we create a . txt file, an annotation file, where:

- w, h is the distance from the center of the bounding box to the left and the top edges of the image.
- Center: is the coordinates of the center of the bounding box (X_center, Y_center).

We collected 4803 images (with sizes from 91×88 to 1622×626) in which the training set includes 3205 images (67%), the validation set, used to evaluate the model's performance and refine the parameters consisting of 926 images (19%), and the testing set includes 672 images (14%). The dataset comprises seven

---

[1] https://www.kaggle.com/datasets/phhasian0710/za-traffic-2020.

classes for types of traffic signs, e.g., "No parking or stopping", "No entry", "No turn", "Speed limit", "Command", "Danger", and other signs.

3.2 Data Augmentation

In this study, we perform a data augmentation process on Roboflow tool to form an enhancement dataset, with the techniques shown in Table 1. After applying the data augmentation process, the number of images from the dataset is doubled, so the enhancement dataset is formed with 8008 images, in which the training set has 6410 images (80%), the valid set has 926 images (12%), and the testing set has 672 images (8%). We augment the data on the training set to help the model learn a wide range of variations and aspects of the data. This improves generalization and reduces overfitting, a phenomenon where a model is too fine-tuned on the training data but does not perform well on the new data.

Table 1. Data augmentation for traffic sign images

Name of enhancement technique	Applied technique	Reasons to apply
Auto-Oriented	Technique to automatically rotate the image so that it faces the right side	The technique is applied to ensure that the image is displayed in the correct orientation
Static Crop	30–100% Horizontal Region, 0–100% Vertical Region	This technique is often applied to focus on important areas of an image and make it more visible.
Resize	Fit (white edges) in 640 × 640	Change the aspect ratio to suit the use of the YOLO algorithm

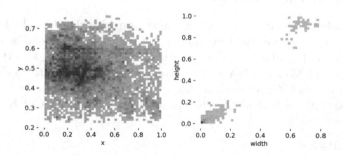

Fig. 1. The chart of resolution distribution (left) and the chart of position of bounding box (right)

3.3   Models for Traffic Sign Recognition

Object detection algorithms can be divided into two groups, including R-CNN and YOLO models. The first category is the family of the R-CNN model, which uses a CNN framework to detect objects in the images. These algorithms split the image into sub-regions and then use a CNN model for classification. The second one is a group of YOLO [14], an object detection algorithm group based on a CNN architecture for real-time object detection. These algorithms help locate the objects faster than the traditional methods and achieve high accuracy. This study considers the recognition speed factor more important than the accuracy factor. Therefore, we decided to use YOLO algorithms to solve this problem. We chose the two versions of YOLO architecture, YOLOv5 [15] and YOLOv8 [16]. YOLOv5 currently has five versions with various sizes, as reported in [17]. YOLOv8 is currently in development and has five versions.

Changes of YOLOv8 in comparison to YOLOv5, as stated in [16], are as follows. Firstly, it replaced module C3 with module C2f. Secondly, it replaced the first 6×6 Conv with the 3×3 Conv in Backbone and the first 1×1 Conv with 3×3 Conv in Bottleneck. Thirdly, it removed two convolutional layers (no. 10 and no. 14 in the YOLOv5 configuration). Lastly, YOLOv8 used the split end and deleted the object branch. A critical feature of YOLOv8 is that it can improve the accuracy and the training time compared to the previous YOLO versions. In addition, the backbone network is updated based on EfficientNet, which improves the ability to capture high-level features of the model. Moreover, YOLOv8 has a new feature consolidation module that integrates features from multiple scales. However, YOLOv8 is only in the development stage; we chose YOLOv5 to develop the proposed model and YOLOv8 to test and recheck the model.

4   Experimental Results

4.1   Environmental Settings

Google Colab is a virtual cloud machine provided by Google for free to researchers. This is the ideal environment for developing small and medium models with a default CPU equipped with an Intel Xeon CPU, two vCPUs (virtual CPUs), and 13 GB of RAM. The process of uploading data to Colab takes time, and after each Colab session (about 5 h), the data can be lost. Therefore, we store the data on Roboflow[2], then connect Roboflow with Colab.

4.2   Results

As shown in Table 2, the YOLOv5 and YOLOv8 models reach the precision accuracy of 0.82 and 0.77, respectively. The low standard deviation, 0.15 and 0.11, indicate the stability of the models in object detection. Comparing the recall, we

---

[2] https://roboflow.com/.

find that the YOLOv5 and YOLOv8 models' recall accuracy is 0.45 and 0.39, respectively. The low standard deviation, 0.08 and 0.07, illustrate stability in detecting real-life objects. Regarding the mAP (mean Average Precision) measure, the YOLOv5 and YOLOv8 models reach the accuracy of mAP@0.5 are 0.50 and 0.44, while those of mAP@0.5:0.95 accuracy are 0.30 and 0.27, respectively. The low standard deviation (from 0.06 to 0.11) shows stability in object detection at different thresholds. The results indicate that the YOLOv5 model takes the longest training time but has the highest accuracy.

Table 2. Results of various versions of YOLO

YOLO version	Precision	Recall	mAP@0.5	mAP@0.5:0.95	Training time
YOLOv5	0.82 ± 0.15	0.45 ± 0.08	0.50 ± 0.11	0.30 ± 0.07	9972 s
YOLOv8	0.77 ± 0.11	0.39 ± 0.07	0.44 ± 0.09	0.27 ± 0.06	9612 s

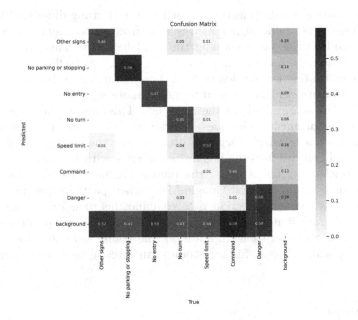

Fig. 2. A confusion matrix of YOLOv5 on various traffic sign classes.

Figure 2 shows that the results are still low across the signs, possibly because the sign is too small for the image to contain the entire context. "Danger", "No parking or stopping", and "Speed limit" signs give better results than the rest. The remaining signs are often confused with "Speed limit" and "No parking or stopping" signs. As shown in Fig. 3, we see that in the same frame, the proposed method can recognize even the minor signs on the truck's trunk. The other signs (blue) are not within the scope of the considered dataset.

(a) The original image   (b) The detected traffic signs

Fig. 3. An illustration on traffic sign recognition containing traffic signs.

5   Conclusion

This study proposes a deep learning-based approach using different YOLO versions to recognize traffic signs. As revealed from experiments, although the YOLOv5 model takes more training time and has fewer parameters than the YOLOv8 model, the former can get better accuracy in traffic sign recognition tasks. Therefore, the new version of YOLO needs further evaluation to improve accuracy. In future studies, we need to collect more signs with different viewing angles, e.g., left side, middle of the road, etc. This may be necessary to learn more data enhancement techniques. In addition, the number of images in the different classes is relatively large. In other words, the data is unbalanced. To solve this problem, we take the images in the class with less data enhancement and data thickening, but at least the images are not precisely the same. As most road users often turn on Google Maps when participating in traffic, the subsequent development direction is to combine this model with Google Maps and then integrate it into autonomous vehicles or vehicle support systems for vehicle control and operation. In addition, the traffic signs may be overlapped or hidden by some objects, further studies can investigate more to improve the performance.

References

1. Abedin, Z., Dhar, P., Hossenand, M.K., Deb, K.: Traffic sign detection and recognition using fuzzy segmentation approach and artificial neural network classifier respectively. In: 2017 International Conference on Electrical, Computer and Communication Engineering (ECCE). IEEE (2017). https://doi.org/10.1109/ecace.2017.7912960
2. Kamal, U., Tonmoy, T.I., Das, S., Hasan, M.K.: Automatic traffic sign detection and recognition using SegU-Net and a modified Tversky loss function with L1-constraint. IEEE Trans. Intell. Transp. Syst. **21**(4), 1467–1479 (2020). https://doi.org/10.1109/tits.2019.2911727

3. Khan, M.A., Park, H., Chae, J.: A lightweight convolutional neural network (CNN) architecture for traffic sign recognition in urban road networks. Electronics **12**(8), 1802 (2023). https://doi.org/10.3390/electronics12081802

4. Megalingam, R.K., Thanigundala, K., Musani, S.R., Nidamanuru, H., Gadde, L.: Indian traffic sign detection and recognition using deep learning. Int. J. Trans. Sci. Technol. **12**(3), 683–699 (2023). https://doi.org/10.1016/j.ijtst.2022.06.002

5. Boujemaa, K.S., Berrada, I., Bouhoute, A., Boubouh, K.: Traffic sign recognition using convolutional neural networks. In: 2017 International Conference on Wireless Networks and Mobile Communications (WINCOM). IEEE (2017). https://doi.org/10.1109/wincom.2017.8238205

6. Li, J., Wang, Z.: Real-time traffic sign recognition based on efficient CNNs in the wild. IEEE Trans. Intell. Transp. Syst. **20**(3), 975–984 (2019). https://doi.org/10.1109/tits.2018.2843815

7. Zhang, Y., Xu, H., Huang, L., Chen, C.: A storage-efficient SNN–CNN hybrid network with RRAM-implemented weights for traffic signs recognition. Eng. Appl. Artif. Intell. **123**, 106232 (2023). https://doi.org/10.1016/j.engappai.2023.106232

8. Garg, P., Chowdhury, D.R., More, V.N.: Traffic sign recognition and classification using YOLOv2, faster RCNN and SSD. In: 2019 10th International Conference on Computing, Communication and Networking Technologies (ICCCNT). IEEE (2019). https://doi.org/10.1109/icccnt45670.2019.8944491

9. Shahud, M., Bajracharya, J., Praneetpolgrang, P., Petcharee, S.: Thai traffic sign detection and recognition using convolutional neural networks. In: 2018 22nd International Computer Science and Engineering Conference (ICSEC). IEEE (2018). https://doi.org/10.1109/icsec.2018.8712662

10. Sichkar, V., Kolyubin, S.: Real time detection and classification of traffic signs based on YOLO version 3 algorithm. Sci. Tech. J. Inf. Technol. Mech. Opt. **20**(3), 418–424 (2020). https://doi.org/10.17586/2226-1494-2020-20-3-418-424

11. Tai, S.K., Dewi, C., Chen, R.C., Liu, Y.T., Jiang, X., Yu, H.: Deep learning for traffic sign recognition based on spatial pyramid pooling with scale analysis. Appl. Sci. **10**(19), 6997 (2020). https://doi.org/10.3390/app10196997

12. Zhu, Y., Yan, W.Q.: Traffic sign recognition based on deep learning. Multimedia Tools Appl. **81**(13), 17779–17791 (2022). https://doi.org/10.1007/s11042-022-12163-0

13. Roboflow (version 1.0) [software]. https://roboflow.com/research

14. Redmon, J., Divvala, S., Girshick, R., Farhadi, A.: You only look once: unified, real-time object detection (2016)

15. Jocher, G.: ultralytics/yolov5: v3.1 - Bug Fixes and Performance Improvements. https://github.com/ultralytics/yolov5 (2020). https://doi.org/10.5281/zenodo.4154370

16. Reis, D., Kupec, J., Hong, J., Daoudi, A.: Real-time flying object detection with YOLOv8. arXiv preprint arXiv:2305.09972 (2023)

17. Dlužnevskij, D., Stefanovič, P., Ramanauskaitė, S.: Investigation of YOLOv5 efficiency in iPhone supported systems. Baltic J. Mod. Comput. **9**(3), 333–344 (2021). https://doi.org/10.22364/bjmc.2021.9.3.07

FaceMask Detection Using Transfer Learning

Nguyen Thai-Nghe, Tran Minh-Tan, Le Minh Hai, and Nguyen Thanh-Hai[(✉)]

Can Tho University, Can Tho, Vietnam
{ntnghe,tmtan,nthai.cit}@ctu.edu.vn

Abstract. Wearing face masks in public places has reduced the global spread of COVID-19 and other bronchi/lung diseases. This paper proposes an approach for the face detection of people from the photo with or without a facemask. This task is implemented based on the features around their eyes, ears, nose, and forehead by using the original masked and unmasked images to form a baseline for face mask detection. In this work, we have used the Caffe-MobileNetV2 model for feature extraction and image classification. First, the convolutional architecture for the fast feature embedding Caffe model is used as a face detector, and then the MobileNetV2 is used for facemask identification. Experimental results revealed that the proposed approach performed well, with an accuracy of 98.54%. The work is expected to be deployed in practical cases.

Keywords: Facemask Detection · FaceMask Classification · Transfer Learning

1 Introduction

According to the WHO's official reports[1], over 500 million people worldwide have been infected with COVID-19, resulting in over 5 million fatalities. The symptoms experienced by COVID-19 patients range widely from mild signs to significant sickness. One is a respiratory issue, such as having trouble breathing or feeling short of breath. Elderly adults with lung disease may get major COVID-19 sickness problems since their risk seems to be increased. A clinical mask is essential to treat some respiratory viral infections, such as COVID-19. The general population has to be informed about whether to wear a mask for source control or to avoid contracting COVID-19 and other diseases. Potential benefits of using masks include lowering vulnerability to danger from infectious individuals during the "pre-symptomatic" stage and stigmatizing specific individuals who use masks to stop the spread of viruses. Thus, facemask detection is an essential task in today's global society.

This work proposed using transfer learning with the Caffe-MobileNetV2 [1] model for facemask detection. After being trained on the two-class image dataset,

---

[1] https://covid19.who.int/, accessed on 30 July 2023.

N. Thai-Nghe et al. (Eds.): ISDS 2023, CCIS 1950, pp. 212–219, 2024.
https://doi.org/10.1007/978-981-99-7666-9_18

the proposed model was tested on the image dataset. Depending on the output and the designated class name, the bounding boxes were painted around the faces in either green or red.

2 Related Work

Recent research on face mask identification during COVID-19 has been conducted and is presented in the literature. Researchers have created methods based on deep learning to investigate the problem of face mask identification, as illustrated in [2]. ResNet50 was created as a dependable approach based on occlusion reduction and deep learning-based features to address the issue of masked face recognition [3]. To construct new models, transfer learning is applied using the VGG16 and AlexNet convolutional neural network architectures [4]. The authors created an automatic system in [5] to determine whether or not someone is wearing a mask. If the individual is not, the system generates an alarm. The authors included CNN's VGG-16 architecture into their system. Their system achieved overall detection accuracy. In the future, the authors decide to make a system that will not only detect whether a person is wearing a mask or not but will also detect the physical distance between each individual and will sound an alert if the physical distancing is not followed correctly.

A model to determine if someone is wearing a mask or not in public spaces was introduced by the authors in [6] using a DL algorithm. They accomplished this using the pre-trained MobileNetV2 picture categorization algorithm. A method for automatically determining whether someone is wearing a mask is presented in another work in [7]. With MobileNetV2, they created a transfer learning-based technique for identifying masks in streaming video and photos. Their algorithm successfully detected 98% of the 4,095 pictures in the dataset. The work in [8] built a method by merging pre-trained DL models like Xception, MobileNetV2, and VGG19 to extract features from the input photos. Following feature extraction, they classified the extracted characteristics of the images using various ML classifiers, including SVM and k-nearest neighbor (k-NN). A total of 1,376 photos from two classes, those with masks and those without were used. According to the experimental findings, MobileNetV2 and SVM had the most significant classification accuracy (97.11%). This work proposes using the Caffe-MobileNetV2, based on the MobileNetV2 [1] with transfer learning.

3 Method

3.1 The Proposed Workflow

The architecture of the proposed system is presented in Fig. 1. The workflow includes two phases, including training and testing. In the training phase, we use MobileNet to learn images, including faces with and without masks. Then, we evaluate the performance with various metrics.

3.2   Dataset

Datasets have been used for experiments downloaded from Kaggle consisting of 14,835 images (Fig. 2), and their countenances are clarified either with a mask or without a mask. In Fig. 3, some face collections are head turn, tilt, slant with multiple faces in the frame, and different types of masks with different colors.

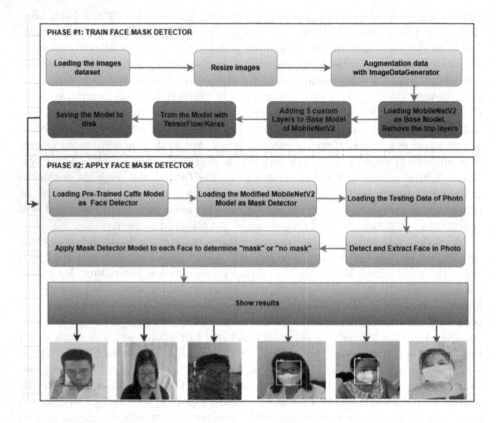

Fig. 1. The flowchart of the facemask detection.

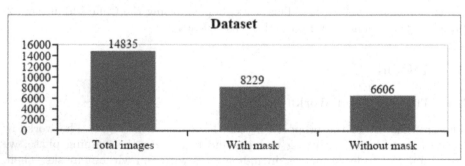

Fig. 2. Data distribution.

Fig. 3. Some samples from the dataset, including faces with and without a mask.

The collected dataset was divided into three parts. The first is a training dataset consisting of 11,868 images, representing 80% of the entire dataset. The second category is the validating data, consisting of the remaining 2,076 images, representing 14% of the total dataset. The third category is the testing data, consisting of the remaining 891 images representing 6% of the entire dataset.

3.3 Data Pre-processing

During pre-processing, all photos from a folder are loaded as input, and the photographs are resized in preparation for use with the suggested model. The data is added after loading and transformed into an array to pre-process the photos. Furthermore, we would like to add more data to train on to improve our model's performance. Scaling, rotating, and flipping the data will artificially generate more data samples, etc., as presented in Fig. 4.

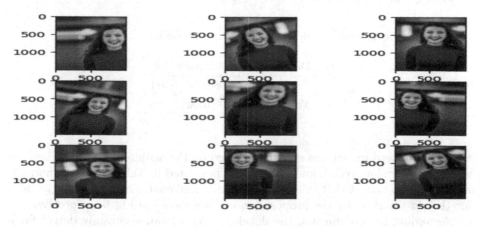

Fig. 4. Result after using augmentation (with Keras ImageDataGenerator).

3.4 Proposed Model

The pre-trained MobileNetV2 model was imported. After loading the base model, We add five layers for the transfer learning. A 7×7 average pooling, flattening, density of 128 neural networks, dropout size of 0.5, and densen_1 of 2 make up the additional layers. Through the use of the average pooling layer, these layers enabled the improved MobileNetV2 to perform better while requiring fewer parameters. Coming into a single dimension is made easier by flattening. A thick layer using an activation function like the rectified linear unit (ReLU), in which every output depends on every input, was utilized to generate fully connected layers. Undoubtedly, adding a dropout layer, followed by adding two thick layers for binary classification, aids in preventing overfitting. The Softmax function was used since there were several output neurons.

4 Experimental Results

Results during the training phase are presented in Table 1, Fig. 5a, and Fig. 5b. The primary layers were frozen to prevent the loss of previously acquired properties. A new set of trainable layers was created to distinguish a face wearing a mask from a face not wearing one, and these layers were trained using the collected dataset. The model was then updated after saving the weights. A learning rate of 0.001 and 50 epochs were added to the suggested model's training, along with a batch size 32. We used the Adam optimizer, which modifies each neural network weight's learning rate by calculating the first and second moments of the gradient. After each batch was received, the appropriate algorithm used transfer learning to change the weights in the neural network. A neural network was trained and ultimately preserved for use in producing predictions on the image dataset. Before training the network, we specify an early stopping condition to prevent repeating epochs after the model converges. With early stopping, the process of training stops at epoch 30. The trained model achieved an accuracy of 98.84% and a loss of 0.031.

Table 1. Result table of accuracy and loss during training phase

No.	Dataset	Accuracy	Loss
1	Training set	98.94%	311
2	Validation set	98.54%	509

The proposed model was evaluated based on the acquired measures, such as accuracy, precision, recall, and f1-score, as presented in Table 2. The unnormalized confusion matrix (left side) and the normalized confusion matrix (right side) for the classification by the proposed model are shown in Fig. 6, respectively.

According to experiments, the developed model can accurately detect faces with and without masks and uses fewer parameters than the existing models.

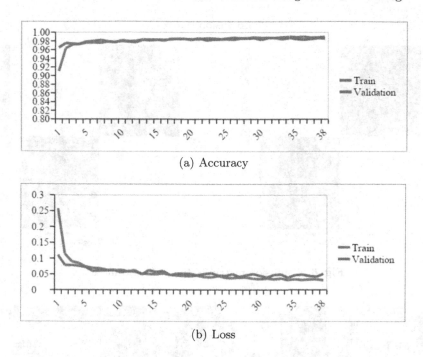

(a) Accuracy

(b) Loss

Fig. 5. Performance during epochs in training and validation set.

According to investigations, the overall model achieved an accuracy of 98.54%. The proposed model makes predictions based on the training dataset's pattern and labels. The visualization of the classification results of the photo image data is shown in Fig. 7. The face with a proper mask is displayed with a green bounding box, and the face without/with an incorrect mask is shown with a red bounding box. Table 3 indicates that the proposed model produces better results in all metrics for classifying faces with and without masks. The performance measure of the several versions of the proposed model using Caffe-MobileNetV2, MobileNetV3 Small, and MobileNetV3 Large are also calculated and compared in terms of various metrics such as accuracy, precision, recall, f1-score, and error rate as shown in Table 3.

Table 2. Summary results table

	Precision	Recall	F1-Score
With_mask	98.00%	99.39%	98.69%
Without_mask	99.23%	97.48%	98.35%
Accuracy			98.54%
Macro avg	98.62%	98.45%	98.52%
Weighted avg	98.55%	98.54%	98.54%

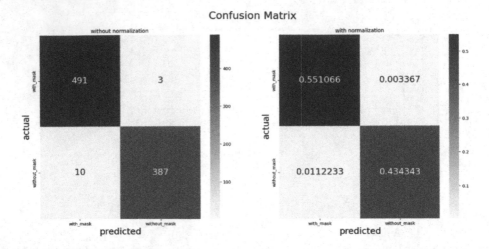

Fig. 6. Confusion matrix of the proposed model.

Fig. 7. Some illustration of Face detection with and without masks.

Table 3. Comparison of the proposed model, MobileNetV3 Small and MobileNetV3 Large.

Model Version (use weighted avg)	Accuracy	Precision	Recall	F1-Score	Error Rate
Proposed model	98.54%	98.55%	98.54%	98.54%	1.46%
MobileNetV3 Large	72.95%	73.63%	72.95%	72.89%	27.05%
MobileNetV3 Small	64.53%	66.98%	64.53%	64.34%	35.47%

5 Conclusion

This work proposed a method for detecting facemasks by modifying the MobileNetV2. The transfer learning technique helped to distinguish between people wearing or not wearing a face mask. The proposed method worked very well on the dataset with an accuracy of 98.54%. A facial mask detection approach in real-time can be used in various places, such as in airports, shopping malls, subway stations, temples, offices, hospitals, educational institutions, etc. The model does, however, have certain drawbacks. For instance, it was mainly tested and trained on pictures of straight faces with and without masks.

In the future, it can be extended to detect if a person is wearing the mask properly. The model can be further improved to detect if the mask is virus-prone or not, i.e., the type of the mask is surgical, N95, or not. The alarm functionality can be added to the system, i.e., an alarm is raised whenever a face mask is violated. The prediction quality can be improved by using a high-definition video camera for video capture.

References

1. Howard, A.G., et al.: Mobilenets: efficient convolutional neural networks for mobile vision applications. arXiv preprint arXiv:1704.04861 (2017)
2. Mandal, B., Okeukwu, A., Theis, Y.: Masked face recognition using ResNet-50. arXiv preprint arXiv:2104.08997 (2021)
3. Hariri, W.: Efficient masked face recognition method during the COVID-19 pandemic. arXiv preprint arXiv:2105.03026 (2021)
4. Koklu, M., Cinar, I., Taspinar, Y.S.: CNN-based bi-directional and directional long-short term memory network for determination of face mask. Biomed. Signal Process. Control **71**, 103216 (2022). https://doi.org/10.1016/j.bspc.2021.103216
5. Militante, S.V., Dionisio, N.V.: Real-time facemask recognition with alarm system using deep learning. In: 2020 11th IEEE Control and System Graduate Research Colloquium (ICSGRC), pp. 106–110 (2020)
6. Sanjaya, S.A., Adi Rakhmawan, S.: Face mask detection using mobilenetv2 in the era of COVID-19 pandemic. In: 2020 International Conference on Data Analytics for Business and Industry: Way Towards a Sustainable Economy (ICDABI), pp. 1–5 (2020)
7. Sandler, M., Howard, A., Zhu, M., Zhmoginov, A., Chen, L.C.: Mobilenetv 2: inverted residuals and linear bottlenecks (2019)
8. Oumina, A., El Makhfi, N., Hamdi, M.: Control the COVID-19 pandemic: face mask detection using transfer learning. In: 2020 IEEE 2nd International Conference on Electronics, Control, Optimization and Computer Science (ICECOCS), pp. 1–5 (2020)

Human Intrusion Detection for Security Cameras Using YOLOv8

Nguyen Thai-Nghe[1], Huu-Hoa Nguyen[1(✉)], Wonhyung Park[2], Quang Thai Ngo[3], and Minh Toan Truong[3]

[1] Can Tho University, Can Tho, Viet Nam
{ntnghe,nhhoa}@ctu.edu.vn
[2] Sungshin Women's University, Seoul, Korea
whpark@sungshin.ac.kr
[3] VNPT Ca Mau, Ca Mau, Viet Nam
{thaim2522024,toanm2522025}@gstudent.ctu.edu.vn

Abstract. Human intrusion detection is a critical concern in the field of security, especially for finding illicit entrance at private or restricted places. Unauthorized intrusions occur when attackers gain illicit access to residential or restricted areas and engage in property theft. Security monitoring is essential in various locations, including residential areas and households, particularly during specified periods, typically from 10 pm to 6 am. In this study, we propose an approach for detecting human intrusion with YOLOv8 for security cameras. The model was identified and tracked objects in the predetermined ROI area, and got over 80% accuracy from human identification data. After training the model, the software was employed with several fundamental functions for detection and tracking of unauthorized human intrusions.

Keywords: Human intrusion detection · Yolov8 · deep learning · security camera

1 Introduction

The topic of human intrusion detection is crucial to the security industry. Intrusion detection is a crucial aspect of ensuring security in various environments. Whether it is protecting residential areas, commercial spaces, or restricted facilities, the ability to identify and respond to unauthorized access is of paramount importance. Traditional security measures often rely on human vigilance, which can be prone to errors and limitations. However, advancements in technology have paved the way for more sophisticated intrusion detection systems.

This article proposed an approach for human intrusion detection for security cameras using the YOLOv8[1] approach. In this approach, by leveraging the power of YOLOv8, security cameras can effectively identify and track various objects within their field of view, including potential intruders. This approach combines object detection, face

---

[1] https://github.com/ultralytics/ultralytics

N. Thai-Nghe et al. (Eds.): ISDS 2023, CCIS 1950, pp. 220–227, 2024.
https://doi.org/10.1007/978-981-99-7666-9_19

detection, and intrusion detection tasks, allowing for comprehensive monitoring and analysis. This multi-task classification model enables security systems to not only detect unauthorized access but also recognize faces and identify specific objects of interest. This integration of tasks enhances the overall surveillance capabilities and provides a more robust security solution. After training the model, we build an application which has two key components: people detection and object tracking.

2 Related Works

Intrusion detection using cameras and the YOLO (You Only Look Once) algorithm is an emerging area in the field of computer vision and security. This part will summarize some recent studies related to intrusion detection using the YOLO algorithm and cameras. In the study [1], it is a crucial resource for researchers and professionals working in the field of computer vision and object identification since the authors provide an in-depth analysis of a real-time object detection system using YOLO and CNN models. The authors of [2] suggest a method that focuses on creating a network structure to balance processing speed and accuracy. According to test results, their trained model can process information at a rate of 2 FPS on the Raspberry PI 3B while retaining a high degree of human detection accuracy. The model's accuracy is 95.05% on the INRIA dataset and 96.81% on the PENN FUDAN dataset, respectively.

For unique circumstances, the authors of the study [3] suggest a small size target detection technique. The method ensures the accuracy of detection for all sizes while demonstrating improved precision for detecting tiny size targets. In order to successfully merge shallow and deep information, the feature fusion network is improved, and a new network topology is suggested in order to increase detection accuracy. The research also provides a new down-sampling technique to retain context feature information. According to the experimental results, the suggested algorithm is more accurate than YOLOX, YOLOXR, scaled YOLOv5, YOLOv7-Tiny, and YOLOv8. For small size targets, minimal size targets, and regular size targets, the proposed approach outperforms YOLOv8 in experiments using reliable public datasets, with improvements ranging from 0.5% to 2.5%.

The study in [4] focuses on the creation of an automatic human intruder detection system that can quickly alert the control room about human intrusions into restricted regions. The model that is being described makes use of the YOLO Model to find human objects in films taken by cameras that have a reasonable level of quality. To assess the model's performance in various circumstances, a variety of experiments were run. The acquired findings demonstrate the model's effectiveness in distinguishing human objects from other items in the recorded videos with an accuracy of roughly 96% in the taken-into-account environmental conditions.

In this study, we propose an approach for detecting human intrusion with YOLOv8 for security cameras. However, other techniques could also be used [5, 6].

3   Proposed Approach

3.1   System Architecture

The proposed system is presented in Fig. 1. In this system, the real-time video of security camera at the restricted place is inputted into the detection engine, e.g., using the
YOLOv8. The engine will analyzes and detects if there is a human in this video to send
an alert for security employee. By using this approach, we do not need many security
employees for manual surveillance, thus it can reduce the cost for companies/units.

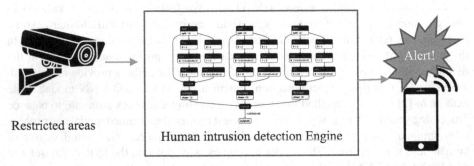

Fig. 1. Human intrusion detection system

In this work, the human intrusion detection engine is YOLOv8. This is the most recent
and cutting-edge YOLO model. It can be utilized for applications including object identification, image categorization, and instance segmentation. The same company that developed the well-known and industry-defining YOLOv5 model also developed YOLOv8.
Compared to YOLOv5, YOLOv8 has a number of architectural updates and enhancements. Its changes compared to the YOLOv5 by[2] replacing the C3 module with the C2f
module, replacing the first 6 × 6 Conv with 3 × 3 Conv in the Backbone, deleting two
Convs (No.10 and No.14 in the YOLOv5 config), replacing the first 1 × 1 Conv with 3 ×
3 Conv in the Bottleneck, and using decoupled head and delete the objectness branch.

3.2   Evaluation Measures

After training an entity recognition model, we used the IoU (Intersection Over Union)
as a metric for evaluation. The IoU is a function that evaluates the accuracy of an object
detector on a particular data set as presented in Fig. 2. In which, the Area of Overlap is
the area of intersection between predicted bounding box and the grouth-truth bounding
box, and the Area of Union is the area of intersection between predicted bounding box
and grouth-truth bounding box. If the IoU $>= 0.5$ then prediction is evaluated as good.
Moreover, the mAP (mean Average Precision) is also used. The mAP for object
detection is the average of the AP calculated for all the classes and the mAP@50 means
that it is the mAP calculated at IOU threshold 0.5.

---

[2] github.com/ultralytics/ultralytics/issues/189.

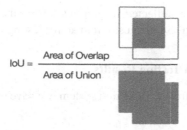

$$IoU = \frac{\text{Area of Overlap}}{\text{Area of Union}}$$

Fig. 2. IoU calculation formula

4 Experimental Results

4.1 Data Sets

We have used two data sets. The first one is a public data set from Kaggle[3]. This data set has 559 images which having one or several humans in each image. We have used the App CVAT[4] for labelling by drawing a bounding box around the human, as presented in Fig. 3. This data set is divided to 80% for train (447 images) and 20% for test (112 images).

Fig. 3. Loss and accuracy during training and validation phases

[3] https://www.kaggle.com/datasets/constantinwerner/human-detection-dataset.
[4] https://app.cvat.ai/

The second dataset was collected from a video from the security camera with the length of 10 min, which has 38 images of human for testing.

4.2 Model Training and Testing Results

For building the human intrusion detection system, we have used the following libraries and environments:

- Libraries used: opencv-python, onnxruntime-gpu, app CVAT.
- Environment: Windows 10, Python 3.11, MySQL, PHP.

The loss and accuracy during training and validation are presented in Fig. 4. In this figure, box_loss displays the quantity of bounding box mistakes that were found, and cls_loss shows the number of errors in the detected feature classes. The level of box_loss and cls_loss drop with each epoch after running 25 epochs, while the accuracy is increased.

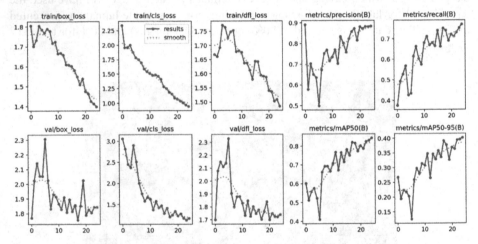

Fig. 4. Loss and accuracy during training and validation phases

Figure 5 presents for some samples of human detection during the test phase. After testing, the MAP@50 achieves 0.839 on the first data set and 0.6 on the second data set, respectively. This shows that the model accuracy is acceptable.

4.3 System Interface

After training and testing the model, we have integrated it into the system. A screenshot of system interface is presented in Fig. 6. In this system, when a human is detected, it send an alert/sound to notify the security employee. Moreover, after detecting human, the system is automatically extract his/her images to store in database with the current date-time for later tracking, as presented in Fig. 7.

Fig. 5. Human detection result in the test set

Fig. 6. Identification results through images extracted from the camera

Fig. 7. Human image extraction by appearing time for later tracking

5 Conclusions

Intrusion detection is an important issue in the security field, as unauthorized intrusions can lead to property theft and other criminal activities. To protect people's property, it is crucial to have effective unauthorized intrusion detection methods in place. In this study, we have proposed an approach utilizing features extracted from three deep learning network architectures to create a human intrusion detection system. This system can provide for businesses and localities with a sense of security in protecting their personal accounts.

In our future research direction, we aim to enhance the results of the predictive model by exploring and applying different data preprocessing techniques. By doing so, we expect to further improve the accuracy and efficiency of our intrusion detection system. Additionally, we will focus on developing applications that can identify and analyze the behavior of subjects, including facial recognition to identify strangers. This expansion of capabilities will enable a more comprehensive approach to intrusion detection. Furthermore, to ensure the best accuracy of the model, it is imperative to gather more high-quality image and video data. By collecting diverse and representative datasets, we can train the model to better handle real-world scenarios and increase its effectiveness in detecting intrusions.

References

1. Viswanatha, V., Chandana, R.K., Ramachandra, A.C.: Real time object detection system with YOLO and CNN models: a review. arxiv preprint: arXiv:2208.00773 (2022)
2. Nguyen, H.H., Ta, T.N., Nguyen, N.C., Bui, V.T., Pham, H.M., Nguyen, D.M.: YOLO based real-time human detection for smart video surveillance at the edge. In: Proceedings of the 2020 IEEE Eighth International Conference on Communications and Electronics (ICCE), Phu Quoc Island, Vietnam, pp. 439–444 (2021). https://doi.org/10.1109/ICCE48956.2021.9352144

3. Lou, H., et al.: DC-YOLOv8: Small-size object detection algorithm based on camera sensor. Electronics **12**(10), 2323 (2023). https://doi.org/10.3390/electronics12102323
4. Chandra, N., Panda, S.P.: A human intruder detection system for restricted sensitive areas. In: Proceedings of the 2021 2nd International Conference on Range Technology (ICORT), Chandipur, Balasore, India, pp. 1–4 (2021). https://doi.org/10.1109/ICORT52730.2021.9582099
5. Thai-Nghe, N., Tri, N.T., Hoa, N.H.: Deep learning for rice leaf disease detection in smart agriculture. In: Dang, N.H.T., Zhang, YD., Tavares, J.M.R.S., Chen, B.H. (eds) Artificial Intelligence in Data and Big Data Processing. ICABDE 2021. Lecture Notes on Data Engineering and Communications Technologies, vol 124. Springer, Cham (2022). https://doi.org/10.1007/978-3-030-97610-1_52
6. Thanh-Du, H., Huk, M., Dung, N.H., Thai-Nghe, N.: An attendance checking system on mobile devices using transfer learning. In: New Trends in Intelligent Software Methodologies, Tools and Techniques. Frontiers in Artificial Intelligence and Applications, vol. 355, pp. 499–506 (2021). https://doi.org/10.3233/FAIA220279

Intelligent Systems

Application of 3-gram to English Statements Posted at Q&A Sites to Obtain Factor Scores

Yuya Yokoyama(✉) [iD]

Advanced Institute of Industrial Technology, 1-10-40, Higashiooi, Shinagawa-ku, Tokyo, Japan
yokoyama-yuya@aiit.ac.jp

Abstract. The widespread diffusion of the Internet has enabled us to communicate on Question and Answer (Q&A) sites. While there have been increasing numbers of registered users and posted Q&A statements, the issues of mismatches between the intentions of questioners and respondents are becoming conspicuous. In order to solve these problems, nine factors for Japanese Q&A statements were obtained through impression evaluation experiments. Syntactic information was then extracted through morphological analysis (MA) and employed as the feature values in performing multiple regression analysis to obtain factor scores for any Q&A statements. Factor scores can be calculated by utilizing N-gram applied to Part-of-Speech extracted through MA. Since most of the methodology so far have mainly depended on Japanese, the validity of the method applied to other languages must also be inspected. So far, in a similar fashion as in Japanese, the nine factors were experimentally obtained from English Q&A statements. The factor scores were then calculated utilizing feature values of English statements extracted through MA. It has also been confirmed that N-gram could be applicable to calculate factor scores instead of syntactic information using mere MA. Therefore, in this paper, 3-gram, as well as MA, was utilized as the feature values of English Q&A statements to obtain factor scores. Through multiple regression analysis, it has been shown that applying 3- gram would lead to as good accuracy as mere MA for all nine factors. Hence it could be concluded that 3-gram would also be applicable to the methodology when applied to English materials.

Keywords: Q&A Site · 3-gram · English · Multiple Regression Analysis

1 Introduction

The prompt expansion of technology has generated the blooming of numerous Question and Answer (Q&A) sites, which are the communities where users can reciprocally post Q&A statements, e.g. Yahoo! Chiebukuro (Y!C) [1]. With increasing numbers of people registering on to them, those Q&A sites are considered as databases with massive amounts of knowledge to resolve a great variety of issues, matters or problems. The fundamental procedure of a Q&A system is as follows: a user posts a question, then others might respond, regardless of their content or accuracy. Among the answer statements posted, the questioner chooses the most agreeable one as the "Best Answer" (BA) and provides the respondent with awards as a token. The BA is the one the questioner subjectively finds most satisfying.

© The Author(s), under exclusive license to Springer Nature Singapore Pte Ltd. 2024
N. Thai-Nghe et al. (Eds.): ISDS 2023, CCIS 1950, pp. 231–246, 2024.
https://doi.org/10.1007/978-981-99-7666-9_20

With all the growing numbers of people using Q&A sites, it is becoming harder for respondents to catch sight of questions that meet their specialty or interests. Thus, qualified respondents would be likely to overlook a question posted by another user. Moreover, even though Q&A sites are becoming so enormous as to be the collective knowledge for society, improper answers can also be accumulated. Hence, not encountering any appropriate respondents could lead to mismatches and the following issues:

- Inappropriate answers may confuse the questioner and spread incorrect knowledge.
- Deficient knowledge might prevent respondents from providing appropriate answers, eventually leaving the question unsolved.
- Abusive words, slander, or statements against public order and standards of decency might offend users.

Therefore, demanding respondents to be users who are likely to post appropriate answers is in-dispensable for compiling appropriate answer statements. In order to solve the issues stated earlier, a number of preceding studies researching Q&A sites [2–4] with the usage of textual features or link analysis have been reported. These prior works, however, have yet to regard the tendencies of the written styles of the users. In addition, it cannot be said that a method to introduce appropriate respondents to a questioner has been settled yet. Thus, by measuring the impressions made by the statements, our work aims to introduce appropriate respondents to a questioner. The promotion and development of this work will lead to the accumulation of appropriate answer statements alone and make Q&A sites more functional for society, thus contributing to rapid and efficient promotion of social activities.

The objective of this work is to present questions to users qualified to properly answer them, resulting in the reduction of the problematic issues described above. Through factor analysis applied to the results of impression evaluation experiment, nine factors describing the impression of Japanese Q&A statements have been obtained [5]. Using the feature values of statements extracted through morphological analysis (MA), factor scores have been estimated by performing multiple regression analysis [6]. This methodology was subsequently developed to find appropriate respondents who would be expected to answer a newly posted question [7].

Nevertheless, most of the feature values have largely depended on the syntactic information (Syn-Info) extracted through MA. In addition, a number of explanatory variables (EVs) with the consideration of quadratic terms might result in obtaining considerably complicated multiple regression equations to estimate factor scores. Therefore, as an alternative syntactic method of MA, N-gram has been regarded and applied to the feature values of Syn-Info [8–10]. In performing multiple regression analysis, the feature values based on N-gram and those other than the Syn-Info were aggregately used as EVs, whereas the factor scores of nine factors were set as respondent variables [8–10]. The analysis results with the cases when N is 2 through 5 have shown that employing N-gram in addition to MA would be as applicable as using mere MA [8–10].

As most of the feature values have been mainly dependent on Japanese, another significant requirement is to investigate the validity of extending this method. Hence, in an analogous fashion as with Japanese, nine factors representing the impression of English Q&A statements were experimentally obtained [11]. Additionally, factor scores were estimated through multiple regression analysis utilizing the feature values extracted through MA. Similar to Japanese, all nine factors also showed good estimation accuracy in English [12].

Likewise, whether N-gram in place of Syn-Info would be applicable has also been investigated. 2-gram was applied to Part-of-Speech (PoS) extracted through MA and utilized as the feature values of English Q&A statements. The analysis results have shown that applying 2-gram on top of MA outperformed MA alone from the viewpoints of estimation accuracy.

So far only 2-gram has been analyzed for English Q&A statements. Therefore, longer units of N-gram must also be applied to verify whether the methodology could be extended in English. In this paper, 3-gram is applied to the feature values of English Q&A statements. Similar to the previous analyses, through multiple regression analysis, the feature values based on 3-gram and appearance of PoS were jointly utilized as EVs, whereas the factor scores were set as respondent variables. As a result of analysis, the 3-gram would be as applicable as 2-gram and MA alone to estimate factors scores.

The remainder of this paper consists of the following: Sect. 2 introduces related works. As with previous works, the method of obtaining factor scores of Japanese/English statements and estimating them are summarized in Sect. 3. In Sect. 4, the previous work on applying 2-gram is explained. Then the application of 3-gram to English Q&A statements is presented in Sect. 5. Considerations toward the analysis results are discussed in Sect. 6. Finally, Sect. 7 concludes the paper.

2 Related Works

A number of studies investigating Q&A sites have been reported in the literature. Due to space limitations, several research studies on introducing users to answer statements will be summarized. Yang *et al.* proposed four challenges through a broad review of the present literature on expert recommendation [2]: (1) Extant recommendation methods disregard the users' willingness to keep contributing within the online knowledge community. (2) Insufficient information in user profiles which hinders identifying potential experts. (3) Recommending experts as a collaborative group rather than looking for familiar individuals could drastically enhance the recommended answer rate. (4) It is vital to regard the self-evolution of present expert recommendation approaches. Zhang *et al.* challenged the problems where the patients' current usage of clinical data is considerably limited due to the technical nature of the clinical report [3]. The analysis results has conveyed users provided both objective and subjective information to the community. This result accentuates the consequence of developing mechanisms to address the problem of the quality of online health information. Haq *et al.* investigated on the Q&A site reputation through Quora, a Q&A platform that aggregates elements of social networks to the traditional Q&A model [3]. They demonstrated that anonymity does not impact the polarity; and that anonymous answers and non-anonymous ones are considerably different in terms of length, subjectivity, and lexical diversity. They have also showed that stronger subjectivity results in more extreme polarity, owing to the self-experience argued in the anonymous content.

Although these earlier findings have developed their research by using textual features, link analysis or user profiles, the tendency of answer statements have not been sufficiently regarded; some users might prefer a polite style, whereas others may tend to post their response in a ruder tone. Some are fond of abstract words, while others are inclined to write in a more concrete style. Therefore, our work focus on impressions in addition to textual features. Furthermore, a method to introduce appropriate respondents to a questioner has yet to be established. Thus, by using the impression of statements, the objective of our work is to introduce appropriate respondents to a questioner.

3   Previous Works

3.1   Obtaining Factors of Statements (Japanese)

Impression evaluation experiment was carried out to evaluate the impressions of Q&A statements. Forty-one evaluators were requested to evaluate the style or content of statements and allot labels from a set of fifty impression words [5]. Twelve sets of sixty Q&A statements were evaluated. These experimental materials were chosen from those actually posted at Y!C in 2005 [1]. Nine factors were obtained by applying factor analysis to experimental results. The factors indicate the nature of statements, as interpreted by multiple impression words assigned to the statement; they were named *accuracy, displeasure, creativity, ease, persistence, ambiguity, moving, effort*, and *hotness*.

3.2   Estimation of Factor Scores (Japanese)

The factor scores were obtained for only the sixty statements utilized as the experimental materials as stated in Sect. 3.1. To be able to estimate the factor scores of other statements, multiple regression analysis was applied to their feature values [6]. Overall, seventy-seven feature values were adopted.

Multiple regression analysis was executed on the sixty Japanese Q&A statements used in the impression evaluation experiment stated in Sect. 3.1. As one of results of multiple regression analysis, multiple correlation coefficients (MCCs) are obtained. MCCs indicate the goodness of the estimation accuracy were above 0.9 for all nine factors [6]. Thus, MCCs for all the nine factors showed very good estimation accuracy.

3.3   Obtaining Factors in English

Most of the feature values in this study mainly depended on Japanese. Thus, it would be required to investigate how far this methodology could be applicable in other languages as well [11]. Similarly, as an initial step, factors that describe impression of English statements were obtained in a similar way as explained in Sect. 3.1.

Most of the English impression words utilized are the same ones as Japanese so that the method applied in English can easily and directly compared with that in Japanese. With employing those fifty impression words, impression evaluation experiment was carried out with the cooperation of four foreign subjects (three males: Mexican, Vietnamese, and French, and one female: French). Experimental materials were six sets of thirty Q&A English statements actually posted to Yahoo! Answers (Y!A) [13]. These materials were extracted from sourceforge [14], where data on Y!A are open to the public so that research on Community Question Answering could be facilitated. Similar to the previous experiment summarized in Sect. 3.1, each set is composed of one question and four answer statements including the "BA."

As a result of factor analysis applied to the experimental results, nine factors were obtained. These factors were named *accuracy, evaluation, disappointment, discomfort, novelty, potency, difficulty, politeness,* and *nostalgia.* Both similarity and difference were observed between the English factors and the Japanese ones shown in Sect. 3.1.

3.4   Obtaining Factors in English

Similar to the case in Japanese, the factor scores obtained were for a mere thirty English Q&A statements used in the experiment. Multiple regression analysis was applied to their feature values so that the factor scores of any statements could be estimated [12]. Overall, twenty-six feature values were adopted with the consideration of multicolin-earity. They are shown in Table 1 and briefly summarized as follows:

- The number of words that each statement consisted of was extracted. The variable is denoted as "Word" (Eg1). Moreover, the ratio of word to sentence was also taken into consideration. This variable is denoted as "Word/SENT" (Eg2). These Eg1 and Eg2 are collectively denoted as "WORDs."
- As with the method in Japanese, Syn-Info was extracted through MA applied to the statements of experimental materials stated in Sect. 3.2. MA was performed through installing TreeTagger [15], resolving a chunk of sentences into each word, its Part-of-Speech (PoS) tag, and its basic form. Additionally, some words can be assigned as more segmentalized tags. These detailed explanations of PoS tags are provided in Appendix [16].
- PoS: The appearance times of base POS tags were adopted. These feature values are jointly denoted as "PoS" and shown in Eg3–Eg10.
- PoS (%): In addition to the words with their base PoS tags, their appearance percentage of each Q&A statement is considered. These feature values are collectively denoted as "PoS (%) and shown in Eg11–Eg26. Taking an example of the words given an "MD"

Table 1. English feature values adopted. [11]

Eg	Variable	Eg	Variable	Eg	Variable	Eg	Variable
Eg1	Word	Eg3	CD	Eg11	CC(%)	Eg19	PP(%)
Eg2	Word/SENT	Eg4	FW	Eg12	EX(%)	Eg20	RB(%)
		Eg5	MD	Eg13	IN(%)	Eg21	TO(%)
		Eg6	PDT	Eg14	JJ(%)	Eg22	VH(%)
		Eg7	RP	Eg15	MD(%)	Eg23	VV(%)
		Eg8	UH	Eg16	NN(%)	Eg24	WH(%)
		Eg9	VD	Eg17	NP(%)	Eg25	GESY(%)
		Eg10	CUSY	Eg18	POS(%)	Eg26	CUSY(%)

tag (Eg5), the ratio of their appearance to all the words in a statement is considered and denoted as "MD(%)" (Eg15).

As with the method in Japanese, multiple regression analysis was executed on the thirty English Q&A statements used in the experiment depicted in Sect. 3.3. Factor scores of nine factors were estimated from the twenty-six feature values summarized in Table 1. MCCs of nine factors are shown in Table 2. From the viewpoint of MCCs, because all these figures are above 0.9, it could be indicated that estimation accuracies were very good.

Table 2. Multiple correlation coefficients (MCCs) with morphological analysis (MA). [12].

Factor	MCC (MA)	Factor	MCC (MA)	Factor	MCC (MA)
1st (Accuracy)	0.905	4th (Discomfort)	0.971	7th (Difficulty)	0.990
2nd (Evaluation)	0.999	5th (Novelty)	0.936	8th (Politeness)	0.919
3rd (Disappointment)	0.953	6th (Potency)	0.954	9th (Nostalgia)	0.992

4   Application of 2-gram to Q&A Statements

4.1   Objective

From the previous works summarized in Sect. 3, these methodologies so far have considerably depended on the syntactic information (Syn-Info) extracted through morphological analysis (MA). Furthermore, taking account of quadratic terms has generated numerous EVs, inducing significantly complicated multiple regression equations estimated for factor scores. Therefore, the principle has been shifted from mere Syn-Info to N-gram in employing feature values to estimate factor scores. The application of N-gram on top of Syn-Info would be expected to contribute to improved estimation accuracy and result in more concise equations used for estimating factor scores.

4.2   N-gram

N-gram is also known as an alternative syntactic analysis method for MA. N-gram represents the contiguous sequence of N units of characters, morphemes, or Part-of-Speeches, where N is an arbitrary integer at least 2 [15]. As a beginning phase of regarding N-gram, N was set to 2 in the analysis using English Q&A statements. In order to explain a 2-gram PoS example, one question out of the thirty English Q&A statements was employed. The original English question statement is shown in Table 3. For convenience, the question is denoted as "EAQ02."

For the example of 2-gram for the statement EAQ02, their Part-of-Speeches, examples, and frequencies are summarized in Table 4.

Table 3. The original English statements of EAQ02 [8, 13].

	Statements
EAQ02	If I sell on eBay, will they take the shipping cost from the amount the buyer pays? So I'm getting ready to post a watch for sale on eBay, and I'm not too clear on the shipping method. So let's say I post the watch for sale at $50. When I'm set up with a buyer, and the money is paid to the PayPal account, if I post free shipping, will eBay automatically take the shipping cost from the balance that the buyer paid? If I post a specific amount for shipping, will eBay take that from the balance the buyer paid?

- The column entitled "2-gram" shows 2-gram with the abbreviations of PoS, whose detailed explanations are summarized and provided in Appendix. Taking an example of the notation [DT - NN] listed in the first row, this 2-gram is made up of a determiner (DT) and a noun (NN).
- The column entitled "Example" provides one example per each 2-gram extracted from EAQ02.
- The column entitled "Frequency" shows the appearance time of respective 2-gram.

Table 4. 2-gram, one example and frequency for EAQ02 [8].

2-gram				Example				Frequency
[	DT	-	NN	] [	the	-	amount	]	11
[	PP	-	VV	] [	they	-	take	]	5
[	IN	-	PP	] [	if	-	I	]	3
[	NN	-	IN	] [	sale	-	at	]	9
[	NN	-	SENT	] [	method	-	.	]	2
[	VV	-	DT	] [	take	-	the	]	5
[	VV	-	PP	] [	say	-	I	]	1
[	IN	-	DT	] [	on	-	the	]	6
[	JJ	-	NN	] [	free	-	shipping	]	3
[	IN	-	NN	] [	for	-	sale	]	6

4.3 Analysis Method with Application of 2-gram (English)

As an initial step to applying the methodology of N-gram, the case when N is set to 2 (namely 2-gram) is applied to the PoS extracted through MA. The analysis result using 2-gram was then compared with those using mere MA so that applying 2-gram to English Q&A statements would be validated.

The feature values based on 2-gram were used in place of PoS (Eg3–Eg10, as explained in Sect. 3.4 and shown in Table 1). These feature values with 2-gram and those besides PoS (namely WORDs and PoS(%), as stated in Sect. 3.4) were collectively utilized as EVs, while the factor scores for the respective nine factors were employed as respondent variables.

Similar to the previous analysis method using MA explained in Sects. 3.3 and 3.4, multiple regression analysis was performed. Likewise, factor scores of the nine English factors were set as the respondent variables. Meanwhile, as for the explanatory variables, among the English feature values shown in Table 1, "WORDs" (Eg1 and Eg2) "PoS (%)" (Eg11–Eg26) were used for the subsequent analyses. On the other hand, instead of adopting "PoS" (Eg3–Eg10), 2-gram was applied and utilized as the feature value. The outline of selecting the English feature values for 2-gram through total amounts for the thirty English Q&A statements is described in Table 5.

Table 5. The outline of sorting and extracting 2-gram for the thirty Q&A statements.

(a) Before sort

	Q&A	...	[IN-NN]	[NN-IN]	[NN-SENT]	...	[DT-NN]	[JJ-NN]	...
					206 combinations				
	EAQ01	...	2	1	4	...	1	2	...
	EAA01-01	...	1	1	0	...	2	3	...
	EAA01-02	...	5	5	5	...	3	3	...
	EAA01-03	...	0	0	2	...	2	0	...
30 Q&A	EAA01-04	...	2	0	0	...	0	0	...
	EAQ02	...	6	9	2	...	11	3	...
	...	...	...	...	...	...	...	...	...
	EPA03-03	...	0	0	1	...	1	0	...
	EPA03-04	...	0	0	0	...	0	0	...
	Sum_2-gram	...	48	58	53	...	93	49	...

(b) After sort

Q&A	En2gr_g1 [DT-NN]	En2gr_g2 [PP-VV]	En2gr_g3 [IN-PP]	En2gr_g4 [NN-IN]	En2gr_g5 [NN-SENT]	En2gr_g6 [VV-DT]	En2gr_g7 [VV-PP]	En2gr_g8 [IN-DT]	En2gr_g9 [JJ-NN]	En2gr_g10 [IN-NN]
EAQ01	1	2	1	1	4	0	3	1	2	2
EAA01-01	2	1	2	1	0	3	2	1	3	1
EAA01-02	3	8	3	5	5	1	5	1	3	5
EAA01-03	2	1	1	0	2	0	2	3	0	0
EAA01-04	0	1	0	0	0	0	0	0	0	2
EAQ02	11	5	3	9	2	5	1	6	3	6
...	...	...	...	...	...	...	...	...	...	...
EPA03-03	1	0	0	0	1	1	0	0	0	0
EPA03-04	0	1	0	0	0	1	0	0	0	0
Sum_2-gram	93	84	59	58	53	53	50	50	49	48

For the thirty English Q&A statements, 206 combinations of 2-gram were generated. The respective appearance numbers of each 2-gram combination were then counted and summed up for the respective thirty English Q&A statements, as shown in Table 5-(a). The row entitled "Sum_2-gram" is the appearance number of the respective 2-gram combinations. These 2-gram combinations were then sorted in the descending order of their total appearance times, as shown in Table 5-(b). The amount of feature values of 2-gram extracted as the feature values instead of "PoS" was experimentally set to ten through trial and error. These feature values of 2-gram are denoted as En2gr_g1, En2gr_g2, ..., and En2gr_g10, as shown in Table 6.

Table 6. English feature values of 2-gram.

Eg	Variable	Eg	Variable	Eg	Variable	Eg	Variable
En2gr_Eg1	[DT-NN]	En2gr_Eg3	[IN-PP]	En2gr_Eg5	[NN-SENT]	En2gr_Eg8	[IN-DT]
En2gr_Eg2	[PP-VV]	En2gr_Eg4	[NN-IN]	En2gr_Eg6	[VV-DT]	En2gr_Eg9	[JJ-NN]
				En2gr_Eg7	[VV-PP]	En2gr_Eg10	[IN-NN]

With the mixtures of WORDs (Eg1, Eg2), POS(%) (Eg11–Eg26) and 2-gram (2gr_Eg1-2gr_Eg10), a total of twenty-eight feature values are utilized as EVs. The detailed descriptions and abbreviations of POS tags are referred to in the Appendix.

As a result of multiple regression analysis, MCCs are shown in Table 7 [8]. Similar to the previous analyses using mere MA shown in Table 1, all the MCCs were over 0.9 for all the nine English factors.

Table 7. Multiple correlation coefficients (MCCs) with 2-gram [8].

Factor	MCC (2-gram)	Factor	MCC (2-gram)	Factor	MCC (2-gram)
1st (Accuracy)	0.966	4th (Discomfort)	0.981	7th (Difficulty)	0.997
2nd (Evaluation)	1.000	5th (Novelty)	0.982	8th (Politeness)	0.994
3rd (Disappointment)	0.997	6th (Potency)	0.999	9th (Nostalgia)	0.998

5   Application of 3-gram to English Q&A Statements

5.1   Aim

In Sect. 4, the former analysis using 2-gram indicated that N-gram along with MA could be more effective than using MA alone in estimating factor scores of English Q&A statements. So far, however, mere 2-gram has been analyzed for this approach. Therefore, for the purpose of strengthening the effectiveness of using N-gram, we sought to validate if the method using N-gram would be effective with the longer unit of N. Thus, in this paper, 3-gram is adopted as the feature values with a view to rein-forcing the approach using N-gram. The analysis result using 3-gram is then com-pared with that utilizing 2-gram or that employing MA in order to inspect the validity of applying 3-gram to English Q&A statements.

Similar to the previous analysis with the usage of feature values based on 2-gram, in this analysis 3-gram is employed in lieu of PoS (Eg3–Eg10). These feature values with 3-gram and those besides WORDs (Eg1 and Eg2) and PoS(%) (Eg11–Eg26) are jointly utilized as EVs, whereas the factor scores for the respective nine factors are used as respondent variables.

5.2 Analysis Method of 3-gram (English)

Similar to the preceding analysis method analyses using 2-gram or MA depicted in Sects. 3.2, 3.4 and 4.3, multiple regression analysis was performed. Similarly, factor scores of the nine English factors are used as the respondent variables. On the other hand, 3-gram as well as WORDs and PoS (%) are set as EVs. Likewise, the amount of feature values of 3-gram is set to ten through trial and error. Therefore, a total of twenty-eight feature values (3-gram: ten, WORDs: two, and POS (%): sixteen) are set as EVs. The feature values of 3-gram are denoted as En3gr_g1, En3gr_g2, ..., and En3gr_g10, as shown in Table 8. Similar to the notation of 2-gram explained in Sect. 4.2, an example of 3-gram denoted as "En3gr_g1," [IN-DT-NN] indicates a 3-gram composed of a complementizer (IN), a determiner (DT) and a noun (NN).

Table 8. English feature values of 3-gram.

Eg	Variable	Eg	Variable	Eg	Variable	Eg	Variable
En3gr_g1	[IN-DT-NN]	En3gr_g3	[IN-PP-VV]	En3gr_g5	[DT-NN-IN]	En3gr_g8	[DT-JJ-NN]
En3gr_g2	[VB-DT-NN]	En3gr_g4	[NN-IN-NN]	En3gr_g6	[NN-IN-DT]	En3gr_g9	[VV-IN-PP]
				En3gr_Eg7	[VV-TO-VV]	En3gr_g10	[VV-RB-VV]

5.3 Estimation Result

Similar to the previous analysis method analyses using 2-gram or MA, MCCs are obtained for the nine factors as a result of multiple regression analysis. These figures are summarized in Table 9. Similarly, because all the MCCs are greater than 0.9, it could be concluded that 3-gram would also be applicable to estimate factor scores.

Table 9. Multiple correlation coefficients (MCCs) with 3-gram.

Factor	MCC (3-gram)	Factor	MCC (3-gram)	Factor	MCC (3-gram)
1st (Accuracy)	0.967	4th (Discomfort)	0.974	7th (Difficulty)	1.00
2nd (Evaluation)	0.978	5th (Novelty)	0.998	8th (Politeness)	0.980
3rd (Disappointment)	0.999	6th (Potency)	0.999	9th (Nostalgia)	0.974

6 Consideration

6.1 Interpretation on Standardized Partial Regression Coefficient (SPRC)

As a result of multiple regression analysis, standardized partial regression coefficients (SPRCs) can be obtained. SPRC plays a vital role in evaluating which explanatory variables (EVs) heavily influence on respondent variables, since the size and unit of EVs can be eliminated and disregarded through standardization. Therefore, the maximum three positive/negative biggest EVs with SPRCs over 1.0 are focused on for each of the nine factors. These EVs are summarized in Table 10. Most of the factors show three biggest positive/negative biggest EVs with SPRCs over 1.0 are focused on for each of the nine factors. These EVs are summarized in Table 10. Most of the factors show the three biggest positive/negative SPRCs, although there are a couple of exceptions; there are only the top two EVs for the negative SPRC on the 3rd factor (disappointment) and

Table 10. Explanatory variable (EV) and feature falue (FV) with three biggest positive/negative standardized partial regression coefficient (SPRC): 3-gram, English.

1st (Accuracy)			2nd (Evaluation)			3rd (Disappointment)		
EV	FV	SPRC	EV	FV	SPRC	EV	FV	SPRC
g1	WORDs	2.85	En3gr_g1	3-gram	2.77	g1	WORDs	1.82
En3gr_g1	3-gram	1.49	g16	PoS (%)	1.34	En3gr_g6	3-gram	1.59
En3gr_g3	3-gram	1.25	En3gr_g3	3-gram	1.20	g13	PoS (%)	1.44
g13	PoS (%)	-1.58	g13	PoS (%)	-1.29	En3gr_g3	3-gram	-1.19
En3gr_g2	3-gram	-2.35	g20	PoS (%)	-1.59	En3gr_g1	3-gram	-2.60
En3gr_g6	3-gram	-2.94	En3gr_g6	3-gram	-3.25			

4th (Discomfort)			5th (Novelty)			6th (Potency)		
EV	FV	SPRC	EV	FV	SPRC	EV	FV	SPRC
En3gr_g6	3-gram	2.30	En3gr_g3	3-gram	4.07	g1	WORDs	1.77
En3gr_g9	3-gram	1.90	En3gr_g9	3-gram	2.96	En3gr_g3	3-gram	1.69
En3gr_g5	3-gram	1.34	En3gr_g1	3-gram	2.96	g16	PoS (%)	1.53
g16	PoS (%)	-1.31	g13	PoS (%)	-3.96	g13	PoS (%)	-2.71
En3gr_g7	3-gram	-1.40	g14	PoS (%)	-5.12	g14	PoS (%)	-3.14
En3gr_g1	3-gram	-1.93	En3gr_g2	3-gram	-5.90	En3gr_g2	3-gram	-3.61

7th (Difficulty)			8th (Politeness)			9th (Nostalgia)		
EV	FV	SPRC	EV	FV	SPRC	EV	FV	SPRC
En3gr_g2	3-gram	3.08	En3gr_g6	3-gram	2.97	En3gr_g2	3-gram	6.40
En3gr_g6	3-gram	2.25	En3gr_g2	3-gram	2.33	En3gr_g6	3-gram	5.01
g13	PoS (%)	2.19	g1	WORDs	-1.55	g14	PoS (%)	4.21
g16	PoS (%)	-1.82	En3gr_g1	3-gram	-1.90	En3gr_g9	3-gram	-2.28
En3gr_g1	3-gram	-2.09	En3gr_g3	3-gram	-2.20	En3gr_g1	3-gram	-3.70
En3gr_g3	3-gram	-2.45				En3gr_g3	3-gram	-4.39

the positive SPRC on the 8th factor (politeness). The column entitled "FV" indicates either "3-gram," "WORDs" or "POS (%)", which are classifications of feature values corresponding to the column entitled "EV."

From the summarizations shown in Table 10, most of the EVs are 3-gram. Therefore, it could be implied that applying 3-gram would be effective in estimating factor scores of English Q&A statements. It would also be suggested that N-gram would be as applicable as MA to estimate them.

6.2 Comparison among N-gram and Morphological Analysis (MA)

For the purpose of directly comparing the analysis results among 3-gram, 2-gram and mere MA, the results shown in Tables 2, 7 and 9 are integrated and shown in Table 11. From these results, MCCs when using 3-gram is almost as good as those when using 2-gram. Nevertheless, both cases show better results than the MCC with only MA. Therefore, it could be suggested that an analysis applying N-gram would outperform one merely utilizing MA.

Table 11. Comparison of MCCs for analyses using 3-gram, 2-gram and MA.

Factor	MA	2-gram	3-gram
1st (Accuracy)	0.905	0.966	0.967
2nd (Evaluation)	0.999	1.000	0.978
3rd (Disappointment)	0.953	0.997	0.999
4th (Discomfort)	0.971	0.981	0.974
5th (Novelty)	0.936	0.982	0.998
6th (Potency)	0.954	0.9990	0.9988
7th (Difficulty)	0.990	0.997	1.000
8th (Politeness)	0.919	0.994	0.980
9th (Nostalgia)	0.992	0.998	0.974

From the viewpoints of MCCs, any of these methods would be applicable. However, it could be suggested that the best method among N-grams and MA should be applied to each factor depending on their respective best. For example, 2-gram would contribute to the best result for the 2nd factor (Evaluation), 4th (Discomfort), 8th (Politeness) and 9th factor (Nostalgia). Meanwhile, 3-gram could be applied to the 1st (Accuracy), 3rd (Disappointment), 5th (Novelty), 6th (Potency) and 7th factor (Difficulty).

Nevertheless, the meanings or contents of Q&A statements have not been regarded orthodoxly yet. Therefore, meaning analysis must be applied to this methodology in the future. Additionally, we would be required to investigate whether the method of employing N-gram could be extended to other languages as well.

7 Conclusion

In this paper, 3-gram was extracted as the feature values of English Q&A statements to obtain factor scores. Similar to the previous analyses using 2-gram or MA, multiple regression analysis was performed. The feature values based on 3-gram, WORDs and PoS (%) were collectively employed as EVs, whereas the factor scores were utilized as respondent variables. The analysis result applying 3-gram on top of MA showed as good estimation accuracy as that with the application of 2-gram. Since either case using 3-gram or 2-gram outperformed that utilizing MA alone, it could also be implied that applying N-gram on top of MA could exceed employing MA alone.

For future work, the bigger dataset (e.g. Stack Overflow, Quora, etc.) will be used for the analysis, since Y!A site operations ceased in 2021. In order to avoid and eliminate over-fitting, cross-validation test should be adopted and performed. Furthermore, as the semantic between Japanese Q&A statements have been analyzed [15], those of English Q&A statements must also be taken into consideration. Moreover, with the Japanese feature values of syntactic information based on MA, the factor scores obtained were subsequently utilized for inspecting the possibility of finding respondents who would be expected to appropriately answer a newly posted question [7]. Thus, whether the feature values based on N-gram could be effective in detecting appropriate respondents must be inspected and compared with the case employing MA alone.

Acknowledgement. This research was partially supported by the Japan Society for the Promotion of Science, Grant Number 26008587, 2014–2015, and Grant-in-Aid for Young Scientists, Grant Number 20K19933, 2020–2023.

Appendix

In performing Treetagger [16], through morphological analysis, a set of sentences is broken down into each word, its Part-of-Speech (PoS) tag, and basic form. Nevertheless, some words are classified into more profoundly subdivided tags. The detailed explanations of PoS tags are provided as follows [17]:

- One type of prepositions or subordinate conjunctions (IN): complementizer
- Comparative/Superlative form of adjective (JJ)/adverb (RB)
- Plural form of noun (NN)/proper noun (NP)
- Different form of pronoun (PP)
- Different tense of verbs (VB, VD, VH, VV)
- Wh-words (WH)

In these cases, the more deeply subdivided tag is equally regarded as its base tag. Each base tag and its subdivided tags are summarized in Table 12. Base (Subdivided, respectively) tags are given in the column entitled "Base." ("Subdivided.") PoS tags available through TreeTagger are shown in Table 13. The meaning of each PoS Tag and its example is summarized in the column entitled "Explanation" and "Example," respectively [17].

Table 12. Base tags and their subdivided tags [16].

Base	Subdivided	Description	Base	Subdivided	Description	Base	Subdivided	Description
IN	IN/that	Complementizer		VDD	Past form of VD		VVD	Past form of VV
JJ	JJR	Comparative of JJ		VDG	Gerund/participle of VD		VVG	Gerund/participle of VV
	JJS	Superlative of JJ	VD	VDN	Past participle of VD	VV	VVN	Past participle of VV
NN	NNS	Plural of NN		VDZ	Pres, 3rd p. sing of VD		VVZ	Pres, 3rd p. sing of VV
NP	NPS	Plural of NP		VDP	Pres non-3rd p. of VD		VVP	Pres non-3rd p. of VV
PP	PP$	Possessive pronoun		VHD	Past form of VH		WDT	Wh-determiner
RB	RBR	Comparative of RB		VHG	Gerund/participle of VH	WH	WP	Wh-pronoun
	RBS	Superlative of RB	VH	VHN	Past participle of VH		WP$	Possessive wh-pronoun
	VBD	Past form of VB		VHZ	Pres, 3rd p. sing of VH		WRB	Wh-adverb
	VBG	Gerund/participle of VB		VHP	Pres non-3rd p. of VH			
VB	VBN	Past participle of VB						
	VBZ	Pres, 3rd p. sing of VB						
	VBP	Pres non-3rd p. of VB						

Table 13. PoS tags, explanations, and examples [17].

PoS Tag	Explanation	Example	PoS Tag	Explanation	Example
CC	coordinating conjunction	and, but, or, &	VB	verb be, base form	be
CD	cardinal number	1, three	VBD	verb be, past	was, were
DT	determiner	the	VBG	verb be, gerund/participle	being
EX	existential there	there is	VBN	verb be, past participle	been
FW	foreign word	d'œuvre	VBZ	verb be, pres, 3rd p. sing	is
IN	preposition/subord. conj.	in,of,like,after,whether	VBP	verb be, pres non-3rd p.	am|are
IN/that	complementizer	that	VD	verb do, base form	do
JJ	adjective	green	VDD	verb do, past	did
JJR	adjective, comparative	greener	VDG	verb do gerund/participle	doing
JJS	adjective, superlative	greenest	VDN	verb do, past participle	done
LS	list marker	(1),	VDZ	verb do, pres, 3rd per.sing	does
MD	modal	could, will	VDP	verb do, pres, non-3rd per.	do
NN	noun, singular or mass	table	VH	verb have, base form	have
NNS	noun plural	tables	VHD	verb have, past	had
NP	proper noun, singular	John	VHG	verb have, gerund/participle	having
NPS	proper noun, plural	Vikings	VHN	verb have, past participle	had
PDT	predeterminer	both the boys	VHZ	verb have, pres 3rd per.sing	has
POS	possessive ending	friend's	VHP	verb have, pres non-3rd per.	have
PP	personal pronoun	I, he, it	VV	verb, base form	take
PP$	possessive pronoun	my, his	VVD	verb, past tense	took
RB	adverb	however, usually, here, not	VVG	verb, gerund/participle	taking
RBR	adverb, comparative	better	VVN	verb, past participle	taken
RBS	adverb, superlative	best	VVP	verb, present, non-3rd p.	take
RP	particle	give up	VVZ	verb, present 3d p. sing.	takes
SENT	end punctuation	?, !, .	WDT	wh-determiner	which
SYM	symbol	@, +, *, ^, |, =	WP	wh-pronoun	who, what
TO	to	to go, to him	WP$	possessive wh-pronoun	whose
UH	interjection	uhhuhhuhh	WRB	wh-adverb	where, when
			GESY	general joiner	;, -, --
			CUSY	currency symbol	$, £

References

1. Yahoo! Chiebukuro (in Japanese). http://chiebukuro.yahoo.co.jp/. Accessed 03 Sep 2023
2. Yang, Z., Liu, Q., Sun, B., Zhao, X.: Expert recommendation in community question answering: a review and future direction. Int. J. Crowd Sci. **3**, 348–372 (2019)
3. Zhang, Z., Lu, Y., Wilson, C., He, Z.: Making sense of clinical laboratory results: an analysis of questions and replies in a social Q&A community. In: Proceedings of the 17th World Congress on Medical and Health Informatics (MEDINFO 2019), Lyon, France, pp. 2009–2010 (2019)

4. Haq, E.U., Braud, T., Hui, P.: Community matters more than anonymity: analysis of user interactions on the Quora Q&A platform. In: Proceedings of the International Conference Series on Advances in Social Network Analysis and Mining (ASONAM 2020), pp. 94–98. Virtual (2020)
5. Yokoyama, Y., Hochin, T., Nomiya, H., Satoh, T.: Obtaining factors describing impression of questions and answers and estimation of their scores from feature values of statements. In: Software and Network Engineering, vol. 413, pp. 1–13. Springer (2013). https://doi.org/10.1007/978-3-642-28670-4_1
6. Yokoyama, Y., Hochin, T., Nomiya, H.: Using feature values of statements to improve the estimation accuracy of factor scores of impressions of question and answer statements. Int. J. Affect. Eng. **13**(1), 19–26 (2016)
7. Yokoyama, Y., Hochin, T., Nomiya, H.: Quantitative evaluation of potential tendency differences between English and Japanese in detecting appropriate respondents at Q&A sites. Int. J. Affect. Eng. **18**(3), 145–154 (2019)
8. Yokoyama, Y., Hochin, T., Nomiya, H.: Application of 2-gram and 3-gram to obtain factor scores of statements posted at Q&A sites. Int. J. Networked Distrib. Comput. **1–2**, 11–20 (2022)
9. Yokoyama, Y., Hochin, T., Nomiya, H.: Using 4-gram to obtain factor scores of japanese statements posted at Q&A site. In: Proceedings of the 13th International Congress on Advanced Applied Informatics (AAI2022-Winter), Phuket, Thailand, pp. 25–31 (2022)
10. Yokoyama, Y., Hochin, T., Nomiya, H.: Application of 5-gram to obtain factor scores of Japanese Q&A statements. In: Proceedings of the 14th IIAI International Congress on Advanced Applied Informatics (AAI 2023), Presented, Koriyama, Japan, 7 p. (2023)
11. Yokoyama, Y., Hochin, T. and Nomiya, H.: Factors describing impression of English question and answer statements. In: Proceedings of the 11th Spring Conference of Japan Society of Kansei Engineering, G12-3, Kobe, Japan (2016)
12. Yokoyama, Y., Hochin, T., Nomiya, H.: Estimation of factor scores from feature values of English question and answer statements. In: Proceedings of the IEEE/ACIS 15th International Conference on Computer and Information Science (ICIS 2016), Okayama, Japan, pp. 741–746 (2016)
13. Yahoo! Answers. https://answers.yahoo.com/. Accessed 04 May 2021
14. sourceforge, Yahoo! Answers Datasets, Summary. http://sourceforge.net/projects/yahoodataset/. Accessed 03 Sep 2023
15. Yokoyama, Y.: Consideration of semantics between Q&A statements to obtain factor score. In: Proceedings of the IEEE/ACIS 21st International Conference on Software Engineering Research, Management and Applications (SERA 2023), Orlando, U.S., pp. 169–175 (2023)
16. TreeTagger - a language independent part-of-speech tagger. http://www.cis.uni-muenchen.de/~schmid/tools/TreeTagger/. Accessed 03 Sep 2023
17. Tree Tagger Tag Set (58 tags). https://courses.washington.edu/hypertxt/csar-v02/penntable.html. Accessed 06 May 2021

Spatial-Temporal Information-Based Littering Action Detection in Natural Environment

Cu Vinh Loc[✉], Le Thi Kim Thoa, Truong Xuan Viet, Tran Hoang Viet, and Le Hoang Thao

Can Tho University, Can Tho, Vietnam
{cvloc,thoam2521014,txviet,thviet,lhthao}@ctu.edu.vn

Abstract. A significant environmental problem in Vietnam and around the world is litter and garbage left on the street. The illegal dumping of garbage can have negative consequences on the environment and the quality of human life, as can all forms of pollution, in addition to having a large financial impact on communities. Thus, in order to reduce the impact, we require an automatic litter detecting method. In this study, we propose a new method for spotting illegal trash disposal in surveillance footage captured by real-world cameras. The illegal littering is recognized by a deep neural network. An ordered series of frames makes up a video. The order of the frames carries the temporal information, and each frame contains spatial information. To model both of these features, convolutional layers are used for spatial processing, and instead of using recurrent layers for obtaining temporal information, we take advantage of transformer encoder for encoding sequential data by evaluating each element in the sequence's relevance. Besides, CNNs is also used to extract important features from images, hence lowering input size without compromising performance. This leads to reduce computational requirements for the transformer. Through testing on actual recordings of various dumping operations, we show that the proposed strategy is effective. Specifically, the validation outcomes from the testing data's prediction reveal an accurate value of 92.5%, and this solution can be implemented in a real-time monitoring system.

Keywords: Deep neural network · action recognition · garbage dumping action · video surveillance · illegal littering action

1 Introduction

Waste is currently a very painful problem for our country in particular and the whole world in general because we often see garbage bags on the side of the road, on the sidewalk or on the surface of a public lake. Therefore, the environment we live in is becoming more and more polluted. Illegal dumping of waste includes both the illegal deposit of e-waste inside and outside of national boundaries as

© The Author(s), under exclusive license to Springer Nature Singapore Pte Ltd. 2024
N. Thai-Nghe et al. (Eds.): ISDS 2023, CCIS 1950, pp. 247–261, 2024.
https://doi.org/10.1007/978-981-99-7666-9_21

well as the unlawful disposal of wastes larger than litter onto land (landfilling) or into water. The illegal dumping of garbage can have negative consequences on the environment and the quality of human life, as can all forms of pollution, in addition to having a large financial impact on communities. As a result of occasionally being larger than the fine for illegal dumping, the fees that must be paid for the legal disposal of waste at appropriate waste disposal facilities don't always serve as a deterrent. Along with pollution in the air and the seas, pollution on land is a serious issue that involves a number of actions. Around the world, illegal dumping has been a persistent issue in many cities. In addition to degrading the appearance of the city, the aromas and toxins brought on by trash dump leftovers, abandoned household objects, and construction debris also pose a health risk to the populace.

In Vietnam, the problem of indiscriminate littering is becoming an urgent problem of society when everywhere we see garbage bags littering the streets on the sidewalks. In public places, people's awareness when throwing garbage is even more unconscious, even though the trash can is not far from them. Some public places after the holidays or fairs or outdoor activities, when everything is over, those places become giant piles of garbage that is scattered. This scene was terrible. Or in the park, where people entertain after eating and drinking, they conveniently throw plastic bags on the ground, causing a loss of beauty of that place. According to statistics [1], in 2022, Can Tho city has more than 238,637 tons of garbage collected, transported and treated, with an average of more than 653 tons per day.

(a)   (b)

(c)   (d)

Fig. 1. Various behaviors of illegal dumping.

Many cities or provinces operate various education initiatives, social media-based community portals, surveillance camera monitoring, and ordinances that levy fines on the dumpers to reduce the unlawful waste disposal. The informed locals proactively report trash dumps close to their homes. City staff monitor illegal dumping hot spots in non-residential areas on a daily basis. Once wastes are discovered, city-sponsored trash trucks are dispatched to the disposal site to collect wastes. The current approach for detecting and removing unlawful dumping is ineffective and expensive since city staff must personally keep an eye on a few hotspots all the time, and reports of illegal dumping by citizens sometimes contain false alarms. Even if the allegation turns out to be false, trash trucks should still be sent to the claimed location because it is difficult to verify individual reports of dumping. Such a manual and voluntary reporting approach wastes government funds and makes it more difficult to effectively clean up waste. Therefore, it is essential to include a system that identifies and reports unlawful dumping actions automatically and with little human intervention.

There is now a pressing need for research to find solutions to the aforementioned issues. We were motivated to suggest a monitoring system that may categorize human activities in a public setting in order to safeguard against improper human behaviors in the environment and currently used public amenities. People will be compelled to follow the rules if this strategy is used, making them less likely to engage in criminal activity. The existence of coercion will make people understand how important maintaining the environment is. They will also start to do it naturally because of it.

Compared to garbage classification [3,6,9,10,18], there are less works involved in recognizing garbage dumping behaviors. Here are some typical studies [4,8,11,12,15,20,21] that have been done on the subject of identifying littering action, and many of them make use of methods including human pose, object detection, object tracking, their combination, and deep learning-based action recognition. We have seen that strategies based on measuring the objects' bounding boxes and estimating their distance from one another to identify whether they're littering give lower generalization than deep learning-based approaches. Figure 1 demonstrates various ways that the people carry out their illegal littering. This shows that it is difficult to identify general littering behavior by measuring the distance from a person's arm to the object they are holding. This is the reason we opt for an approach to recognize littering actions using deep neural network. With this approach, we use video data to train the neural networks in order that the model knows how to recognize littering actions in videos. In other words, we feed video data into a trained model, and the model tells us what action is happening in the video. The hybrid learning framework that can model a number of significant characteristics of the video data is the primary contribution of this work. We also demonstrate that (1) the CNN architecture is combined with the transformer architecture that replaces the traditional combination of CNN and RNN for processing temporal information extracted from the video frames. The combination of CNN and transformer architecture leverages the strengths of these approaches providing a strong solution for this difficult task, (2) before feeding into the transformer, the deep neural network is utilized to extract key features from images, hence reducing input size without impacting performance. As a result, the transformer's computing requirements are reduced, and (3) in place of the conventional RNN, the transformer encoder is employed to extract temporal information.

The paper is set up as follows. Section 2 and Sect. 3 describe relevant works and background information, respectively. The proposed technique is described in Sect. 4. The experimental findings are highlighted in Sect. 5. In the section titled 6, the conclusion and potential changes are covered.

2   Related Work

To identify diverse sorts of regularly discharged garbage, the authors [5] suggest using a deep learning approach. They investigated different strategies to increase accuracy and show how accuracy varies depending on the number of classes, baseline models, and input image characteristics. Additionally, they suggest using edge computing to cut down on pointless image transfers to the servers. When a deep learning model detects frequently discharged waste in a taken image of a specific dumping hot point, the edge computing station sends the image to the server. The outcomes of their experiments demonstrate that the suggested methods offer high recognition accuracy with little memory footprint. Yun *et al.* [21] have proposed a new method for spotting illegal trash disposal on real-world security cameras. A person's relationship to the object they are holding may alter, which allowed the authors to identify the dumping action. A background subtraction method and human joint estimation were utilized to locate the person-held object of undefined form. Next, a relationship model between the joints and the object was constructed while the thing was being held by a person. The voting-based decision module, in the end, was able to detect the dumping activity. By testing on real-world videos of different dumping behaviors, the authors demonstrate the efficacy of the proposed methodology.

The work presented in [8] consists of two methods like using CNN only, and using CNN-Long Short-Term Memory (LSTM). The authors show that only about 67.7–75% of the time did the algorithm accurately identify the activity while using CNN. This is because the CNN's training procedure could only attain 56% accuracy and had a high loss value of 70%. However, the system could operate more effectively if it used the CNN-LSTM. It indicated a 10% loss and 97.7% accuracy. When used in the actual studies, this strategy likewise yielded positive results, with a percentage of accurate categorization of about 97.2%. Through multi-task learning, Bae *et al.* [2] train the efficient model to comprehend the person using several datasets, including human poses, human coarse action (e.g., upright, bent), and human fine action (e.g., pulling a cart). Their method does away with the requirement for scene-by-scene tuning and gives the performance of behavior understanding in a visual surveillance system more stability. Additionally, the authors propose a new object identification network that is well-suited for identifying people and carryable goods. By limiting the list of possible suspects to those who are carrying an object, the proposed detection network lowers the computational cost. On a real-world surveillance video dataset, their system surpasses cutting-edge techniques in detecting garbage dumping activity.

Another work [20] is based on image processing in Open Automatic License Plate Recognition (OpenALPR). To record any cars accessing the unlawful dumping site, a Raspberry Pi camera module with a microwave radar sensor is interfaced to the Raspberry Pi, which serves as the microprocessor and is programmed in Python. To identify the vehicle's license plate, the picture is taken. The study's results include the automatic real-time email notification of cases of illegal dumping and the identification of vehicle license plate numbers. Due to the fact that the plate number recognition is done in real time, the detection system can be used for case monitoring. Kim and Cho [11] proposed a system for the automatic detection and reporting of unlawful waste dumping. The authors propose a system for tracking illegal garbage disposal based on deep neural networks to accomplish this. With the help of OpenPose and the object detection model You Only Look Once (YOLO), the proposed monitoring method obtains the articulation points (joints) of a dumper in order to measure the distance between his wrist and the garbage bag and determine whether it is illegal to dump or not. In order to decrease the likelihood of falsely detecting unlawful dumping, they also implemented a technique for tracking the IDs assigned to the garbage bags using the multi-object tracking (MOT) model. The authors show that the proposed method was more accurate and produced fewer false alarms, making it suitable for a range of forthcoming applications. Malik [15] et al. have provided a CNN-based automated model for categorizing urban rubbish into various groups. They have portrayed the model being used for litter classification utilizing fine tuning of pretrained neural network model with new datasets. As it is the issue linked with healthy living provisioning across cities, this paradigm enables software and hardware to both be produced using low-cost resources and to be deployed at a big scale. The use of pre-trained models and the application of transfer learning for optimizing a pre-trained model for a particular job are the two most important factors for the creation of such models.

Additionally, we have recently discovered a number of publications [3, 6, 9, 10, 18] that are pertinent to the classification and identification of garbage.

3 Background

Deep Learning. Processing of natural data like images, voices, etc. was a restriction of traditional machine learning systems. The creation of a machine learning algorithm to process natural data needs precise engineering abilities. Large-scale, intricate computing techniques were needed for natural language processing, yet they were difficult to implement. Representational learning, a novel approach, was developed to address the complexity issue mentioned above. A collection of techniques known as "representation learning enables" a computer to be fed with unstructured data and automatically find the representations required for detection or classification. Deep learning is a branch of machine learning that relies on approaches for representational learning. Using straightforward but non-linear functions, deep learning is a multi-layered, back propagating, representational, self-learning technique.

Each layer of deep learning converts the data representation from one form to another, extracting information from the initial raw data. This method is used from the first layer, which receives raw input, to the last layer, which yields the final representation. Each layer takes a small portion of the data, preventing irrelevant data from being processed further and giving the subsequent layer abstracted data. Multiple layers working together can quickly pick up complex functionalities. Deep learning employs the back propagation method to fix faults and self-learn in order to provide the desired results. Only the top and bottom layers of a multilayered construction are visible. All more layers are concealed and inaccessible on the surface. By computing the derivative of the error with respect to the function's output, the backpropagation algorithm propagates errors backward towards the first layer. The hidden layers can be managed by using this approach. When we are aware of the error derivative of one layer, we may utilize that knowledge to determine the error derivative of the layers behind it. The error derivative of the final layer of the learning process, which is visible to the outside world, can be calculated. The error derivative of the hidden layers can be calculated using the back propagation gradient algorithm. The error at each level is reduced using the results from the error derivative, which leads to a faster learning method.

Convolutional Neural Networks. By reducing complexity and addressing overfitting issues from deep learning, CNN [13] is a realistically valuable technique, particularly in computer vision tasks. The combination of convolution and pooling layers with the fundamental concepts of sparse weight, parameter sharing, and equivariant representation gives it its effectiveness. In image processing, a feature map was extracted using a filter as part of the convolution operation, which was a preprocessing step for machine learning. The fact that all convolution filters are now learning the simpler model has improved the situation. Following the convolution layer, the algorithm uses subsampling to further compress the feature map by making use of the neighbors pixels from a related featured image. Forward inference pass is constructed by the deep concatenated structure of convolution and pooling layers. By calculating the gradient for the provided loss or cost function from the forward pass, the backward propagation technique is used to learn a model. CNN's internal structure consists of straightforward, repeating matrix multiplications without branching. As a result, it offers parallel computations that are highly efficient on GPUs.

Transfomer Architecture. A paper titled "Attention Is All You Need" was published by Vaswani *et al.* [17] in 2017 for the NeurIPS conference. In place of the widely used RNN encoder-decoder models, they created the original transformer architecture for machine translation, which was more effective and quick. Originally intended for sequence-to-sequence activities like machine translation, question-answering, etc., a transformer is composed of an encoder and a decoder. The encoder takes the input embeddings, which are the sum of the regular embeddings and the positional embeddings, and outputs tensors of the same shape, but these tensors include a great deal of information, such as contextual meaning, part of speech, position of the word, etc., encoded in them. Often referred to as the hidden state, the tensors produced by the encoder. According to the job at hand and the pre-training goal, these tensors are supplied to the decoder, which then decodes the hidden state. The first to incorporate understandable information into those numbers, or vectors would be a more accurate term, was an embedding. Thus, embeddings of it are sent to the encoder rather than the tokens themselves. But one significant issue was encoding the data regarding the placement of the words in the embeddings. The encoder receives static embeddings that contain some information encoded in them. However, the self-attention layer also encodes that information in the context of each embedding, taking into account the words or tokens inside a sentence or collection of sentences.

Basically, several self-attention units make up multi-head attention. This is so that each attention head can pay attention to a different aspect of the text. One head may choose to concentrate on the subject-verb agreement, the other on the text's tense, and so on. This is comparable to utilizing several filters in a single convolutional layer, and as we've seen with ensembling, using multiple models almost always yields successful results.

4   Proposed Method

Action recognition from video is more difficult than image classification because we have to process multiple images at the same time. For binary image classification, multi-class image classification, two-dimensional convolutional neural network (2D-CNN)-based modeling was employed. One straightforward method would be to use a 2D-CNN model to process each image in a video individually before averaging the results. The temporal correlation between frames is not taken into account in this method, though. In order to extract temporal correlation, we would rather utilize a model that analyses many images from a video. To achieve this, we will employ a combination of CNN and Transformer architecture for recognition of littering action in videos using temporal input. The architecture is made up of a number of parts, such as a CNN-based feature extraction module, a positional embedding layer, several transformer encoder blocks, and modules for recognition and aggregation. Figure 2 provides a general representation of the architecture. In order to conduct precise littering action recognition, the architecture must successfully capture the temporal information included in the video sequences. By combining CNN with Transformers, an effective solution to this difficult issue is provided.

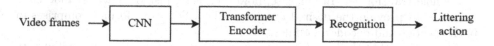

Fig. 2. The CNN-Transformer architecture.

Feature Extraction Using CNN and Positional Embedding. When the transformer was first employed for visual categorization, particularly for images, the frames were frequently broken into smaller patches and used as the main input [14,23]. The transformer needed a lot of processing power because these characteristics were frequently extremely huge. CNN can be used to extract important features from images, allowing for a reduction in input size without compromising performance while maintaining accuracy and efficiency. We assume that the size of the CNN features retrieved for each frame is (w, h, c), where w and h represent the width and height of a $2D$ feature, respectively, and c represents the number of filters. Global average pooling is used to further reduce the size of the features, changing their size from $w \times h \times c$ to c. The transformer uses a positional encoding approach to save the position of each frame in the sequence. The following formulas are used to compute the values of the positional encoding vector, which is added to the feature and has the same size as it. This is different from the RNN block's sequential data processing since it enables simultaneous handling of each entity in the sequence.

$$pos\_en_{(p,2i)} = sin(p/10.000^{2i/l_{model}}) \tag{1}$$

$$pos\_en_{(p,2i+1)} = cos(p/10.000^{2i/l_{model}}) \tag{2}$$

where p, i, pos_en are the input vector's time step index, dimension, and positional encoding matrix, respectively. l_{model} is the positional encoding vector's length.

Transformer Encoder. One essential element of the transformer architecture, which combines CNN technology, is the transformer encoder. It consists of a stack of N identical layers, each with multiple heads of self-attention and positionally fully connected feed-forward network sub-layers. Prior to each operation, residual connections are used, followed by layer normalization, to guarantee the retention of crucial input data. The multi-head self-attention mechanism, which is made up of many self-attention blocks, forms the basis of the module. This technique, which encodes sequential data by evaluating the relationship between each element in the sequence, is comparable to RNN. To give a more realistic representation, it takes advantage of the built-in relationships between frames in a video. Furthermore, because the calculation can be parallelized utilizing contemporary GPUs, the self-attention operates on the full sequence at once, yielding notable performance improvements. Since a classification label rather than a sequence is sought after, our architecture just uses the encoder portion of a whole transformer. In our case, however, using just the encoder module is sufficient to provide the desired outcome. The whole transformer comprises of both encoder and decoder modules.

We assume that the input sequence $(X = x_1, x_2, ..., x_n)$ is first projected onto these weight matrices $Q = XW_Q$, $K = XW_K$, and $V = XW_V$ with W_Q, W_K, and W_V being three trainable weights, the query $(Q = q_1, q_2, ..., q_n)$ and the key

$(K = k_1, k_2, ..., k_n)$ of the dimension d_k, and the value $(V = v_1, v_2, ..., v_n)$ of the dimension d_v, the output of self-attention is calculated as follows.

$$Attention(Q, K, V) = Softmax(\frac{QK^T}{\sqrt{d_k}})V \qquad (3)$$

The multi-head attention is made up of a number of heads, which are all combined and input into another linear projection to create the following final outputs.

$$MultiHead(Q, K, V) = Concat(head_1, head_2, ..., head_n)W^0 \qquad (4)$$

$$head_i = Attention(QW_i^Q, KW_i^K, VW_i^V) \qquad (5)$$

where $i = 1, 2, ..., n$, n depicts the number of heads. W_i^Q, W_i^K, W_i^V, and W^0 are parameter matrices.

Video Frame Pre-processing. The model, which needs a fixed number of inputs, may have difficulty with input videos with changing numbers of frames. Simply put, we included a time distributed layer that needs a specific number of frames in order to process a video sequence. We use a variety of methods for choosing a smaller subset of frames to overcome this issue. The method that establishes a maximum sequence duration for each video. Although this method is simple, it may cause lengthy recordings to be chopped short and information to be lost, especially if the intended length is not obtained. The second approach reduces the amount of frames used while still achieving the entire duration of the video by using a step size to skip some frames. The images are also center-cropped to produce square images.

Littering Recognition. We assume that we have a collection of videos $Vi = \{vi_1, vi_2, ..., vi_n\}$ and matching labels $y = \{y_1, y_2, ..., y_n\}$, where n is the number of samples. We select fr frames from the videos and obtain fe features from the global average pooling $2D$ layer. Each transformer encoder generates a set of representations by consuming the output from the previous block. After N transformer encoder blocks, we can obtain the multi-level representation $H^N = \{h_1^N, h_2^N, ..., h_{fr}^N\}$ where each representation is $1D$ vector with the length of fe. The littering recognition module uses global max pooling in addition to well-known layers like fully connected and softmax to shrink the network. We use Gaussian noise and dropout layers into the architecture to avoid overfitting. The categorical cross entropy loss is employed as the optimization criterion, and stochastic gradient descent is used to train the CNN-transformer model.

5 Experimental Results

5.1 Dataset

To demonstrate the effectiveness of the proposed model compared to existing models, we will be utilizing a subsampled version of the original UCF101 dataset

[16] in this work to make the runtime of the program reasonably short. It includes 101 types of human motions, such as punching, boxing, and walking, plus 13,320 realistic video clips pulled from YouTube. The UCF101 dataset is an extension of UCF50 dataset which has 50 categories. The subsampling dataset consists of 594 and 224 videos for training and testing respectively. To speed up the calculation, we reduce the image resolution from 224×224 to 128×128. Additionally, we train the proposed approach using our own dataset of garbage behavior. This dataset comprises of 320 videos for training, and 200 videos for testing the model. The videos are divided into four categories like walking, running, standing, and littering. We tested our work using GPU (T4) on Google Colab environment. To evaluate the results, we use the common classification accuracy statistic, which is defined as follows: the proportion of correctly predicted samples to the total number of predictions.

Since our collaborative models demand real-time information flow, training them for littering recognition requires developing a new, specialized framework. We believe that this is the first system of its kind ever created for this purpose. Each model is trained in a separate environment to account for the significant hardware resources needed. Each model updates its location, prior location, estimate of the loss function's gradient, and any other pertinent data after one training epoch. Neighboring models are then informed of these modifications. Figure 3 demonstrates a snapshot of action samples from UCF-101 dataset and our dataset (littering actions).

Fig. 3. The samples of actions from UCF-101 dataset (a), and our dataset has the four categories of standing, running, walking, and littering (b).

5.2   Results

We used both the original UCF101 dataset and the modified UCF101 dataset to test our proposed architecture. 15 classes are included in the updated UCF101 dataset: 1. Baseball Pitch, 2. Basketball Shooting, 3. Bench Press, 4. Biking, 5. Billiards Shot, 6. Breaststroke, 7. Clean and Jerk, 8. Diving, 9. Drumming, 10. Fencing, 11. Golf Swing, 12. High Jump, 13. Horse Race, 14. Horse Riding, and 15. Hula Hoop. Random selection was used to choose these classes from the whole UCF101 dataset. Due to its size, the entire UCF101 dataset requires extra work. Using a modified UCF101 dataset, the suggested approach was able to attain an overall test accuracy of 82.65%. However, utilizing the entire UCF101 dataset, a test accuracy score of 87.5% was obtained overall. Our findings demonstrate the suggested architecture's applicability to various datasets that include human activity.

The proposed neural network's evaluation metrics (P, R, and F1) were calculated using the full UCF101 dataset as well as the modified UCF101 dataset presented in Table 1. The accuracy attained was calculated by adding the total of samples that were accurately predicted to the sums of all samples, 82.65% using modified UCF101 dataset, 87.95% using full UCF50 dataset, and 87.5% using full UCF101 dataset.

Table 1. The evaluation of the proposed approach using various datasets.

Evaluation metrics	Full UCF101 dataset	Modified UCF101 dataset	UCF50 dataset
Precision (P)	87.14	89.31	94.14
Recall (R)	81.36	87.05	85.02
F1 score (F1)	84.15	88.17	89.35

Table 2. The accuracy and loss function model after 50 iterations of training and testing.

Parameters	Full UCF101 dataset	Modified UCF101 dataset	UCF50 dataset
Train loss	0.66	0.82	0.71
Train accuracy	89.65	86.95	91.08
Test loss	0.75	0.91	0.47
Test accuracy	87.5	82.65	87.95

The training and testing strategy for the proposed neural network is provided in Table 2, which compares accuracy and loss function for 50 epochs. In this instance, the value of the objective function that was minimized is the train loss. The training loss value was computed using the whole training dataset. The test accuracy, on the other hand, denotes that the trained model correctly detects independent images that were not used in training, despite the fact that the same images were used for both training and testing.

Table 3. A comparison of the accuracy achieved by various neural network architectures.

Algorithm for Recognition	Accuracy (%)
Motion feature architecture [22]	65.4
Static feature architecture [22]	63.1
Hybrid feature architecture [22]	71.2
3DCNN architecture [7]	29
3DCNN Architecture [19]	85.2
Our approach	88.56

Comparing the proposed architecture to previous neural network architectures based on motion, static, and hybrid features, it produced the best experimental results. The figures presented in Table 3 were examined using the same dataset. Overfitting might have contributed to this substantial drop in accuracy. The architecture was subsequently changed. We deleted two layers based on the observation of the change in accuracy during layer addition and removal. Additionally, we saw a decrease in accuracy as the number of filters for a particular layer increased. We have also decreased the number of filters for various layers in this instance through observation. We were able to get a simpler model that performed better on the datasets as a result. The complexity of the used datasets could be to blame for such a quick change in the outcomes.

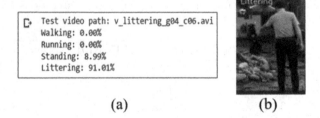

(a)   (b)

Fig. 4. The samples of littering action recognition: (a) the result of classifying the behavior into the corresponding class (a), and (b) a snapshot illustrating the recognition of littering taken from a video.

Additionally, we use our dataset of videos of actions divided into four types to train the proposed neural network. Figure 4 depicts the recognition of littering action, and a snapshot of littering action extracted from the video is presented in

Fig. 4(b). After 75 epochs we obtain the train loss function decreased to 0.0031 and the train accuracy increased to 96.02%. Meanwhile, the validation loss and validation accuracy are 0.0064 and 92.50% respectively. We may conclude that the proposed architecture handled the issue of time continuity across frames well because it was able to analyze both spatial and temporal data. We may conclude from analyzing the confusion matrix that the process of recognizing actions into groups was successful and mistake rates were low. Figure 5(a) shows the accuracy during training increasing positively with each subsequent session. We observe a sharp rise in accuracy in the initial epochs. The change in accuracy slows but still increases at a value of about 80. In training, 96% accuracy is the maximum possible.

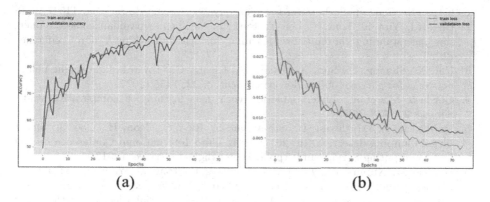

(a)   (b)

Fig. 5. The accuracy and loss during training process.

The reduction of the loss function during the training phase is depicted in Fig. 5(b). Good recognition results are indicated by the loss function's decreasing trend, which is directly inversely proportional to accuracy. In the initial training epochs, a sharp decline in values is seen. The loss function changes less quickly around the value of 0.010, but it still drops until it reaches its minimum value of about 0.0031. A directly proportionate rise in accuracy reflects the minimizing of errors, which is indicated by a decline in the loss function. Most of the time, we would see an increase in accuracy as loss decreased. The accuracy and loss function are two parameters that measure different things and have different definitions. There is no mathematical connection between these two measurements, despite the fact that they frequently appear to be inversely proportionate.

6   Conclusion

The primary value of this contribution to the community is in the area of recognizing people's out-of-the-ordinary behavior in general and littering action in particular in public areas. This monitoring can be applied to a variety of real-world situations. The proposed neural network's overall accuracy is adequate. It is particularly difficult to monitor and categorize human non-standard behavior in public spaces like city parks, sporting arenas, training facility campus, or city squares. The output of this work will be the warning in case of abnormal human behavior to be carried out in the next study. Besides, collecting a variety of data on garbage behavior, and training the model on a larger dataset are also issues that need to be improved.

Acknowledgements. This study is funded in part by the Can Tho University, Code: T2023-92.

References

1. https://laodong.vn/xa-hoi/can-tho-kien-quyet-xu-ly-van-de-rac-thai-1154773.ldo
2. Bae, K.I., Yun, K., Kim, H., Lee, Y., Park, J.: Anti-litter surveillance based on person understanding via multi-task learning. In: British Machine Vision Conference (2020)
3. Cai, X., Shuang, F., Sun, X., Duan, Y., Cheng, G.: Towards lightweight neural networks for garbage object detection. Sensors **22**, 7455 (2022)
4. Cordova, M., et al.: Litter detection with deep learning: a comparative study. Sensors **22**, 548 (2022)
5. Dabholkar, A., Muthiyan, B., Srinivasan, S., Ravi, S., Jeon, H., Gao, J.: Smart illegal dumping detection. In: 2017 IEEE Third International Conference on Big Data Computing Service and Applications (BigDataService) (2017)
6. Haohao, M., Xuping, W., As'Arry, A., Weiliang, H., Tong, M., Yanwei, F.: Domestic garbage target detection based on improved YOLOv5 algorithm. In: 2023 IEEE 13th Symposium on Computer Applications & Industrial Electronics (ISCAIE) (2023)
7. Hlavatá, R., Hudec, R., Sykora, P., Kamencay, P., Radilova, M.: Education of video classification based by neural networks (2020)
8. Husni, N.L., et al.: Real-time littering activity monitoring based on image classification method. Smart Cities **4**, 1496–1518 (2021)
9. Jiang, X., Hu, H., Qin, Y., Hu, Y., Ding, R.: A real-time rural domestic garbage detection algorithm with an improved yolov5s network model. Sci. Rep. **12**, 16802 (2022)
10. Jin, S., Yang, Z., Królczykg, G., Liu, X., Gardoni, P., Li, Z.: Garbage detection and classification using a new deep learning-based machine vision system as a tool for sustainable waste recycling. Waste Manage. **162**, 123–130 (2023)
11. Kim, Y., Cho, J.: AIDM-Strat: augmented illegal dumping monitoring strategy through deep neural network-based spatial separation attention of garbage. Sensors **22**, 8819 (2022)
12. Kulbacki, M., et al.: Intelligent video analytics for human action recognition: the state of knowledge. Sensors **23**, 4258 (2023)
13. LeCun, Y., et al.: Backpropagation applied to handwritten zip code recognition. Neural Comput. **1**, 541–551 (1989)
14. Liu, Z., et al.: Video swin transformer. In: Proceedings of the IEEE/CVF Conference on Computer Vision and Pattern Recognition (CVPR) (2022)
15. Malik, M., et al.: Machine learning-based automatic litter detection and classification using neural networks in smart cities. Int. J. Semant. Web Inf. Syst. **19**, 1–20 (2023)

16. Soomro, K., Zamir, A.R., Shah, M.: Ucf101: a dataset of 101 human actions classes from videos in the wild. arXiv:1212.0402 (2012)
17. Vaswani, A., et al.: Attention is all you need. In: Advances in Neural Information Processing Systems. Curran Associates, Inc. (2017)
18. Verma, V., et al.: A deep learning-based intelligent garbage detection system using an unmanned aerial vehicle. Symmetry **8**, 960 (2022)
19. Vrskova, R., Hudec, R., Kamencay, P., Sykora, P.: Human activity classification using the 3DCNN architecture. Appl. Sci. **12**, 931 (2022)
20. Wan Ismail, W.Z.: An illegal dumping detection system based on image processing in OpenALPR. ASM Sci. J. (2021)
21. Yun, K., Kwon, Y., Oh, S., Moon, J., Park, J.: Vision-based garbage dumping action detection for real-world surveillance platform. ETRI J. **41**, 494–505 (2019)
22. Zhang, X., Yao, L., Huang, C., Sheng, Q.Z., Wang, X.: Intent recognition in smart living through deep recurrent neural networks. In: Liu, D., Xie, S., Li, Y., Zhao, D., El-Alfy, E.S. (eds.) Neural Information Processing. Lecture Notes in Computer Science(), vol. 10635, pp. 748–758. Springer, Cham (2017). https://doi.org/10.1007/978-3-319-70096-0_76
23. Zhang, Y., et al.: VidTR: video transformer without convolutions. In: 2021 IEEE/CVF International Conference on Computer Vision (ICCV) (2021)

A Novel Ensemble K-Nearest Neighbours Classifier with Attribute Bagging

Niful Islam, Humaira Noor, and Dewan Md. Farid$^{(\boxtimes)}$

Department of Computer Science and Engineering, United International University,
United City, Madani Avenue, Badda, Dhaka 1212, Bangladesh
dewanfarid@cse.uiu.ac.bd
https://cse.uiu.ac.bd/profiles/dewanfarid/

Abstract. Classification of supervised learning in machine learning is a challenging task. K-Nearest Neighbours (kNN) is one of the simple classifiers that is commonly used in many real-life pattern recognition applications. The kNN is a lazy learner as it does not build any decision line from historical training data. In this paper, we have proposed a novel ensemble based K-Nearest Neighbours classifier with attribute bagging technique. The proposed method is designed to reduce the time complexity of the traditional kNN classifier, which also ameliorates the classification accuracy. It employs the attribute bagging technique and selects a sorted sub-set of instances from the original training data to apply kNN classifier. We have applied binary search technique to effectively reduce the search space to find the nearest neighbours. The proposed approach reduces the time consumption of kNN classifier from linear to logarithmic making the algorithm feasible for Big Data. We have tested the performance of the proposed algorithm with traditional machine learning classifiers e.g. kNN, Random Forest, AdaBoost on 10 benchmark datasets. The datasets are taken from UCI Machine Learning Repository. The results show that the proposed method outperformed the basic kNN and AdaBoost classifiers, and is compatible with Random Forest classifier.

Keywords: Attribute Bagging · Classification · K-Nearest Neighbours

1 Introduction

The human brain's inventiveness in completing tasks rapidly resulted in machine learning (ML) algorithms. In 1959 Arthur Samuel coined the term machine learning and defined it as a field of study that gives computers the ability to learn without being explicitly programmed [11]. These days, machine learning is considered a subfield of Artificial Intelligence that blends computer science and statistical techniques. In addition, machine learning teaches computers how to handle data more effectively. Sometimes, even after examining the data, we are unable to evaluate or extrapolate the information. We then use machine learning

N. Thai-Nghe et al. (Eds.): ISDS 2023, CCIS 1950, pp. 262–276, 2024.
https://doi.org/10.1007/978-981-99-7666-9_22

in such a situation. In recent years, the availability of Big Data has increased the demand for machine learning. ML algorithms may be roughly categorised into three groups, supervisor learning, unsupervised learning, and reinforced learning. The supervised method analyses data that has been fully class-labeled and determines the connection between the data and its class [16].

The k-Nearest Neighbour (kNN) algorithm is a supervised machine learning technique mainly used for classification problems. kNN categorises unlabelled data by figuring out how far apart each labeled data point is from the test instance. Finally, using patterns in the dataset, it assigns each unlabelled data point to the class label with the majority instances of the k nearest neighbours [2,12]. The selection of the hyper-parameter k is one of the numerous factors that influence how well the kNN algorithm performs. The method would be more susceptible to overfitting data points if k were too small. The model won't be flexible if the value of k is too high [14]. Among several drawbacks of kNN, the laziness of this algorithm is the most prominent one [4]. Since KNN does not draw any decision boundary at the training phase, and only performs computation at the test phase, it is commonly known as lazy learner. Due to it's laziness, it has a very high computational complexity at the testing phase that makes KNN infeasible for large datasets.

This paper introduces a novel ensemble based K-Nearest Neighbours classifier that is designed to reduce the time complexity of the traditional approach by a significant margin. Unlike the naive algorithm, the proposed approach does some computation by sorting sub-datasets at the training phase to so that while inference, binary search can be applied. This approach reduces the time consumption of kNN classifier from linear to logarithmic making the algorithm feasible for Big Data. Furthermore, the proposed method is tested on 10 benchmark datasets from UCI Machine Learning Repository. The result shows a noteworthy improvement in classification performance over basic kNN and AdaBoost classifiers, and compatible with Random Forest classifier. To summarise, the paper's main contribution is to proposed a novel approach for kNN classifier that has logarithmic time complexity, and the proposed classifier outperforms the traditional ML algorithms.

The remainder of this paper is structured as follows. Section 2 holds the related works along with their brief descriptions and drawbacks. Section 3 and Sect. 4 contain the proposed method and the results obtained from that method respectively. The article concludes in Sect. 5.

2   Related Work

For solving the resource consumption of kNN, researchers have proposed several solutions. Among them, tree based solutions are the most dominant ones. The idea is to pick splitting criteria and construct tree (mostly binary tree) to hold the dataset so that nearest neighbour searching becomes fast. The tree based approaches mainly differ in the splitting condition. Gupta et al. [7] proposed a new compression technique that creates the Combi tree from a binary search tree. This technique divides the data points into clusters and then uses a hash table to compress each cluster. The compressed clusters are then combined to form the Combi tree. The primary limitation of this paper is it operates solely in Hamming space, a type of high-dimensional space used for binary data. Therefore, it may not be effective or applicable for other types of data or similarity measures that operate outside of Hamming space. Hassanat [8] presented a method that constructs binary search tree (BST) based on the norms of the data points. The main concept is to use a partitioning scheme based on norms to ensure that the data points are distributed in the BST. This approach expects the data points to be uniformly distributed. In the real world, there might be skewness in the dataset. Pappula [13] proposed a novel BST approach that leverages a scaling factor to enhance the speed of search. This method is specifically devised for dealing with large datasets, as conventional binary search trees tend to be cumbersome and ineffective in such cases. The fundamental concept of the approach involves dynamically modifying the search intervals' sizes by utilising the scaling factor as the search proceeds. As a result, the method can swiftly narrow down the search space and discover the required item with logarithmic time complexity. The experiment, however, was conducted on synthetic data. Hence, the algorithm's performance on real world data is missing.

Another popular approach for reducing the time consumption of kNN is dimensionality reduction. Shokrzade et al. [18] reduced dimensionality using Extreme Learning Machine (ELM) to transform high dimensional data into a set of feature space. ELM is a supervised machine learning technique that has a single hidden layer. Since ELMs are sensitivity to noise and their dependence on the choice of random weights and biases, the drawbacks of ELMs apply to the proposed algorithm as well. Shekhar et al. [17] proposed to reduce dimensionality using Mutual Information (MI) and parallelise the process of nearest neighbour search using General Purpose Graphics Processing Units. This approach costs extra hardware resources.

One more very popular approach is to run KNN on a small set of data. Typically this small subset is selected through clustering. Saadatfar et al. [15] clustered the dataset using k-means clustering algorithm. Then it identifies and eliminates the data points that do not have a significant impact on the classification accuracy. On classification, for any given instance, it finds the cluster where the test instance belongs and runs KNN on that sub-dataset. Due to the pruning technique used in this algorithm, it lacks optimality where the decision boundary is non-linear or the dataset is noisy. Wang et al. [19] used clustering to reduce the number of data points that need to be compared for each query. After that, the algorithm uses region division to further reduce the search space. The search space is divided into several smaller regions, and only the points within the region that contain the query point need to be considered for the KNN search. Since, clustering can be expensive for high dimensional data, both the previous algorithms lack scalability as the number of features increases. Gul et al. [6] proposed an ensemble method based on KNN. The algorithm is composed of some kNN base classifiers that are constructed through random feature and instance selection. Although the approach increases classification accuracy, for ideal values of the hyper-parameters, the algorithm consumes more computation than the traditional one. Islam et al. [9] presented a novel algorithm named KNNTree that builds a decision tree (DT) to a certain depth and runs KNN on the rest of the sub-dataset. This algorithm allows DT to effectively reduce the number of samples where KNN is to be run. The proposed hybrid algorithm outperforms both the DT and KNN by a noteworthy margin. Table 2 holds a comparison between different approaches along with their drawbacks that our algorithm addresses. The tree based approaches mainly lack scalability which combi tree addresses by hashing the dataset. However, the hashing only applies to hamming space. Another limitation of these algorithms is that they expect the dataset to be uniformly distributed, which applies to most of the tree based kNN algorithms. Other algorithms, however, fail to reduce the computational cost by a significant margin (Table 1).

Table 1. Comparison between different methods

Algorithm	Advantage	Drawback
Combi tree [7]	Logarithmic	Limited to Hamming space
Norm based [8]	Logarithmic	Expects uniform distribution only
BST based [13]	Logarithmic	Experimented on synthetic data
ELM based [18]	Faster	Sensitive to noise
Pknn-mifs [17]	Faster	Requires hardware
EDP [15]	Faster	Drops performance on non-linear data
SRBC [19]	Faster	Not scaleable for higher dimension
KNNTree [9]	Logarithmic	Performance highly varies on the selection of hyper-parameters

3   Methodology

This section presents the detailed algorithm along with the asymptotic analysis.

3.1   Proposed Method

Among various techniques of lessening the time complexity of kNN, approximation based search techniques are one of the most efficient ones. However, sometimes this leads to a reduction in classification efficacy. Ensemble techniques are found to escalate the performance of any classifier by a noteworthy margin. In this algorithm, we employ these two concepts (approximate search and ensemble technique) to propose a new algorithm called Ensemble of Sorted Subset KNN (ESSK) that minimises the computation without decreasing the performance. The searching process in KNN is linear that requires $O(nd)$ time complexity where n is the number of instances and d is the dimension. If a monotonic nature is found on the dataset, a more efficient searching technique, binary search, can be applied. Therefore, the algorithm sorts a subset of data to make binary search possible. Let's consider some data points in a three dimensional space as described in Fig. 1. In the algorithm, we randomly select some dimensions. Let the randomly selected dimensions be D1 and D3. In our example, assume, D3 has more correlation with the class label than D1. Therefore, we sort the dataset according to D3 as shown in Fig. 2. At inference, it uses the searching algorithm to efficiently search a group of instances that are more likely to be the nearest neighbour. This process is repeated m number of times for higher accuracy.

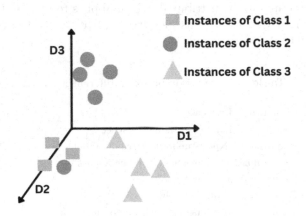

Fig. 1. Data points in three dimensions.

Let the independent variables be X=$\{x_1, x_2, x_3....x_N\}$ and dependent variables be y=$\{y_1, y_2, y_3...., y_N\}$. Other than the training data (X and y), the model takes their parameters as input. The first parameter K represents the number of instances to be considered as nearest neighbours while inference. Since this is an ensemble algorithm, it requires a certain number of iterations. Therefore, the variable m is used. Lastly, the variable g represents grace. Further details about the parameter is mentioned in Sect. 3.3. In the training phase, described in Algorithm 1, we first take a set (SA) that stores sets of four elements. The main task initiates by randomly selecting a subset of features. Among those features, the best feature is selected using Mutual Information (MI). Details of MI is present in Sect. 3.2. Using the best feature as comparator, the sub-dataset along with the class column is sorted. Finally the sorted sub-dataset, class column, randomly selected attributes and the best attribute is stored in SA. This procedure is carried out for a total of m iterations.

In the testing phase, for any given instance, X_i, we use binary search to find its position in a sub-dataset using binary search with the best attribute as comparator. The next step involves selecting a maximum of $K + g$ instances from both sides, which are then used to run the KNN algorithm. This process

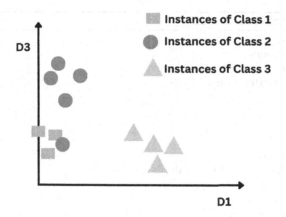

Fig. 2. Data points in randomly selected two dimensions.

is also iterated m times. Finally, the majority of these predictions is returned as output. Algorithm 2 outlines the detailed process.

3.2 Mutual Information

Mutual Information (MI) is a metric that measures the extent of shared information between two variables [5]. The metric calculates how much the presence of one variable increases the certainty of the other variable. The range of MI score between two variables is 0 to 1. A higher MI score implies strong dependency between the two variables, whereas a lower score indicates their independence. The formula for calculating MI is mentioned in Eq. 1. Where $p(x, y)$ is the joint probability distribution of variable X and Y and $p(x)$ and $p(y)$ are the respective marginal probability distributions.

$$MI(X, Y) = \sum_{y \in Y} \sum_{x \in X} p(x, y) \log \left(\frac{p(x, y)}{p(x)p(y)} \right) \tag{1}$$

3.3 Grace

The proposed algorithm is based on approximate search. In this algorithm, the search space is reduced such that only a few instances are left. However, if the features have a very small range, there could lie a large number of highly potential nearest neighbor candidates in a small span. In that scenario, high deduction of search space might lead to a substantial decrease in model's performance. Given that context, it is necessary to have a larger number of instances on which the KNN classifier should run. The variable grace (g) is utilized under those circumstances.

Algorithm 1 ESSK on Training

1: **Input:** $X = \{x_1, x_2, x_3 .... x_N\}$,
   $y = \{y_1, y_2, y_3 .... y_N\}$,
   K = Number of Nearest Neighbours,
   m = Number of Iterations,
   g = Grace
2: **Output:** $\emptyset$
3: **Method :**
4: $SA \leftarrow \emptyset$
5: **for** $i = 1 ... m$ **do**
6:    $attr$= randomly select some attributes from X
7:    $best\_attr \leftarrow \emptyset$
8:    $MI \leftarrow$ -1
9:    **for each** $a \in attr$ **do**
10:      mi = mutual information of (a, y)
11:      **if** $mi > MI$ **then**
12:        $MI = mi$
13:        $best\_attr = a$
14:      **end if**
15:    **end for**
16:    $X' = X[attr]$
17:    $y' = y$
18:    sort X' and y' with $best\_attr$ as comparator
19:    $val = \{X', y', attr, best\_attr\}$
20:    $SA = SA \cup val$
21: **end for**

3.4 Asymptotic Analysis

Let n be the number of instances, d be the number of dimensions and K, m, g be the number of nearest neighbor, iterations and grace respectively. At the training phase, since some number of features are randomly selected, at the worst case, all the features can be selected. The next step is to find the best attribute using MI. There are several techniques for calculating MI and the best approach costs $O(n)$ complexity. Therefore, the complexity of finding the best attribute is $O(nd)$. The sorting consumes $O(ndlog(nd))$ with some promising sorting algorithms (like marge sort) at the worst case [3]. Other calculations are insignificant and not considered in the overall complexity. Hence, at each iteration the complexity is $O(nd + ndlog(nd))$ which can also be written as $O(ndlog(nd))$. Finally, after m iterations, the asymptotic complexity at training becomes $O(mndlog(nd))$. Although traditional approach costs $O(1)$ due to it's laziness, the main difference is in the testing phase.

In testing, it uses binary search that costs $O(log(n))$. Because the searching operation takes place in one dimension, the effect of having a higher dimension is missing. After that, it runs KNN on a small set which costs $O(Kd+dg+K^2+Kg)$. According to asymptotic theory, it is similar to $O(K^2)$. Thus, the complexity of the test phase becomes $O(m(log(n) + K^2)$ which is a huge improvement com-

---

Algorithm 2 ESSK on Testiing

1: **Input:** X_i = Any given test instance
2: **Output:**
3: C_i = Predicted class label
4: **Method :**
5:   $preds \leftarrow \emptyset$
6: **for** $i =1 \dots m$ **do**
7:     $low=1$
8:     $high = N$
9:     $X'=SA[m][1]$
10:    $y'=SA[m][2]$
11:    $att=SA[m][4]$
12:    $X_i'=X_i[SA[m][3]]$
13:    **while** $low < high$ **do**
14:       $mid= low + (high - low)/2$
15:       **if** $X'[att][mid] < X_i'[att]$ **then**
16:         $low= mid+1$
17:       **else**
18:         $high=mid\text{-}1$
19:       **end if**
20:    **end while**
21:    $left = \max(0,low\text{-}K\text{-}g)$
22:    $right = \min(N,low+K+g)$
23:    $pred$ = predict X_i' from $\{X'[left{:}right], y'[left{:}right]\}$ using traditional KNN
24:    $preds = preds \cup pred$
25: **end for**
26: **return** majority element on $preds$

---

pared to the complexity of the naive approach that is $O(nd + Kd)$. For space complexity, the algorithm takes $O(mnd)$ auxiliary space. Compared to the primitive approach which is $O(nd)$, this has slightly more consumption in the worst case. For the average case, however, the proposed ESSK has less complexity than kNN. To summarize, the proposed ESSK algorithm has a training time complexity of $O(mndlog(nd))$ and testing time complexity of $O(m(log(n) + K^2))$. Lastly, the space complexity is $O(mnd)$.

4 Experimental Results

This section presents the results obtained from the algorithm along with a comparison with some state-of-the-art algorithms.

4.1 Dataset

Ten benchmark datasets were selected mostly from UCI machine learning repository for evaluating the model. Medical data is often complex, diverse, imbalanced and high-dimensional, which poses significant challenges for machine learning models. Evaluating a model using medical data allows to test the effectiveness in handling these challenges, as well as its ability to accurately predict outcomes or diagnose diseases [1]. Therefore, five medical datasets (ECG, Diabetes, Lymphography, Fertility and Breast Cancer) were chosen for evaluating the performance all of which are imbalanced. Producing high classification performance with less amount of data is always a challenging task for machine learning models. That is why the collection contains six datasets with less than one thousand samples. The model is evaluated on both binary class and multiclass datasets. The datasets also differs in dimensionality. Such as the high dimensional MNIST dataset that is a collection of ten categories of handwritten digits ranging from 0 to 9 [10]. However, the Iris dataset, with a low dimensionality, comprises descriptions of three different types of Iris flowers. Table 2 contains the detailed description of the datasets. For training 80% of the samples were randomly selected and rest of them were left for evaluation.

4.2 Experimental Setup

The experiment was conducted on a machine having CPU Ryzen 5 5600 with 16 GB RAM. The programming language selected for the experiment was Pyhton. Three python libraries named Numpy, Pandas and scikit-learn were also used.

Table 2. Dataset Description.

No.	Dataset Name	Instances	Features	Classes
1	Iris	150	4	3
2	ECG	109446	187	5
3	Glass	214	9	6
4	Diabetes	768	8	2
5	Spambase	110201	4	2
6	MNIST	70000	784	10
7	Magic	19020	10	2
8	Lymphography	148	18	4
9	Fertility	100	8	2
10	Breast Cancer	569	30	2

Table 3. Comparison of accuracy on different datasets.

No.	Dataset Name	KNN	RF	AdaBoost	ESSK
1	Iris	0.95	0.96	0.94	**0.97**
2	ECG	0.96	**0.97**	0.86	0.91
3	Glass	0.70	0.80	0.49	**0.77**
4	Diabetes	0.69	0.76	0.76	**0.79**
5	Spambase	0.81	**0.96**	0.94	0.90
6	MNIST	0.55	0.75	0.35	**0.80**
7	Magic	0.80	**0.88**	0.84	0.85
8	Lymphography	0.76	0.84	0.66	**0.86**
9	Fertility	0.85	0.88	0.80	**0.95**
10	Breast Cancer	0.94	**0.96**	**0.96**	0.94
	Average	0.80	0.88	0.76	0.87

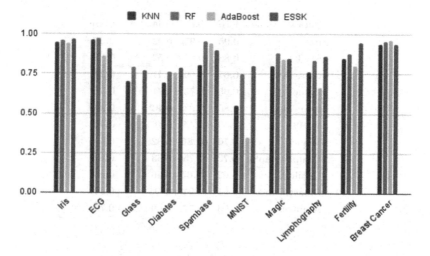

Fig. 3. Accuracy comparison on datasets.

4.3 Models and Hyperparameters

The model was tested against three popular classification algorithms (KNN, Random Forest and AdaBoost). For both the KNN and proposed ESSK algorithm, the number of nearest neighbors (i.e. K) was 3. For ESSK, there are two additional parameters. Number of iterations (m) was chosen as 3 and the grace (g) was 0. For Random Forest (RF), hundred full grown decision trees were used as base classifier. For AdaBoost, however, although the number of decision trees were same, the trees were pruned at level one.

Table 4. Comparison of precision on datasets.

No.	Dataset Name	KNN	RF	AdaBoost	ESSK
1	Iris	0.95	0.95	0.94	**0.97**
2	ECG	0.96	**0.97**	0.85	0.91
3	Glass	0.69	**0.80**	0.48	0.77
4	Diabetes	0.70	0.77	0.76	**0.79**
5	Spambase	0.81	**0.96**	0.94	0.90
6	MNIST	0.45	0.44	0.30	**0.66**
7	Magic	0.81	**0.87**	0.84	0.85
8	Lymphography	0.75	0.85	0.66	**0.86**
9	Fertility	0.87	0.88	0.78	**0.95**
10	Breast Cancer	0.92	0.95	**0.95**	0.94
	Average:	0.79	0.84	0.75	0.86

Table 5. Comparison of recall on datasets.

No.	Dataset Name	KNN	RF	AdaBoost	ESSK
1	Iris	0.95	0.95	0.92	**0.97**
2	ECG	0.83	0.80	0.52	**0.85**
3	Glass	0.61	0.69	0.41	**0.79**
4	Diabetes	0.65	0.73	0.71	**0.76**
5	Spambase	0.80	0.93	**0.92**	0.90
6	MNIST	0.50	0.45	0.42	**0.74**
7	Magic	0.76	**0.86**	0.82	0.63
8	Lymphography	0.72	0.75	0.63	**0.84**
9	Fertility	0.62	**0.65**	0.56	0.48
10	Breast Cancer	0.91	0.94	**0.95**	0.94
	Average:	0.73	0.77	0.69	0.79

4.4   Evaluation Matrix

All the models were tested on four evaluation matrices. The first matrix is accuracy. It refers to the percentage of instances correctly classified. Other there matrices are precision, recall and F1 socre. Precision is to measures the contribution of true positive results among the total positive results. Recall, on the other hand, specifies the proportion of actual positive instances that were correctly identified by the model. F1 score is the harmonic mean of precision and recall. This matrix is useful especially for imbalance data [20].

Table 6. Comparison of F1 on datasets.

No.	Dataset Name	KNN	RF	AdaBoost	ESSK
1	Iris	0.94	0.95	0.93	**0.97**
2	ECG	**0.87**	**0.87**	0.51	0.54
3	Glass	0.61	0.69	0.35	**0.78**
4	Diabetes	0.66	0.72	0.73	**0.77**
5	Spambase	0.80	**0.95**	0.92	0.90
6	MNIST	0.46	0.44	0.29	**0.66**
7	Magic	0.76	**0.87**	0.82	0.66
8	Lymphography	0.72	0.74	0.62	**0.86**
9	Fertility	0.59	**0.64**	0.56	0.49
10	Breast Cancer	0.90	0.94	**0.95**	0.93
	Average:	0.73	0.78	0.67	0.76

4.5 Results

The findings demonstrate a substantial enhancement compared to the conventional KNN approach. As shown in Table 3, on average, the proposed model has 7% higher accuracy than KNN and 11% higher than AdaBoost algorithms. Although RF was found to have a higher accuracy, the difference is only 1%. Figure 3 contains a diagrammatic overview of accuracy on benchmark datasets. Since, accuracy is not a perfect matrix, especially for imbalanced datasets, we have compared performance on F1 score, precision and recall matrices that is presented on Table 6, Table 4 and Table 5 respectively. ESSK outperforms KNN and AdaBoost on these matrices also. While for RF classifier, it has higher precision and recall score.

Figure 4 shows a comparative analysis of F1 score of different models. On imbalanced medical datasets, the average F1 score, precision and recall is 71.8%, 89% and 77.4% each. These results imply, that the model is not overfitted and can be a reliable option for medical data analysis. On MNIST, the dataset with the 784 dimensions, ESSK achieves the highest performance among other models compared to all matrices. In short, the proposed ESSK is a high potential machine learning option for data classification.

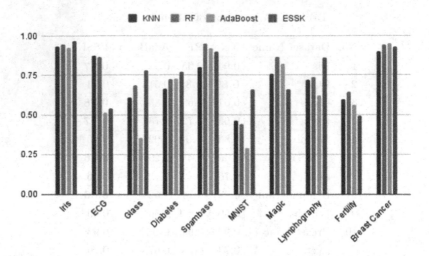

Fig. 4. F1 score comparison on various datasets.

4.6   Comparison with Existing Solutions

An asymptotic comparison with existing solutions of the training and testing time has been presented in Table 7. Let the number of instances in a dataset be represented with n, the number of features with d, for tree based approaches, the depth of the tree with L, the number of iterations with m and the number of nearest neighbors with K. The table illustrates that there are some tree based solutions that consume slightly less computation than the proposed solution. However, they come with some reasonable concern that has been discussed in Table 2. Moreover, some of the solutions compromised classification performance to reduce complexity which, nonetheless, is significantly increased in the proposed ESSK algorithm. Since different devices have different configurations that can hinder the running time, only the asymptotic complexity has been compared.

Table 7. Asymptotic complexity comparison with existing methods.

Algorithm	Training Time	Testing Time
Combi tree [7]	$O(nd * log(nd))$	$O(log(n))$
BST based [13]	$O(nd * log(nd))$	$O(log(n))$
KNNTree [9]	$O(Ln * log(n))$	$O(L + dlog(n))$
Pknn-mifs [17]	$O(nd)$	$O(nd)$
ESSK	$O(mndlog(nd))$	$O(m(log(n) + K^2)$

4.7 Discussion

The proposed ESSK algorithm is found to outperform traditional KNN and AdaBoost in all four evaluation matrices and RF in two matrices. There are several reasons for this performance. Firstly, ensemble models tend to outperform single model classifiers. Therefore, a higher accuracy over KNN classifier is expected. Moreover, since KNN uses distance as matrix for finding nearest neighbors, a noisy column can largely affect the performance. As ESSK considers only a subset of features at every iteration, there is less impact of noisy data if only a few columns have noise. Furthermore, with each successive iteration, the most informative feature among a subset of features gets prioritized. Dissimilar to the traditional approach, where each feature gets a different priority depending on its scale. These improvements make the proposed algorithm superior to the classic KNN algorithm.

5 Conclusion

In this paper, we present a novel ensemble approach for classifying instances with KNN classifier. The proposed approach picks a subset of attributes and finds the best attribute among that subset using mutual information. After that it sorts the sub-dataset using the best attribute as comparator. At the testing phase, since the data is sorted, it runs binary search to find approximate nearest neighbours. This process is repeated some number of times for higher accuracy. We have compared the proposed algorithm against various state-of-the-art machine learning algorithms and it shows a noticeable performance on ten benchmark datasets. Nonetheless, the high performance and computational efficiency come with a cost of more memory consumption. Thus, future work includes developing a more memory efficient version and investigating the robustness of this algorithm on various datasets with an exploration of ideal values of the parameters. Moreover, the proposed solution introduced some new hyper-parameters. The best configurations of these hyper-parameters are yet to be discovered.

References

1. Alanazi, A.: Using machine learning for healthcare challenges and opportunities. Inf. Med. Unlocked **30**, 100924 (2022)
2. Almomany, A., Ayyad, W.R., Jarrah, A.: Optimized implementation of an improved KNN classification algorithm using intel FPGA platform: COVID-19 case study. J. King Saud Univ.-Comput. Inf. Sci. **34**(6), 3815–3827 (2022)
3. Choudhury, M., Dutta, A.: Establishing pertinence between sorting algorithms prevailing in n log (n) time. J. Robot. Auto.Res. **3**(2), 220–226 (2022)
4. Gallego, A.J., Rico-Juan, J.R., Valero-Mas, J.J.: Efficient k-nearest neighbor search based on clustering and adaptive k values. Pattern Recogn. **122**, 108356 (2022)
5. Gu, X., Guo, J., Xiao, L., Li, C.: Conditional mutual information-based feature selection algorithm for maximal relevance minimal redundancy. Appl. Intell. **52**(2), 1436–1447 (2022)

6. Gul, A., et al.: Ensemble of a subset of k NN classifiers. Adv. Data Anal. Classif. **12**, 827–840 (2018)
7. Gupta, P., Jindal, A., Sengupta, D., et al.: ComBi: compressed binary search tree for approximate k-NN searches in hamming space. Big Data Res. **25**, 100223 (2021)
8. Hassanat, A.B.: Norm-based binary search trees for speeding up KNN big data classification. Computers **7**(4), 54 (2018)
9. Islam, N., Fatema-Tuj-Jahra, M., Hasan, M.T., Farid, D.M.: KNNTree: a new method to ameliorate k-nearest neighbour classification using decision tree. In: 2023 International Conference on Electrical, Computer and Communication Engineering (ECCE), pp. 1–6. IEEE (2023)
10. Kaplun, V., Shevlyakov, A.: Contour pattern recognition with MNIST dataset. In: 2022 Dynamics of Systems, Mechanisms and Machines (Dynamics), pp. 1–3. IEEE (2022)
11. Mahesh, B.: Machine learning algorithms-a review. Int. J. Sci. Res. (IJSR).[Internet] **9**, 381–386 (2020)
12. Memiş, S., Enginoğlu, S., Erkan, U.: Fuzzy parameterized fuzzy soft k-nearest neighbor classifier. Neurocomputing **500**, 351–378 (2022)
13. Pappula, P.: A novel binary search tree method to find an item using scaling. Int. Arab J. Inf. Tech. **19**(5), 713–720 (2022)
14. Rattanasak, A., et al.: Real-time gait phase detection using wearable sensors for transtibial prosthesis based on a kNN algorithm. Sensors **22**(11), 4242 (2022)
15. Saadatfar, H., Khosravi, S., Joloudari, J.H., Mosavi, A., Shamshirband, S.: A new k-nearest neighbors classifier for big data based on efficient data pruning. Mathematics **8**(2), 286 (2020)
16. Saranya, T., Sridevi, S., Deisy, C., Chung, T.D., Khan, M.A.: Performance analysis of machine learning algorithms in intrusion detection system: a review. Procedia Comput. Sci. **171**, 1251–1260 (2020)
17. Shekhar, S., Hoque, N., Bhattacharyya, D.K.: PKNN-MIFS: a parallel KNN classifier over an optimal subset of features. Intell. Syst. Appl. **14**, 200073 (2022)
18. Shokrzade, A., Ramezani, M., Tab, F.A., Mohammad, M.A.: A novel extreme learning machine based KNN classification method for dealing with big data. Expert Syst. Appl. **183**, 115293 (2021)
19. Wang, H., Xu, P., Zhao, J.: Improved KNN algorithms of spherical regions based on clustering and region division. Alex. Eng. J. **61**(5), 3571–3585 (2022)
20. Yacouby, R., Axman, D.: Probabilistic extension of precision, recall, and f1 score for more thorough evaluation of classification models. In: Proceedings of the First Workshop on Evaluation and Comparison of NLP Systems, pp. 79–91 (2020)

Legar: A Legal Statute Identification System for Vietnamese Users on Land Law Matters

Thien Huynh[1]([✉]), Ty Nguyen[1], Duc Nguyen[1], Phuong Thai[1], Thang Phung[1], Long Huynh[1], Thu Bui[1], An Nguyen[1], Huu Pham[1], and Tho Quan[2,3]

[1] ADAI LAB, 2nd Floor, Mitech Center, No. 75 Road 2/4, Nha Trang City, Vietnam
hnthien.1190@indivisys.jp
[2] Faculty of Computer Science and Engineering, Ho Chi Minh City University of Technology (HCMUT), Ho Chi Minh City 700000, Vietnam
[3] Vietnam National University, Ho Chi Minh City 700000, Vietnam
https://adai-lab.com

Abstract. This paper introduces Legar, an innovative Legal Statute Identification system designed to cater to Vietnamese users seeking legal guidance on land law matters. Legar focuses on practical efficiency by harnessing the digital capabilities of the Vietnam Landlaw 2013. This is achieved through the utilization of a specialized legal-masked language model, based on the RoBERTa architecture, leading to the development of the law-oriented language model, LegaRBERT. Subsequently, LegaRBERT is incorporated into a multi-label classification model, XGBoots, to offer consultation services based on user-generated questions. As a result, Legar effectively addresses prevailing limitations found in comparable LSI research and prototypes. Empirical evaluations conducted using authentic datasets from Vietnamese legal consultations demonstrate that Legar surpasses existing baselines, particularly when considering our proposed K-Utility metric, which reflects the practical expectations of LSI users. The initial version of Legar is now publicly accessible and has garnered positive feedback from users.

Keywords: PhoBERT · LegaRBERT · Legal Stature Identification · Multi-Label Classification · Law-oriented Contextual Embedding · Machine Learning · Deep Learning

1 Introduction

Legal Statute Identification (LSI) is a specialized task whose objective is to discern the pertinent legal statutes associated with a provided description of facts or evidence [1]. These kinds of documents can represent in different length and size which poses considerable challenge for the model to capture the full context, to which many approaches have been researched to address [2]. To a certain extent, LSI can be enhanced with the support of a knowledge system [3]. Let us consider a motivating example that an LSI can help generic user in the Vietnamese context as follows.

N. Thai-Nghe et al. (Eds.): ISDS 2023, CCIS 1950, pp. 277–291, 2024.
https://doi.org/10.1007/978-981-99-7666-9_23

Example 1. In a real conversation concerning legal matter of land business, a user expresses a concern as. '*Tôi đang thực hiện giao dịch mua đất của Ông A, hiện tại đã ký xong hợp đồng chuyển nhượng có công ch´ứng. Nay tôi được công ch´ứng viên báo là ông A đang bị khởi tố vì tội l`ưa đảo chiếm đoạt tài sản và có thể giao dịch mua đất của tôi sẽ bị tạm d`ừng do tài sản của ông A sẽ bị phong toả? Xin hỏi công ch´ứng viên nói như vậy có đúng không? Xin cảm ơn!*' ('I am currently executing the purchase transaction of land from Mr. A. The transfer contract has been signed and notarized. Today, the notary public informed me that Mr. A is under investigation for the offense of fraudulent misappropriation of assets, and as a result, the land purchase transaction may be temporarily suspended due to the possibility of asset freezing against Mr. A. May I inquire if the notary public's statement is accurate? Thank you!').

Ideally, to assist users in self-exloring their case, an LSI system would provide legal consultation by retrieving relevant articles within the Vietnamese Land Law 2013. Specifically, the system will address the question above by retrieving the content of Article 188 (Nói về điều kiện thực hiện quyền chuyển nhượng quyền sử dụng đất - Regarding the conditions for executing the right to transfer the land use rights), Article 168 (Nói về th`ời điểm được thực hiện các quyền của ngư`ời sử dụng đất - Regarding the timing of executing the rights of land users), and Article 186 (Nói về quyền và nghĩa vụ sử dụng đất của ngư`ời Việt Nam định cư ở nư´ớc ngoài và ngư`ời nư´ớc ngoài - Regarding the rights and obligations of land use for Vietnamese nationals residing abroad and foreigners) from the Vietnamese Land Law 2013 (please refer to sectIion 2 for the structural contents of a legal document). All three articles (188, 168, and 186) from the Vietnamese Land Law 2013 have relevance to the user's question, with Article 188 and Article 168 directly addressing specific aspects of land transactions and Article 186 providing a broader context regarding the rights and obligations of individuals involved in land use. ∎

Researchers have extensively explored LSI-like approaches aiming to automate the process. Initially, researchers used statistical algorithms combined with manually crafted rules [4–6]. Subsequently, LSI was treated as a text classification problem, relying on manually engineered features [7, 8]. More recently, attention-based neural models for LSI have emerged, aiming to extract more informative features from text through techniques such as dynamic context vectors [9], dynamic thresholding [10], hierarchical classification [11], and pretrained BERT-based [12], and RoBERTa-based [13], or Longformer-based [14] contextualizers. Xu et al. [15] proposed segmenting statutes into communities and employed a novel graph distillation operator to identify intra-community features. However, existing approaches to address the LSI problem primarily utilize the textual contents of facts and legal articles, without giving sufficient attention to law-specific terms in legal texts. Moreover, when handling real situation of law consultation, the previously proposed prototypes face the following notable obstacles.

- All of the approaches so far the LSI problem as the single-class classification, meaning they classify the user's concerns into a single legal article. However, in most real-world cases, users' concerns often involve more than one legal article as illustrated in Example 1. As one can see, this popular concern would involve 3 articles of Article 188, Article 168, and Article 186 to be fully addressed.

- As such, an LSI is not merely a Question-Answering (QA) system, but it would be furnished with a knowledge base of law, such as the Vietnamese Land Law 2013, properly labeled for multi-label training. To the best of our knowledge, such law-related data is currently not available in Vietnam.
- More specifically, to leverage the emerging deep learning models for the classification task, the knowledge base should not be treated as a generic one, but should be embedded in a manner that captures the domain knowledge of law effectively.

To address these limitations, this paper introduces the LegaR system[1], which aims to overcome the aforementioned drawbacks. The key contributions of this paper are as follows

- We introduce the VN-LandLaw-2013 corpus, which has been digitalized and structurally processed. The corpus has been multi-labeled specifically for the classification-based consultation task.
- We present the LegaRBERT language model, specifically designed for the Vietnamese Land Law 2013 domain. LegaRBERT's architecture is founded upon the architecture of RoBERTa [16], but it does not utilize BERT's default masked language modeling mechanism [17]. Instead, a *legal-masked language model* strategy is introduced to enhance the model's attention towards legal domain-specific terms in the text.
- LegaRBERT employs multi-label classification model based on XGBoost [18]. Moreover, we introduce the *K-Utility* (K-U) metric for evaluating the multi-label classification problem. When applied in practical scenarios, this metric demonstrates its usefulness in retrieving answers for users, especially regarded in the professional expert view.

The structure of this paper includes Sect. 1, which introduces the LSI problem in the Vietnamese law domain, the Legar system, related work, and the contributions of this paper. Section 2 describes the VN-LandLaw-2013 corpus. Section 3 outlines the general architecture of the Legar system. Section 4 explains the training and evaluation processes of the LegaRBERT model with the *K-U* metric. Section 5 presents experiments, their results, and corresponding analysis. Finally, Sect. 6 concludes the paper.

2   Related Works

The VN-LandLaw-2013 corpus is a collection of questions, answers, and corresponding labels related to the Vietnamese Land Law 2013. The process of preparing this dataset involved two main steps: data collection and data labeling. In the data collection step, we acquired a digitalized version of the Vietnamese Land Law 2013. Then, we gathered conversations from landlaw-related e-forums, which reflect real situations of legal consultation by experts concerning the applications of the Vietnamese Land Law 2013. The collected dataset of conversations was labeled by our team of legal experts to extract relevant information such as doctype, legislation, article, clause, and point. Ultimately,

---

[1] Interested readers can try Legar at https://legar.vn.

a total of 5910 data samples were collected for this corpus. The conversation given in Example 1 is, in fact, from a real sample extracted for the VN-LandLaw-2013 corpus. Listing 1 provides the full information of this data item when stored in the corpus. As can be observed, a data item is annotated with a substantial amount of legal information. In the context of this paper, we focus on the annotated Articles, specifically Articles 168, 188, and 186. These Articles are used as labels for this item. Thus, our consultation problem is realized as a multi-label classification task, where the Legar system will classify a user's concern into corresponding Articles (Table 1).

Table 1. An Example of Data Item in VN-Landlaw-2013 Corpus.

{"id":"92",
*　　"question":"Tôi đang thực hiện giao dịch mua đất của Ông A, hiện tại đã ký xong hợp đồng chuyển nhượng có công chứng. Nay tôi được công chứng viên báo là ông A đang bị khởi tố vì tội lừa đảo chiếm đoạt tài sản và có thể giao dịch mua đất của tôi sẽ bị tạm dừng do tài sản của ông A sẽ bị phong toả? Xin hỏi công chứng viên nói như vậy có đúng không? Xin cảm ơn! ",*
*　　"answer":" Điều 188 Luật Đất đai 2013 quy định điều kiện thực hiện quyền chuyển nhượng quyền sử dụng đất như sau: - Có Giấy chứng nhận, trừ trường hợp quy định tại khoản 3 Điều 186 và trường hợp nhận thừa kế quy định tại khoản 1 Điều 168 của Luật đất đai 2013; - Đất không có tranh chấp; - Quyền sử dụng đất không bị kê biên để bảo đảm thi hành án; - Trong thời hạn sử dụng đất. Căn cứ quy định nêu trên, nếu mảnh đất mà bạn đang thực hiện giao dịch mua bán, chuyển nhượng đáp ứng các điều kiện trên thì người bán có quyền chuyển nhượng mảnh đất này. Hợp đồng chuyển nhượng được công chứng theo đúng quy định pháp luật. Căn cứ quy định tại Điều 82 Luật Công chứng 2014, tổ chức hành nghề công chứng có thể cập nhật thông tin về biện pháp ngăn chặn được áp dụng đối với tài sản có liên quan đến hợp đồng, giao dịch đã được công chứng, cụ thể như sau: "Điều 62. Cơ sở dữ liệu công chứng 1. Cơ sở dữ liệu công chứng bao gồm các thông tin về nguồn gốc tài sản, tình trạng giao dịch của tài sản và các thông tin về biện pháp ngăn chặn được áp dụng đối với tài sản có liên quan đến hợp đồng, giao dịch đã được công chứng." Căn cứ quy định trên, văn phòng công chứng nơi bạn đã công chứng hợp đồng chuyển nhượng sẽ cập nhật được những thông tin ngăn chặn đối với tài sản mà bạn đang giao dịch. Do đó, nếu những thông tin ngăn chặn đối với tài sản bạn đang giao dịch đã được cập nhật trên cơ sở dữ liệu công chứng thì nội dung văn phòng công chứng đưa ra như bạn nêu trên là hoàn toàn có cơ sở. Tuy hợp đồng chuyển nhượng quyền sử dụng đất của bạn đã được công chứng nhưng theo quy định tại Điều 11a Thông tư 24/2014/TT-BTNMT (được bổ sung bởi Thông tư 33/2017/TT-BTNMT), cơ quan có thẩm quyền sẽ dừng giải quyết thủ tục cấp Giấy chứng nhận quyền sử dụng đất đối với hồ sơ xin cấp Giấy chứng nhận quyền sử dụng đất của bạn. Xem thêm: Trường hợp nào bị phong tỏa tài khoản ngân hàng? Trên đây là nội dung tư vấn dựa trên những thông tin mà luật sư đã nhận được. Nếu còn bất kỳ thắc mắc nào liên quan, vui lòng liên hệ 19006199 để được hỗ trợ kịp thời. Xin cảm ơn!",*
*　　"summary":"Nếu những thông tin ngăn chặn đối với tài sản bạn đang giao dịch đã được cập nhật trên cơ sở dữ liệu công chứng thì nội dung văn phòng công chứng đưa ra như bạn nêu trên là hoàn toàn có cơ sở. Tuy hợp đồng chuyển nhượng quyền sử dụng đất của bạn đã được công chứng nhưng theo quy định tại Điều 11a Thông tư 24/2014/TT-BTNMT (được bổ sung bởi Thông tư 33/2017/TT-BTNMT), cơ quan có thẩm quyền sẽ dừng giải quyết thủ tục cấp Giấy chứng nhận quyền sử dụng đất đối với hồ sơ xin cấp Giấy chứng nhận quyền sử dụng đất của bạn.",*
*　　"lagels":[{"doctype":"Luật","legislation":"Luật Đất đai 2013","**article":"188**","clause":"Không xác định","point":""},{"doctype":"Luật","legislation":"Luật Đất đai 2013","**article":"168**","clause":"1","point":""},{"doctype":"Luật","legislation":"Luật Đất đai 2013","**article":"186**","clause":"3","point":""},{"doctype":"Luật","legislation":"Luật Công chứng 2014","article":"82","clause":"Không xác định","point":""},{"doctype":"Luật","legislation":"Luật Công chứng 2014","article":"62","clause":"Không xác định","point":""},{"doctype":"Thông tư","legislation":"Thông tư 24/2014/TT-BTNMT (được bổ sung bởi Thông tư 33/2017/TT-BTNMT)","article":"11","clause":"Không xác định","point":""}]}*

3　Legar System

3.1　Legal-Masked Fine-Tuning for LegaRBERT Generation

During the legal-masked fine-tuning process for LegaRBERT generation, we use the pre-trained MLM model loaded from the model checkpoint 'vinai/phobert-base,' which corresponds to the PhoBERT model [19], a Vietnamese language model. The PhoBERT model's configuration is based on that of RoBERTa (Robustly optimized BERT app-roach) [16]. We fine-tune this pre-trained MLM model on the VN-LandLaw-2013 cor-pus. The same pre-trained PhoBERT model's parameters are used to initialize the model for the downstream task of masked language model (MLM). However, instead of the default MLM mechanism supported by BERT, we leverage our novel strategy of legal masked language modeling as subsequently discussed (Fig. 1).

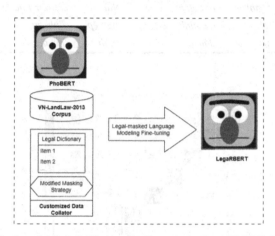

Fig. 1. The Legal-Masked Fine-tuning Process for LegaRBERT Generation.

3.2 Legal Dictionary

Similar to words in other domains, law terminologies are highly specific, so we need a dataset (i.e., legal dictionary) to capture them in this field. We relied on law words from "Từ điển Pháp luật Việt Nam" ("Vietnamese Legal Dictionary") [20], and then used TF-IDF (Term Frequency - Inverse Document Frequency) to learn from the VN-LandLaw-2013 corpus and filter out important words in the field of land law. Finally, we had legal experts perform the final filtering to create a legal dictionary comprising 1255 terms. The developed legal dictionary not only allows us to identify the most important law terms in the Vietnamese Land Law 2013 but also enables us to observe the semantic similarity between articles within this law. Table 2 illustrates the most important terms in the field of land law. Figure 2 illustrates a heatmap of the cosine similarity [21] between legal articles based on the extracted terms. We can observe that Article 186 (Regarding the rights and obligations of land use for Vietnamese nationals residing abroad and foreigners) and Article 188 (Regarding the conditions for executing the right to transfer the land use rights) are determined to have relatively high similarity. In addition, Table 3 illustrates some typical cases of similar laws automatically extracted by the system.

3.3 Legal-Masked Fine-Tuning Strategy

We recall the default MLM mechanism in the original BERT [17] as follows. From the input *sequence of tokens*, the training data generator chooses 15% of the token positions

Table 2. Some Important Terms in the Field of Land Law. (Order: top to bottom, left to right)

đất	sử_dụng	quyền	đất_đai	quy_định	quy_hoạch
tổ_chức	thu_hồi	xây_dựng	nông_nghiệp	bồi_thường	quản_lý
mục đích	đầu_tư	tài_sản	nghĩa_vụ	sở_hữu	quyết_định
cơ_quan	công_trình	sản_xuất	chứng_nhận	pháp_luật	thẩm_quyền
hành_chính	trách_nhiệm	thời_hạn	chính_phủ	chuyển_nhượng	quốc_phòng
huyện	phê_duyệt	an_ninh	doanh_nghiệp	định_cư	hạn_mức
quốc_gia	tái_định_cư	dân_cư	kinh_doanh	thủ_tục	hệ_thống

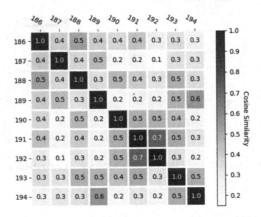

Fig. 2. Heatmap of the Cosine Similarity between Legal Articles based on the Extracted Terms.

at random for prediction. If the ith token is chosen, we replace the ith token in three ways: (1) with the [MASK] token 80% of the time, (2) with a random token 10% of the time, and (3) keep the unchanged i^{th} token 10% of the time.

To enhance LegaRBERT's ability to encode the semantics of legal terminologies, we introduce a modified Legal-Masked Fine-Tuning strategy. In this strategy, the tokens to be masked are not chosen randomly. In fact, the law terminologies constructed from the legal dictionary are given considerably higher priority to be masked due to their occurrences in the input sequence. Example 2 gives detailed information for an example of this strategy.

Example 2. We revisit the question in Example 1, which is then tokenized and masked as follows, with the law terminologies from the legal dictionary boldfaced.

Table 3. Some Typical Cases of Similar Articles.

Cases of Similar Laws	Common Content of Articles
Article 100 and 101	The content of these two articles is related to "Granting certificates of land use/ownership rights" (*Cấp giấy chứng nhận quyền sử dụng/sở hữu đất đai*).
Article 168, 186 and 188	1. Article 168: Regarding the timing of executing the rights of land users (*Nói về thời điểm được thực hiện các quyền của người sử dụng đất*). 2. Article 186: Regarding the rights and obligations of land use for Vietnamese nationals residing abroad and foreigners (*Nói về quyền và nghĩa vụ sử dụng đất của người Việt Nam định cư ở nước ngoài và người nước ngoài*). 3. Article 188: Regarding conditions for executing the right to transfer the land use rights *(Nói về điều kiện thực hiện quyền chuyển nhượng quyền sử dụng đất)*.

Tokenized form: *<s> tôi đang thực_hiện giao_dịch mua đất của ông a hiện_tại đã ký xong hợp_đồng chuyển_nhượng có công_chứng nay tôi được công_chứng_viên báo là ông a đang bị khởi_tố vì tội lừa_đảo chiếm_đoạt tài_sản và có_thể giao_dịch mua đất của tôi sẽ bị tạm dừng do tài_sản của ông a sẽ bị phong_toả xin hỏi công_chứng_viên nói như_vậy có đúng không xin cảm_ơn </s>*

Masked form: *<s> tôi đang <**mask**> giao_dịch mua đất của ông a hiện_tại đã ký xong hợp_đồng chuyển_nhượng <**mask**> công_chứng nay <**mask**> <**mask**> <**mask**> báo <**mask**> ông a đang <**mask**> khởi_tố vì tội lừa_đảo chiếm_đoạt <**mask**> <**mask**> có_thể giao_dịch mua <**mask**> <**mask**> tôi <**mask**> bị tạm <**mask**> do tài_sản của ông a sẽ bị <**mask**> <**mask**> hỏi công_chứng_viên nói như_vậy có đúng <**mask**> xin cảm_ơn </s>*

The words "prioritized" for masking are: *"thực_hiện", "có", "tôi", "được", "công_chứng_viên", "là", "bị", "tài sản", "và", "đất", "của", "sẽ", "dừng", "phong tỏa", "xin", "không".* With the masking strategy mentioned above, LegaRBERT model will learn the relationship between legal terms *"tài_sản"* (property), *"phong_tỏa"* (seizure), and the normal term *"đất"* (land). This is important because in a normal context, *"đất"* is simply a noun, but in the context of *"tài sản"* (property) *"bị"* (being) *"phong tỏa"* (seized), *"đất"* becomes a type of *"tài sản"* (property), relevant and in need of consideration. The embedding vectors of the words *"tài sản" "bị" "phong tỏa"* will be encoded with the semantic meaning of the word *"đất"* (land) (Table 4). ■

Table 4. Presents the *legal_mask_function*, which reflects our masking strategy

---

Algorithm 1. legal_mask_function

Input:
-    Input sequence $S = < s_1, s_2, ..., s_n >$
-    Legal dictionary D
-    Masking rate r

Output:
-    Masked output sequence $S' = < s_1', s_2', ..., s_n' >$

Operations:
    $t \leftarrow$ Percentage of law tokens in S
    $count = 0$
    for each $s_i \in S$
       if $s_i \in D$
          $count \mathrel{+}= 1$
    $t = count/len(S)$

    $M \leftarrow$ List of masked tokens in S
    $L \leftarrow$ List of available law tokens in S
    if $t = r$
$M \leftarrow L$
    elif $t > r$
while $t > r$
        remove random $s_i \in L$
     $M \leftarrow L$
  else
while $t < r$
    add random $s_i \notin L$ to L
$M \leftarrow L$

    for each $s_i \in S$
if $s_i \in M$
    $s'_i \leftarrow$ [MASK]
else
    $s'_i \leftarrow s_i$
      S'.append(s'_i)
    return S'

3.4 LegaR Answering Engine

With the LegaRBERT fully legal-fine-tuned, we present the answering engine of Legar, as depicted in Fig. 3. As observed, we integrate the LegaRBERT and the VN-LandLaw-2013 corpus into the Legar system, which receives users' legal questions and returns relevant articles. The main components of this system are as follows.

- Question Embedding: We use LegarBERT to embed the input question, which is used as the input for *Classification Model*.
- *Classification Model*: In LegaR, the XGBoost classifier is empirically chosen as the classification model. This model serves as multi-label classifier, which determines the articles related to the question illustrated in Example 1 and Example 2

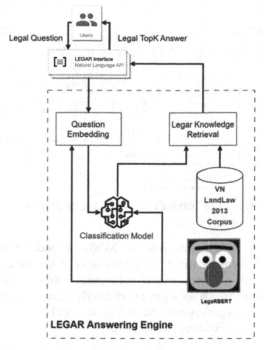

Fig. 3. The Answering Engine of the LegaR System.

- *Legar Knowledge Retrieval:* Based on the relevant articles determined the classification model, we retrieve the full context information from VN-Landlaw-2013 corpus and return it to the user via Legar UI.

4 Training Processes in Legar System

4.1 Training Process Description

In Sect. 3, we have described the functioning of Legar system in responding to user queries. In this section, we detail the process of training our XGBoost Classification Model, as depicted in Fig. 4, which comprises several essential steps as follows.

- *Question Representation:* As mentioned, the user questions, along with their corresponding labels are represented using embedding vectors inferred by LegaRBERT. These vectors result in 5213-dimensional representations for each question.
- *XGBoost Classifier Training:* Our annotated, is used to train the XGBoost classifier. This model employs a histogram-based algorithm to construct decision trees. It learns from the training data to create multiple decision trees, which together form the final ensemble model. Then the number of trees are adjusted to 5000, and the maximum depth of the trees is set to 6 in our system. The final output of this model is a 212-dimentional vector, corresponding to 212 articles in Vietnamese Land Law 2013 of which we need to carry out the multi-label classification task.

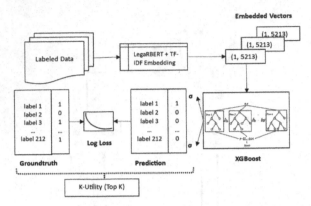

Fig. 4. The Training Process of Classification Model.

- *Evaluation Metric:* The output from the XGBoost model is evaluated using the groundtruth made from the annotated information of Vietnamese Land Law 2013. Especially, alongside of other conventional metrics in supervised learning, we introduce the *K-U* metric, which serves as a particularly meaningful performance evaluation measure for the Legar system's classification model in real scenarios. The details of this *K-U* metric are explained in Sect. 4.2.

4.2   K-Utility Metric

Definition 1 provides the formal definition of the aforementioned *K-U* metric. The *K-U* metric reflects practical information from the operational perspectives of the Legal system: among the top K articles suggested by Legar, the consultation results will be considered successful if there are at least U articles deemed relevant. According to legal experts, in this situation, the system is considered to be effective in providing advice to users as long as $U >= 1$ (i.e., at least one article is considered relevant). Algorithm 2 outlines the algorithm for *K-U* evaluation.

Definition 1 (K-U metric). *Given two values of* K *and* U, *the* K-Utility *index means that at least* U *of the* K *predicted labels must be the same as the ground truth one.* ∎

Example 3.   Table 3 records some illustrated results regarding precision, recall and *K-U* metrics (Table 5). ∎

Table 5. K-U Results from Testing Samples.

Question	Ground Truth	LSI-Prediction(TOP)	Precision	Recall	K-U (5, 1)	K-U (5, 2)
Năm 2005, tôi có ký kết hợp đồng chuyển nhượng quyền sử dụng đất với anh A thì nên lưu ý các điểm gì để tránh bị toàn án vô hiệu khi có tranh chấp?	106, 127, 146, 167, 188	**188**, 168, 3, 186, 100	1/5	1/5	Yes	No
Năm 2013, vợ chồng tôi có mua 1 mảnh đất tại Hương Sơn, Hà Tĩnh thì nếu con trai chủ đất sau khi giao dịch mà không đưa số thì chúng tôi có thể làm gì? Xin cảm ơn!	95, 168, 186, 188, 189, 190, 191, 192, 193, 194	**188**, 168, **186**, 100, 3	2/5	2/10	Yes	Yes
Tôi đang là chủ sử dụng một thửa đất nuôi, trồng thủy sản được cấp số đỏ vào năm 2003. Tôi có thể chuyển thửa đất này sang đất ở thì được không và làm như thế nào?	10, 52, 57, 58, 59	**57**, 100, **10**, 6, 3	2/5	2/5	Yes	Yes
Nhà đất tôi mua giấy viết tay từ năm 2008, đến nay 2021 vẫn chưa được cấp số đi lần đầu thì khi nhà nước quyết định thu hồi thì tôi có được nhà nước bồi thường và nếu có thì ai sẽ là người được hưởng?	64, 65, 75, 77, 100	**75**, **77**, **100**, 101, 3	3/5	3/5	Yes	Yes

---

Algorithm 2. k_utility_function

Input:
- Classification results from the model (probability matrix of labels) *Ypred*
- True label of the sample (multi-hot matrix representing labels) *Ytrue*
- The highest number of results to take *K*
- The minimum length (number) of results that must match the label *U*

Ouput:
- *KUtilityValue* for *K* and *U* (Portion of *Yes* for *True Predictions* over *Total Predictions*)

Operationns:

results ← Boolean list to represent the *Yes/No* status when checking relation of each label

for y in Ypred:

   TopK ← Set of the position of the top *K* with the highest probability from *y*
   target ← Sets of the position of the values to 1 based on *Ytrue* (ground truth)
   TruePredictions ← List of intersection results of *TopK* and *target*
   G ← Length of *target*
      results ← Append *Yes* if length of *TruePredictions* >= *min(G, U)* else append *No*
return *KUtilityValue (Number of Yes / Total results number)*

5   Experimental Results

5.1   Baselines

Based on related works, we gather various suitable techniques to construct comparable experiment baselines to evaluate with our proposed LegaRBERT model. The baselines are represented as follow:

- **Baseline SVM-TF-IDF:** Embedded data using **TF-IDF** and train through **Multi-Ouput SVM Classifier (LinearSVC)** for predicting the relation of each label articles to each input.
- **Baseline FC-PhoBERT:** Embedded data using PhoBERT and train through a Fullly Connected Layer with Sigmoid Activation Function and BCE (Binary Cross Entropy) Loss for predicting multilabel articles.
- **Baseline FC-LegaRBERT:** Embedded data using LegaRBERT and train through a Fullly Connected Layer with Sigmoid Activation Function and BCE (Binary Cross Entropy) Loss for predicting multilabel articles.
- **Baseline XGBoost-TF-IDF-LegaRBERT:** Concating two separate embedded data, one using **TF-IDF** and the other using **LegaRBERT** into final embedded data to train through a **XGBoost** for predicting multilabel articles.

5.2 Experiments and Ablation Study

Table 6. Conventional Metrics.

Baseline Methods	Macro Average Precision	Macro Average Recall	Macro Average F1-Score	Weighted Average Precision	Weighted Average Recall	Weighted Average F1-Score
SVM-TF-IDF	0.45	0.27	0.32	0.51	0.33	**0.39**
FC-PhoBERT	0.20	0.27	0.21	0.20	0.47	0.27
FC-LegaRBERT	0.26	0.34	0.27	0.24	0.54	0.32
XGBoost-TF-IDF-LegaRBERT	0.47	0.14	0.20	**0.83**	0.21	0.31

From Table 6, we can evaluate that the highest **Weighted Average F1-Score** is **0.39** (**SVM-TF-IDF**) and the highest **Weighted Average Precision** is **0.83** (**XGBoost-TF-IDF-LegaRBERT**). We have an imbalanced class distribution in the dataset, where some classes have a significantly larger number of instances (questions) compared to others. Since our models in the four proposed experiments are trained in a supervised manner, classes with more instances tend to have a greater influence on the model evaluation. To address this, we use the weighted average metric as the main evaluation metric for our models in the experiments. In the context of law information retrieval related to input questions in the LSI problem, we aim to demonstrate to customers that our LegaR system can minimize false predictions. Therefore, we prioritize the precision value over the recall value in our evaluation. Overall, for this paper, we focus on the weighted precision score as the primary metric. The weighted precision scores of our models in **Experiment 1 (SVM-TF-IDF)** and **Experiment 4 (XGBoost-TF-IDF-LegaRBERT)** are 0.51 (the lowest score) and 0.83 (the highest score), respectively. The model in Experiment 1 performs moderately in identifying relevant articles, while the model in Experiment 4 shows a substantial improvement in precision compared to the model in Experiment 1, thanks to the inclusion of **LegaRBERT embeddings** and the use of the **XGBoost classifier**. Even though **XGBoost-TF-IDF-LegaRBERT** do not achieve the best Weighted Average F1-Score, its highest Weighted Average Precision propose that in every return prediction results from the model, it most likely will contain the correct label articles that the input related to. However, to address this conclusion more specific, we must use our K-U metric:

Table 7. K-U Metrics (K = 5).

Baseline Methods	U = 1	U = 2	U = 3	U = 4	U = 5
FC-PhoBERT	0.66	0.52	0.45	0.43	0.42
FC-LegaRBERT	0.73	0.59	0.53	0.50	0.49
XGBoost-TF-IDF-LegaRBERT	**0.77**	**0.64**	**0.58**	**0.55**	**0.54**

With our K-U metric that require picking **TopK** from prediction results, only the three methods from Table 7 are applicable. Looking at the results of **XGBoost-TF-IDF-LegaRBERT**, expert in legal domain can immediately guarantee 77% of the prediction results have at least one valuable label articles for exploring more into clauses and points while other labels can be considered as additional information for reference. This result is very practical for legal experts since we can scope down the searching range for them and by looking at the 5 options which will most likely has the option they need, they can easily to pick it up fast compares to checking every option in every article of a law type.

6 Conclusion

People frequently focus on specific articles on people's lives and activities (e.g., the issuance of land use right licenses and home ownership are regulated by Article 100 of the 2013 Land Law, which is the subject of numerous questions in the LSI dataset). In contrast, various executive-specific laws for the state have received little attention (e.g., Article 211. Enforceability implementation, Article 212. Detailed implementation rules have no data). Therefore, the LSI dataset compiled from frequently asked legal queries will have to be a long-tail phenomenon as a matter of course. Extreme Multi-Label Learning (XML) problems with long-tail data present a double challenge. Nonetheless, the experimental results confirmed our initial hypothesis that it is necessary to overcome the above obstacles with a delicately constructed model capable of comprehending Vietnamese and Vietnamese law. By the breakout model LegaRBERT, we increased Precision from 0.51 (Machine Learning Approach) to 0.83 (Bert-based Approach).

However, the low recall indicated that there are still many obstacles to overcome. To surmount this challenge, it is ideally necessary to embed "Legal Inference" into the models; that is, this model must be integrated by "Vietnamese contextual embedding," "Legal contextual embedding," and "Legal reasoning Embedding." LegaRBERT only incorporates the first two embeddings and lacks the third embedding, so the training outcomes highly depend on the data. The model could perform more efficiently with the Long-tail dataset's tail. Different from computer experts' view of a legal document as just a linguistic document, when a lawyer reads a legal document, the lawyer first pays attention to key phrases (e.g., Legal terminology), extracting legal entities, legal relations, the object of impact, and the time of the legal event, thereby identifying which laws and provisions can be used to process a legal document. A heterogeneous graph built by legal entity nodes (e.g., person, legal entity, organization) and legal relationship nodes by extracting verbs (Example buy_sell: legal relationship commercial agent) is a great solution to pursue. Therefore, we will continue to perform legal entity extraction, extract legal relations from the existing legal corpus, and then construct a heterogeneous legal graph, improving the results.

References

1. Paul, S.: LeSICiN: a heterogeneous graph-based approach for automatic legal statute identification from Indian legal documents. In: Proceedings of the AAAI Conference on Artificial Intelligence, vol. 36, pp 11139–11146 (2022)
2. Zhu, G.: Design of knowledge graph retrieval system for legal and regulatory framework of multilevel latent semantic indexing. Comput. Intell. Neurosci. **2022**, 6781043 (2022)
3. Harris-Hellal, J.: Special Metabolites Isolated from the leaves and Roots of Brachiaria humidicola (2010)
4. Kort, F.: Predicting Supreme Court decisions mathematically: a quantitative analysis of the "right to counsel" cases. Am. Polit. Sci. Rev. **51**, 1–12 (1957)
5. Ulmer, S.S.: Quantitative analysis of judicial processes: some practical and theoretical applications. Law Contemp. Prob. **28**, 164–184 (1963)
6. Segal, J.A.: Predicting Supreme Court cases probabilistically: the search and seizure cases, 1962–1981. Am. Polit. Sci. Rev. **78**, 891–900 (1984)
7. Liu, C.-L., Hsieh, C.-D.: Exploring phrase-based classification of judicial documents for criminal charges in Chinese. In: Esposito, F., Raś, Z.W., Malerba, D., Semeraro, G. (eds.) ISMIS 2006. LNCS (LNAI), vol. 4203, pp. 681–690. Springer, Heidelberg (2006). https://doi.org/10.1007/11875604_75
8. Aletras, N.: Predicting judicial decisions of the European Court of Human Rights: a natural language processing perspective. PeerJ Comput. Sci. **2**, e93 (2016)
9. Luo, B.: Learning to predict charges for criminal cases with legal basis. In: Proceedings of the 2017 Conference on Empirical Methods in Natural Language Processing, pp 2727–2736 (2017)
10. Wang, P.: Modeling dynamic pairwise attention for crime classification over legal articles. In: The 41st International ACM SIGIR Conference on Research and Development in Information Retrieval, pp 485–494 (2018)
11. Wang, P.: Hierarchical matching network for crime classification. In: Proceedings of the 42nd International ACM SIGIR Conference on Research and Development in Information Retrieval, pp 325–334 (2019)
12. Chalkidis, I.: Neural legal judgment prediction in English. In: Annual Meeting of the Association for Computational Linguistics (2019)
13. Vold, A.: Using transformers to improve answer retrieval for legal questions. In: Proceedings of the Eighteenth International Conference on Artificial Intelligence and Law, pp 245–249 (2021)
14. Xiao, C.: Lawformer: a pre-trained language model for Chinese legal long documents. AI Open **2**, 79–84 (2021)
15. Xu, N.: Distinguish confusing law articles for legal judgment prediction. In: Proceedings of the 58th Annual Meeting of the Association for Computational Linguistics, pp. 3086–3095 (2020)

16. Liu, Y.: RoBERTa: a robustly optimized BERT pretraining approach. In: Proceedings of the 20th Chinese National Conference on Computational Linguistics, pp. 1218–1227 (2019)
17. Devlin, J.: BERT: pre-training of Deep Bidirectional Transformers for Language Understanding. In: NAACL -HLT(2), vol. 2 (2019)
18. Brownlee, J.: XGBoost With Python: Gradient Boosted Trees with XGBoost and scikit-learn, 2nd edn. Machine Learning Mastery. Google Books (2016)
19. Nguyen, D.Q.: PhoBERT: Pre-trained language models for Vietnamese. arXiv arXiv:2003.00744 (2020)
20. Từ điển Pháp luật Việt Nam. Nhà Xuất bản Thế Giới. https://www.sachluat.com.vn/sach-luat-d0n/tu-dien-phap-luat-viet-nam-luat-gia-nguyen-ngoc-diep/. Accessed 20 Jul 2023
21. Cheng, H.: A text similarity measurement combining word semantic information with TF-IDF method. Chin. J. Comput. **34**, 856–864 (2011)
22. Kingma, D. P., Ba, J.: Adam: a method for stochastic optimization. In ICLR. OpenReview.net (2015)
23. Brownlee, J.: XGBoost With python: Gradient boosted trees with XGBoost and scikit-learn. Machine Learning Mastery (2016)
24. Yao, L., Mao, C., Luo, Y.: Graph convolutional networks for text classification. In: Proceedings of the AAAI Conference on Artificial Intelligence, July 2019, vol. 33, no. 01, pp. 7370–7377 (2019)

An Investigation of a Touch-Based Eye-Tracking System with Smart Feedback and Its Influences on Learning - Simulation of Covid-19 Rapid Test System

Wu-Yuin Hwang[1]([⊠]), Tien-Cheng Wang[1], and Thao Pham[1,2]

[1] National Central University, Taoyuan, Taiwan
wyhwang@cc.ncu.edu.tw
[2] National Economics University, Hanoi, Vietnam

Abstract. This study investigates the impact of a touch-based simulation system (TBSS) with smart feedback on learners' eye movements, behaviors, and performance in Covid-19 rapid testing. The participants included sixty graduate students divided into experimental (TBSS with smart feedback) and control (TBSS with basic feedback) groups. Results revealed that smart feedback improved learning outcomes, leading to higher performance and more efficient task completion compared to the control group. Participants showed positive attitudes toward the usefulness and ease of use of the Covid-19 simulation system, expressing high satisfaction with its functionality. The smart feedback mechanism was found to facilitate efficient decision-making by guiding attention and providing immediate, detailed feedback. This research provides valuable insights into the influence of feedback mechanisms on user behavior and performance in simulation systems, particularly in the context of Covid-19 training and testing procedures.

Keywords: Eye-tracking · Touch-based learning · Smart feedback mechanism · Hand-eye coordination

1 Introduction

Hand-eye coordination plays a pivotal role in performance, as demonstrated by Desmurget et al. (1998), who found that the eyes typically lead the hand in goal-directed movements. To measure the interaction between eye-tracking and touch-based manipulation, Paek et al. (2013) explored the potential of touch-input to improve student learning outcomes in digital learning environments designed to take advantage of such interactions, and the results show that when using the touch-input method, students took less time to complete the tests while making fewer attempts to answer questions.

In intensive human-computer iteration activities, feedback mechanisms have long been acknowledged as powerful tools for improving learning performance (Wu et al. 2012). Immediate and personalized feedback helps learners reflect on their performance, identify weaknesses, and adjust their strategies accordingly (Shute 2008). By incorporating immediate and personalized feedback mechanisms, we can enhance the learning experience and promote skill development in learners.

According to Mayer and Moreno (1998) Cognitive Theory of Multimedia Learning (CTML), deeper learning can be achieved by presenting information in text and graphics rather than relying solely on text. Furthermore, Hwang et al. (2023) expanded upon CTML by introducing the kinesthetic channel, which involves learning through physical movements and manipulation. This concept aligns perfectly with our investigation into touch-based manipulation in the simulation learning system.

In the recent context of the Covid-19 pandemic, which has highlighted the urgent need for effective education and training in testing procedures. This study aimed to explore the potential of integrating hand-eye coordination with smart feedback mechanisms to enhance learning outcomes.

Based on the aforementioned above, we proposed a touched-based simulation system with a smart feedback mechanism to help learners facilitate the procedure of Covid-19 rapid testing. The research questions of this study were shown below:

1) What is the difference in the Learning Performance between two groups with smart feedback mechanism?
2) What is the difference between the visual attention heat map between two groups?
3) What is the difference in the interaction between the gaze plot and touch plot from two groups?

2 Literature Review

2.1 Eye-Tracking in Learning with Simulation System

Eye-tracking technology has emerged as a crucial instrument in the understanding of learners' cognitive processes, especially within the domain of simulation-based learning. This technology provides a unique window into cognitive processes by capturing and analyzing eye movement data, thereby giving researchers access to learners' visual attention, information processing, and engagement levels (Duchowski 2017; Holmqvist et al. 2011).

Furthermore, in the domain of skill acquisition, eye-tracking data has been used to compare experts' and novices' visual search strategies, thereby informing the design of instructional materials (Jarodzka et al., 2010).

2.2 Eye-Tracking with Touch-Based Manipulation for Procedure Knowledge

Touch-based manipulation has been found to facilitate intuitive interactions and enhance the learning experience (Johansson et al. 2001). It is a critical aspect of simulation learning, particularly for procedures that require precise hand-eye coordination. In contrast, Lee (2015) found that students in the touch-based interaction condition significantly outperformed those in the mouse-based condition. With the emergence of advanced touch-based devices, there is a plausible opportunity to integrate touch-based manipulation with eye-tracking in simulation systems to achieve improved learning outcomes.

Unlike traditional mouse-based interactions, touch-based manipulation provides a more intuitive and immersive user experience. When learners directly interact with the system through touch, the physical connection between their gestures and the on-screen actions enhances engagement and motor-skill learning (Lohse et al. 2016). By utilizing touch-based interactions, learners can interact directly with the content, eliminating the distractions associated with cursor movements and enabling more focused visual attention on relevant elements of the simulation system.

Eye-tracking in conjunction with touch-based manipulation has been emerging as a promising approach for teaching procedural knowledge, such as that required for Covid-19 testing. By monitoring eye movement and touch interactions concurrently, educators can comprehensively understand learners' behavior and cognitive processes (Johansson et al. 2001). Recent studies have begun to explore the potential of such multimodal learning environments in healthcare education, revealing that they can support the acquisition of complex procedural skills (Bogossian et al. 2015).

2.3 The Effect of Smart Feedback for Eye-Tracking with Touch-Based Learning

Feedback mechanisms that provide learners with tailored and informative responses based on their performance. By leveraging eye-tracking data, the system can deliver feedback specific to learners' attention and gaze patterns (Conati et al. 2013). This personalized feedback enhances learners' self-awareness, facilitates error correction, and promotes a deeper understanding of the subject matter (Graesser et al., 2004).

The role of smart feedback mechanisms in learning contexts has been well-documented. Shute (2008) contends that immediate, personalized feedback helps learners reflect on their performance, identify their weaknesses, and adjust their strategies.

In conclusion, integrating eye-tracking technology with simulation systems, particularly in touch-based learning, holds great promise for enhancing the acquisition of procedural knowledge related to Covid-19 testing. The combination of touch-based manipulation and eye-tracking enables intuitive interactions and reduces distractions, while smart feedback mechanisms tailored to learners' gaze behavior promote self-reflection and deeper understanding. Educators and researchers can effectively leverage these technologies to improve learning outcomes, engagement, and long-term retention.

3 System Design

In order to facilitate learners' skills in the Covid-19 rapid test, we designed and developed a Covid-19 simulation system. This system supports hand operation, and the user operates on a 21.5-inch touchscreen. The simulation offers a variety of test types, including multiple-choice questions, operating questions, and match questions, based on the Covid-19 rapid test process. Then, we use the open-source software Gaze Pointer to record the user's eye movement. The gaze pointer records the user's eye movement through the webcam above the screen.

3.1 Touch-Based Covid-19 Simulation System

The system is visually divided into three distinct areas, as shown in Fig. 1. Firstly, the simulation area is enclosed within the red rectangle, where users can engage with Covid-19 rapid test simulations, respond to multiple-choice and match questions, and access guidance. This area also provides real-time information, such as the total operation time, question descriptions, and tools for rapid testing of Covid-19. Secondly, the feedback area is located inside the yellow rectangle. If the experimental group (EG) answers a question correctly, the system will display a "good job" message in the feedback area, reinforcing the correct response and enhancing the user's understanding. Conversely, if the EG answers incorrectly, the system will offer detailed explanations based on the user's mistakes, encouraging reflection, and guiding them toward the correct answer. However, if the control group (CG) answers incorrectly, they will receive basic feedback indicating whether their response was right or wrong. Lastly, the procedure area is within the green rectangle, representing the step-by-step progression of the simulation process. There are a total of six buttons in this area. When a user clicks Guidance, the system will show the information about Covid-19 rapid test. Users can view the guidance at any time during the experiment. The remaining buttons correspond to steps 1 to step 5, respectively. After completing the current step, the user can go to the next step.

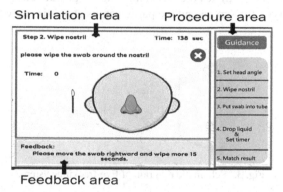

Fig. 1. Interface of Covid-19 simulation system. In step2, the learner needs to move the cotton swab into the nostril and wiping around for 15 s. (Color figure online)

Step 1 presents multiple-choice questions for participants. Learner needs to choose which options are correct. This question was designed to assess participants' understanding of the required head tilt angle for the Covid-19 rapid test. Steps 2 and 3 are hands-on questions; the learner must use his fingertip to move the cotton swab into the right position. Step 4 is also a multiple-choice question. Learners are required to determine the number of drops they need to place on the test strip. Last, Step 5 of the simulation involves a matching question where participants need to match pictures with corresponding words (Fig. 2).

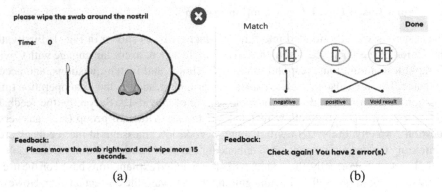

Fig. 2. (a) Step 2: Wipe nostril; (b) Step 5: Match the result.

3.2 Eye-Tracking System

We utilize the Gaze Pointer, a free and open-source software with no time limit to track eye movements. One advantage of this software is its compatibility with participants wearing glasses, its high accuracy rate was a factor in our selection. The eye-tracking system used in our setup utilizes the webcam, which is positioned on top of the screen. The distance between the screen and the participant is approximately 50 cm (Fig. 3).

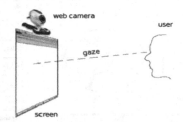

Fig. 3. Schematic diagram of Eye-tracking system.

Initially, a dot appears on the screen, and participants must focus their gaze on it. Once the fixation is achieved, the dot moves to the next position. This process is repeated for a total of sixteen calibration points. Next, arrows appear on the screen, and participants are instructed to rotate their heads according to the direction indicated by the arrows. This calibration method ensures that head movements during the experiment do not affect the accuracy of eye-tracking data. Upon completion of the calibration steps mentioned above, a small gaze pointer appears on the screen, representing the location where the participant is currently fixating. If the accuracy is confirmed, the experiment can proceed. The calibration process ensures accurate and reliable eye-tracking data by aligning the system with the participant's eye movements and compensating for head movements.

4 Methodology

In this study, sixty learners from the National Central University participated in the experiment. They were divided into two distinct groups. One was the experimental group (n = 30, male = 16, female = 14) and the other was the control group (n = 30, male = 18, female = 12). EG learners used a system with a smart feedback mechanism. The system will inform the learner whether he answered the question correctly or incorrectly and provide a detailed explanation. The way of smart feedback is text and voice. Otherwise, CG learners used the system with a basic feedback mechanism. The system will only inform the learner whether he answered the question incorrectly and will not provide a detailed explanation. The way of feedback is text only.

The experimental procedure shown on Fig. 4, includes five stages: pre-test, calibration for Eye-tracking system, operating simulation system, post-test, and questionnaire. The experiment was about 30 min.

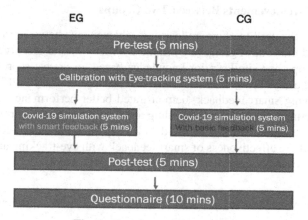

Fig. 4. Experimental Procedure.

We collected the learning behaviors when learners used the Covid-19 simulation system. The variables are listed in Table 1. This learning achievement of this research is the learner's post-test score.

In this study, we used several analysis mechanisms, such as ANOVA analysis was used to investigate the difference between EG and CG in the pretest, ANCOVA analysis was used to examine the difference between EG and CG in the post-test, Independent T-test analysis to compare the means of both groups, and Lag Sequential Analysis was used to analyze whether there is a significant transition in the learner's eye movement and hand manipulation.

Table 1. The research variables.

Variables	Description
(1) watch guidance number	The number of times the learner watched the guidance in the experiment
(2) watch guidance time	The amount of time for the learner to watch the guidance in the experiment
(3) total manipulation time	The amount of time for the learner to complete all steps
(4) step [1, 2, 3, 4, 5] time	The amount of time for the learner to complete step [1, 2, 3, 4, 5]
(5) step [1, 2, 3, 4, 5] error	The number of errors made by the learner in step [1, 2, 3, 4, 5]

5 Results and Discussions

5.1 Learning Achievements Between Two Groups

The ANCOVA was used to compare learning achievements between the experimental group (EG) and the control group (CG) in the COVID-19 rapid test procedural knowledge. Both groups had similar prior knowledge based on pretest scores. The posttest results revealed a significant difference between the EG and CG (F = 15.050, p < .05). The EG, receiving smart feedback, demonstrated better performance due to external stimuli that facilitated deeper understanding and reinforcement of correct behavior. In contrast, the CG's basic feedback lacked detailed explanations for self-reflection. These findings support the effectiveness of smart feedback with eye-tracking and touch-based learning in improving learners' procedural knowledge (Table 2).

Table 2. The ANCOVA analysis results of the learning achievement between two groups.

Group	N	Pre-test		ANOVA	Post-test		ANCOVA
		Mean	SD	F	Mean	SD	F
EG	30	43.67	11.290	1.335	93.33	12.130	15.050***
CG	30	43.00	13.933		82.33	14.782	

5.2 Comparison of Learning Behaviors between Groups

The section investigates learning behavior differences between the two groups. The independent sample t-test analysis revealed significant differences in various aspects. The CG referred to guidance more frequently due to the lack of detailed error explanations in their basic feedback (t = −3.957, p < .01). The EG spent less time and made fewer errors in complex steps, thanks to the smart feedback assisting with swab manipulation (step2_time: $t = -2.329, p < .05$; step2_errors: $t = -2.141, p < .05$) and matching the result (step5_time: $t = -2.090, p < .05$; step5_errors: t = −2.068, p < .05). The total manipulation time of the EG was significantly shorter due to their prompt error correction with smart feedback ($t = -3.587$, $p < .01$). However, no significant differences were found in simpler steps, as both groups were able to complete them easily, irrespective of the feedback system. Overall, integrating smart feedback in eye-tracking with touch-based learning significantly impacted learners' behavior, enhancing task efficiency and accuracy (Table 3).

Table 3. The independent T-test analysis results of the learning behavior between two groups.

Variables	Group	N	Mean	SD	t
watch guidance number	EG	30	1.03	0.183	3.957**
	CG	30	1.47	0.571	
watch guidance time	EG	30	76.13	35.456	0.214
	CG	30	74.23	33.231	
step 1_time	EG	30	9.30	5.200	−1.872
	CG	30	12.67	8.368	
step 2_time	EG	30	28.90	5.241	−2.329*
	CG	30	32.47	6.548	
step 3_time	EG	30	26.33	3.960	−1.579
	CG	30	28.27	5.413	
step 4_time	EG	30	22.40	6.971	-1.031
	CG	30	24.97	3.891	
step 5_time	EG	30	25.50	9.402	−2.090*
	CG	30	36.47	26.485	
step 1_error	EG	30	0.37	0.490	−1.034
	CG	30	0.5	0.509	
step 2_error	EG	30	0.47	0.681	−2.141*
	CG	30	0.93	0.980	
step 3_error	EG	30	0.20	0.407	0.328
	CG	30	0.17	0.379	
step 4_error	EG	30	0.03	0.183	−0.584
	CG	30	0.07	0.254	
step 5_error	EG	30	0.33	0.711	−2.068*
	CG	30	0.97	1.520	
total manipulation time	EG	30	113.43	19.967	3.587**
	CG	30	138.50	32.652	

Note. * p < .05, ** p < .01,

5.3 Comparison Visual Attention Heap Map Between Two Groups for Step 2 and 5

The screen was divided into a 5 × 6 grid, each section labeled with a unique code from A1 to E6. Eye movement data was converted to these coded areas. We designated the rectangular area from A1 to D5 as the simulation area, A6 to D6 as the feedback area, and E1 to E6 as the procedure and guidance button area.

In Step 2, CG focused on the timer areas (A3 & A2) due to their reliance on basic feedback for correct swab positioning and timing. They also paid attention to the nostril areas (B4 & C4) to verify swab insertion. However, they showed minimal engagement with the feedback areas (B6 & C6). In contrast, EG frequently checked the smart feedback area (B6) for crucial swab positioning and timing information. They also verified swab insertion in the nostril areas (B4 & C4) but relied less on the timer (A2 & A3) due to the comprehensive feedback available (Fig. 5).

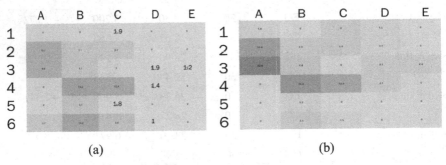

(a)   (b)

Fig. 5. Visual Attention in Step 2. (a) EG; (b) CG.

In Step 5, CG and EG focused on the pictorial representations and options (B3, B4, C4, and C3) for matching questions. CG paid little attention to the feedback areas (B6 & C6), as basic feedback only showed correctness upon completing all connections. However, EG dedicated significant attention to the feedback areas, frequently referring to smart feedback for immediate validation of each connection (Fig. 6).

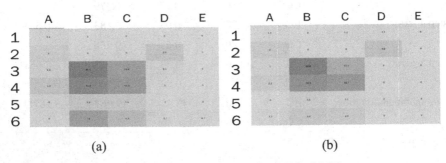

(a)   (b)

Fig. 6. Visual Attention in Step 5. (a) EG; (b) CG.

Overall, EG's attention to the smart feedback areas, reduced reliance on timers in Step 2, and engagement with feedback areas in Step 5, indicates the value and impact of smart feedback on their learning behavior. Conversely, CG's focus on timers and limited engagement with feedback areas highlight the limitations of basic feedback in guiding their actions effectively.

5.4 The Interaction of Eye Movement and Manipulation

The study combines eye movement and manipulation data in this section and uses lag sequential analysis to identify significant transitions. The EG and CG both demonstrated similar attention patterns in hand-eye coordination during the interaction with the system. However, there were distinct differences in gaze behavior regarding the feedback areas. The EG frequently checked the smart feedback, leading to more efficient task completion and fewer errors compared to the CG, which relied on basic feedback provided only after completing all connections.

The interaction process for both groups involve analytical observation, active response, and task completion. Users first visually focus on the question, using cognitive processing to formulate solutions. They then physically interact with the interface to select an answer. For the EG, smart feedback provides immediate validation or error-specific explanations, influencing their subsequent approach and promoting critical thinking and self-correction. This iterative process enhances the EG's learning outcomes and fosters deeper comprehension of the subject matter (Fig. 7).

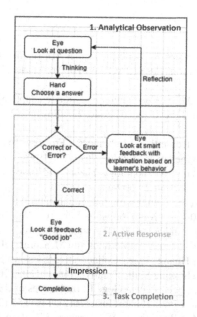

Fig. 7. The interaction between eye and hand with smart feedback

The study highlights the significance of smart feedback in influencing learners' eye gaze and behaviors, leading to differences in learning achievement. Integrating smart feedback in eye-tracking with touch-based learning enhances task efficiency, accuracy, and engagement, contributing to more effective learning outcomes.

6 Conclusion

Integrating touch-based eye-tracking and smart feedback into the Covid-19 Simulation System significantly impacts learners' experience and performance. The experimental group (EG) outperformed the control group (CG) in learning performance, with fewer errors and faster completion times. Smart feedback influenced the correlation between learning behavior and achievement, promoting efficient task completion and active engagement. The heat map analysis revealed differences in focus areas between the two groups, with the EG showing increased engagement with the feedback area, benefiting from immediate, detailed information. Both groups displayed similar hand-eye coordination patterns, but the EG interacted more dynamically with the feedback feature. Smart feedback facilitated an iterative learning process, enhancing comprehension and decision-making. Integrating hand-eye coordination and smart feedback creates a more immersive and interactive learning environment, fostering deeper engagement and better learning outcomes. This research provides valuable insights into the influence of feedback mechanisms on user behavior and performance, highlighting the significance of incorporating eye-tracking and smart feedback in learning systems. Further research is needed to optimize these technologies for educational purposes.

References

Bogossian, F.E., Cooper, S.J., Cant, R., Porter, J., Forbes, H., Team, F.A.R.: A trial of e-simulation of sudden patient deterioration (FIRST2ACT WEB™) on student learning. Nurse Educ. Today **35**(10), e36–e42 (2015)

Conati, C., Jaques, N., Muir, M.: Understanding attention to adaptive hints in educational games: an eye-tracking study. Int. J. Artif. Intell. Educ. **23**, 136–161 (2013). https://doi.org/10.1007/s40593-013-0002-8

Desmurget, M., Pélisson, D., Rossetti, Y., Prablanc, C.: From eye to hand: planning goal-directed movements. Neurosci. Biobehav. Rev. **22**(6), 761–788 (1998)

Duchowski, T.A.: Eye Tracking: Methodology Theory and Practice. Springer, London (2017). https://doi.org/10.1007/978-1-84628-609-4

Graesser, A.C., et al.: AutoTutor: a tutor with dialogue in natural language. Behav. Res. Methods Instrum. Comput. **36**, 180–192 (2004). https://doi.org/10.3758/BF03195563

Holmqvist, K., Nyström, M., Andersson, R., Dewhurst, R., Jarodzka, H., Van de Weijer, J.: Eye Tracking: A Comprehensive Guide to Methods and Measures. OUP Oxford, Oxford (2011)

Hwang, W.-Y., Manabe, K., Huang, T.-H.: Collaborative guessing game for EFL learning with kinesthetic recognition. Thinking Skills Creativity **48**, 101297 (2023)

Jarodzka, H., Scheiter, K., Gerjets, P., Van Gog, T.: In the eyes of the beholder: how experts and novices interpret dynamic stimuli. Learn. Instr. **20**(2), 146–154 (2010)

Johansson, R.S., Westling, G., Bäckström, A., Flanagan, J.R.: Eye–hand coordination in object manipulation. J. Neurosci. **21**(17), 6917–6932 (2001)

Lee, H.W.: Does touch-based interaction in learning with interactive images improve students' learning? Asia-Pacific Educ. Res. **24**, 731–735 (2015). https://doi.org/10.1007/s40299-014-0197-y

Lohse, K.R., Boyd, L.A., Hodges, N.J.: Engaging environments enhance motor skill learning in a computer gaming task. J. Motor Behav. **48**(2), 172–182 (2016). https://doi.org/10.1080/002 22895.2015.1068158

Mayer, R.E., Moreno, R.: A cognitive theory of multimedia learning: implications for design principles. J. Educ. Psychol. **91**(2), 358–368 (1998)

Paek, S., Hoffman, D.L., Black, J.B.: Using touch-based input to promote student math learning in a multimedia learning environment. EdMedia+ Innovate Learning (2013)

Shute, V.J.: Focus on formative feedback. Rev. Educ. Res. **78**(1), 153–189 (2008)

Wu, P.H., Hwang, G.J., Milrad, M., Ke, H.R., Huang, Y.M.: An innovative concept map approach for improving students' learning performance with an instant feedback mechanism. Br. J. Edu. Technol. **43**(2), 217–232 (2012)

Improve Medicine Prescribing Performance Using Recommendation Systems

Thanh Nhan Dinh[1], Ba Duy Nguyen[1], Xuan Dung Vu[2], and Quoc Dinh Truong[3]([✉])

[1] Can Tho University of Technology, Cantho, Vietnam
{dtnhan,nbduy}@ctuet.edu.vn
[2] Vietnam Posts and Telecommunications Group, Camau, Vietnam
[3] Can Tho University, Cantho, Vietnam
tqdinh@cit.ctu.edu.vn

Abstract. In this study, we introduce a decision support system for prescribing medication based on the application of association rules and analysis of prescription history data. Additionally, the research proposes techniques for analyzing prescription data, constructing a storage database, selecting rule sets for information extraction, data visualization, and improving the recommendation system's effectiveness. The experimental results on a real data set, which includes 76,028 health insurance prescriptions, 363,697 data lines, 71 prescribing doctors, 521 medications, and 1,129 disease codes according to the ICD (International Statistical Classification of Diseases and Related Health Problems). Experimental results show that the Apriori algorithm gives good results with high accuracy, improves the efficiency and accuracy of drug prescriptions, support for doctors in making more accurate prescribing decisions and reducing patient risk.

Keywords: Prescription · Clinical decision support systems · Association rule · Apriori algorithm

1 Introduction

The Fourth Industrial Revolution has been creating many drastic changes in the health sectors. The application of information technology advancements can help improve the quality of medical operations. Currently, the number of drug names and brand names is very diverse. In which many drug names are easily confused with each other such as Celebrex and Cerebyx, Losec and Loxen, Levonor and Lovenox,... and many drugs interact, contraindicated with each other. Therefore, if you are careless, the doctor can prescribe the wrong medicine, leading to serious consequences. Based on this reality, this study proposes the development of an electronic prescription support system. The system is built on the combination of analysis of prescription history data and the Apriori algorithm.

There have been many researches on decision support systems serving the health sector in the country and around the world. Reis WC and colleagues [5, 6] have assessed the impact of CDSS applied to medication usage in healthcare for patients by conducting a systematic review of relevant evaluations using PRISMA and Cochran methods. The figures showed positive results in improving the quality of physician prescribing (14/30–46.6%) and reducing prescription errors (5/30–16.6%).

Komal Kumar and colleagues [4] applied machine learning algorithms to generate recommendations about medication use. With a sample data set created solely for testing purposes and not sourced from any specific origin, This experimental evaluation shows that the Random Forest classifier achieves a good recommended accuracy of 96.87% compared to the other classification methods such as Support Vector Machine (SVM), Random Forest, Decision Tree, and K-nearest neighbors.

Tong and colleagues [7] have presented the Apriori algorithm to prevent pharmaceutical conflicts in clinics. However, the system is slow due to the mechanism of the classical Apriori algorithm. To solve this problem, the authors improved the Apriori algorithm based on Linkedlist.

Different from the studies mentioned above, in this paper, we propose to build a prescription recommendation system that solves two main problems. The first problem, selecting an attribute based on the influence of that attribute on the recommender system. The second problem, limiting the omission of drugs with low prescribing frequency and using conflicting drugs in the recommendation system.

2 Proposed System

In this study, we propose a system model with two subsystems. The first subsystem is used to collect, classify, extract, and store the necessary data for the system. The second subsystem utilizes the Apriori algorithm to mine the data and generate rule sets for the recommendation system. The detailed description of the system model is illustrated in Fig. 1.

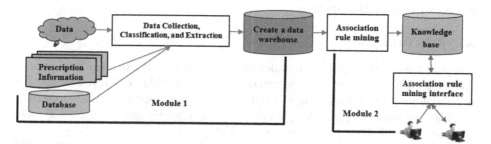

Fig. 1. System Model for the Solution

Module 1: Data Collection, Classification, Extraction, and Preprocessing. With a complex and multi-field data structure, we performed preprocessing and analyzed the value of the data attributes. That, support for identify important attributes, enabling us to select and focus on the attributes that have a significant impact on the recommender systems and eliminating irrelevant attributes. The raw data in CSV format is outpatient prescriptions extracted from VNPT HIS that has the structure described in Fig. 2.

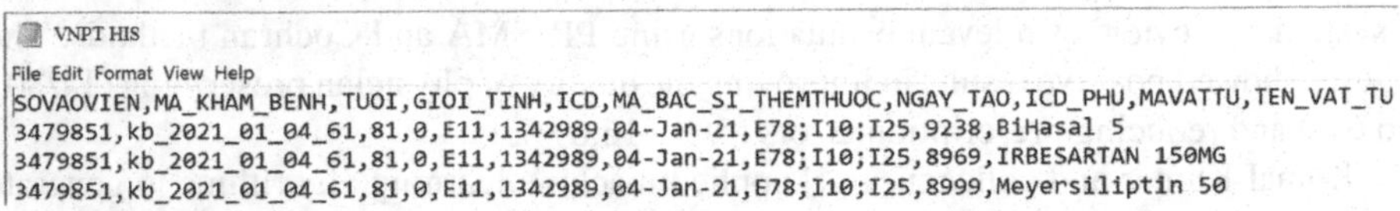

```
VNPT HIS
File Edit Format View Help
SOVAOVIEN,MA_KHAM_BENH,TUOI,GIOI_TINH,ICD,MA_BAC_SI_THEMTHUOC,NGAY_TAO,ICD_PHU,MAVATTU,TEN_VAT_TU
3479851,kb_2021_01_04_61,81,0,E11,1342989,04-Jan-21,E78;I10;I25,9238,BiHasal 5
3479851,kb_2021_01_04_61,81,0,E11,1342989,04-Jan-21,E78;I10;I25,8969,IRBESARTAN 150MG
3479851,kb_2021_01_04_61,81,0,E11,1342989,04-Jan-21,E78;I10;I25,8999,Meyersiliptin 50
```

Fig. 2. Structure of the collected data

Module 2, the subsystem for building the drug recommendation model for the system. In the recommendation subsystem, we propose building two models to compare and assess feasibility in the rule generation process. The first model includes five data fields: age group, gender, main ICD disease codes, prescribing doctor, and the list of medications prescribed for the main ICD disease group. The second model consists of four fields created by removing the prescribing doctor field from the first model. ICD is an abbreviation for the International Statistical Classification of Diseases and Related Health Problems [3, 8]. Each ICD code describes a specific diagnosis in detail.

In this module, we use association rules [2] to build a recommender system. In this study, we compare two algorithms (Apriori and FP-Growth). From the above comparison results, we decided to choose the Apriori algorithm [1] to build the recommender module for our system. Firstly, regarding the search strategy, the Apriori algorithm follows a depth-first search, while the FP-Growth algorithm employs a divide-and-conquer strategy. Secondly, in terms of efficiency on different data sets, the Apriori algorithm works well with large databases, while the FP-Growth algorithm performs better with small databases and generates short frequent itemsets.

Finally in this module, we will build a Web-based application that allows mining of rules that have been built. Based on the above extraction, the doctor will be provided with suggestions from the system.

3 Results and Discussion

3.1 Rules Generation

The data used as the input for the rules generation process is extracted from patients receiving medical treatment at a provincial hospital in Ca Mau in 2021. This data has several attributes including 76,028 health insurance prescriptions, 363,697 prescribed medication entries, 71 prescribing doctors, 521 medications, and 1,129 disease codes according to the ICD-10 [3] that were prescribed.

The first attribute is medications. With the collected data, there are a total of 521 prescribed medications. The most prescribed medication is Partamol 500, with a count of 10,966 out of the total 76,028 prescriptions, accounting for 15%. Additionally, there are still many medications that are less frequently prescribed. The data for the top six most prescribed medications is described in Fig. 3.

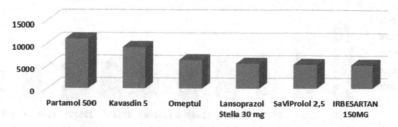

Fig. 3. Analysis of the Most Prescribed Medications

The second attribute is the age of the patient. According to the original data, the prescribed medications are commonly distributed among individuals aged 30 to 90. In addition to analyzing the age attribute, we also preprocess this attribute into target groups to enhance the effectiveness of the recommendation system. The target groups are divided according to medical standards, identified in the system from NT_1 to NT_6. The statistical data on age and the number of prescriptions after preprocessing are described as shown in Fig. 4.

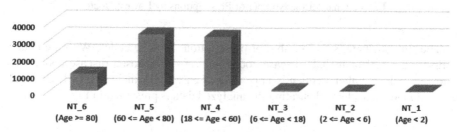

Fig. 4. Chart of the Number of Prescriptions by Age Group

The third attribute that we focused on is gender. In the system, we assigned Male as 1 and Female as 0. The number of prescriptions is roughly balanced between genders, with 40,451 prescriptions for males (53%) and 35,577 prescriptions for females (47%).

The fourth attribute that we are interested in is the number of prescriptions of doctors. In this attribute, the number of prescriptions by doctor code is highest at over 2000, concentrated among 10 doctors out of 71 doctors, while the lowest number is less than 100 prescriptions, also concentrated among 10 doctors out of 71 doctors. The analyzed data is described in Fig. 5 and Fig. 6.

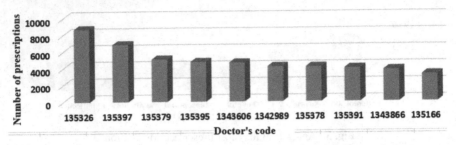

Fig. 5. Chart of the Number of Prescriptions by Highest Doctors

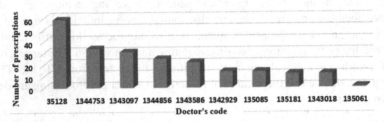

Fig. 6. Chart of the Number of Prescriptions by Lowest Doctors

The fifth attribute is the number of prescriptions by ICD codes. With the data set comprising 1,129 main ICD disease codes, the most diagnosed disease code is I10, followed by E11. Other frequently appearing disease codes include B18.1, C50, K22, I25, and more. The chart illustrating the analyzed data is presented in Fig. 7.

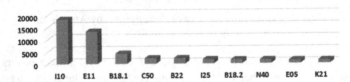

Fig. 7. Chart of the most prescribed main disease codes

The sixth attribute in the analyzed data set is the number of prescriptions by disease chapter [3]. We present the statistical data in Fig. 8.

The seventh attribute is the attribute of sub-ICD disease codes. Which formatted as concatenated strings of sub-ICD codes separated by ";". We analyzed the impact of this attribute on the recommender system and found that there were 31,247 (8.6%) data lines with nulls out of a total of 363,697 data lines. With the analyzed figures, we decided to remove two attributes: sub-ICD codes and disease chapters. These two attributes have a significant amount of noise and can greatly affect the recommendation model. Based on the data collection, statistics, and analysis of attribute values in the overall collected data set, we chose five attributes as input data for the second subsystem: age group, gender, primary ICD disease codes, prescribing doctor, and the list of medications prescribed for the main ICD disease code.

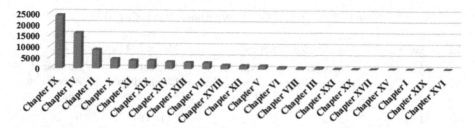

Fig. 8. Description of the Number of Prescriptions by Disease Chapter

To enhance the efficiency of building and updating a set of rules in the recommendation system, we designed a database to improve data accessibility. The shortened model used in the recommendation system is presented in Fig. 9.

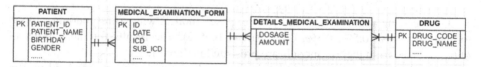

Fig. 9. Entity Relationship Diagram of the Recommendation System

With four input parameters for the first model and five input parameters for the second model, the components of the generated rule sets are described in Fig. 10.

antecedents	consequents	antecedent support	consequent support	support	confidence	lift
(10590)	(MB_E11)	0.012074	0.174633	0.010746	0.889978	5.096277
(10574)	(GT_0)	0.024491	0.467946	0.016007	0.653598	1.396739

Fig. 10. Components of the Generated Rule Sets

The important data fields used in the recommendation subsystem such as Antecedents (Inputted medicine of the rule), Consequents (Suggested medicine of the rule), Support (Support thresholds of the rule), and Confidence (Confidence in the rule).

3.2 System Evaluation

In this study, we use two terms for the number of rule sets to differentiate stages of the system. The first term, "Number of Obtained Rules," describes the total number of rules generated from the recommendation model. The second term, "Number of Appropriate Rules," describes the total number of rules obtained after removing rules where the "Antecedents" attribute does not correspond to medications. The number of appropriate rules will be used in the association rule mining.

Table 1. The number of rule sets obtained from both models

No.	Support Threshold	Model 1 (5 parameters)		Model 2 (4 parameters)	
		Number of Obtained Rules	Number of Appropriate Rules	Number of Obtained Rules	Number of Appropriate Rules
1	1.0%	2.538	0	1.818	0
2	0.5%	12.164	0	8.508	0
3	0.3%	38.548	25	25.116	222
4	0.2%	92.572	87	57.516	509
5	0.1%	400.526	584	225.116	2.100
6	0.05%	829.896	2.923	829.896	7.753

On the same data set, we build two proposed models with corresponding support thresholds (1.0%, 0.5%, 0.3%, 0.2%, 0.1%, 0.05%). The results for the number of obtained rules from the two models are presented in Table 1.

From the description in Table 1, we can see that with a support threshold of 1%, and at a support threshold of 0.5%, no rules are obtained by either model. However, at support thresholds of 0.1% and 0.05%, the first model yields 584 and 2,923 obtained rules, respectively. Similarly, the second model generates 2,100 and 7,753 obtained rules with the same support thresholds is 0.1% and 0.05%. So we can conclude that the second model is more effective in generating a greater number of applicable rules. Furthermore, by removing the "Doctor" parameter from the input data for the rule generation process using the Apriori algorithm, we can limit cases where doctors prescribe based on personal preferences and propose medication based on their habits.

We divide the data into two parts: training data and testing data, using the train_test_split function from the sklearn.model_selection library. The test results show that 75% of the transactions in the testing set are predicted by the generated rules. From the results, it can be seen that many prescription recommendations prioritize based on decreasing confidence ratio, but there are still cases with low accuracy due to the dependency of prescription on test results and diagnostic imaging conclusions.

After finding the association rules, they are evaluated based on support, confidence, and lift measures to select meaningful and high-quality rules. Using the lift measure is a method to evaluate the association between two items in a data set. The lift measure is calculated as the ratio of the joint probability of the two items to the product of their probabilities. A higher lift value indicates a stronger association between the two items. If the lift value is greater than one, it means that the rule has a high likelihood of occurrence. If the lift value is equal to one, it means that the association between the items is random and there is no special relationship. If the lift value is less than one, it means that the association between the items is negative and the likelihood of occurrence of the rule is low.

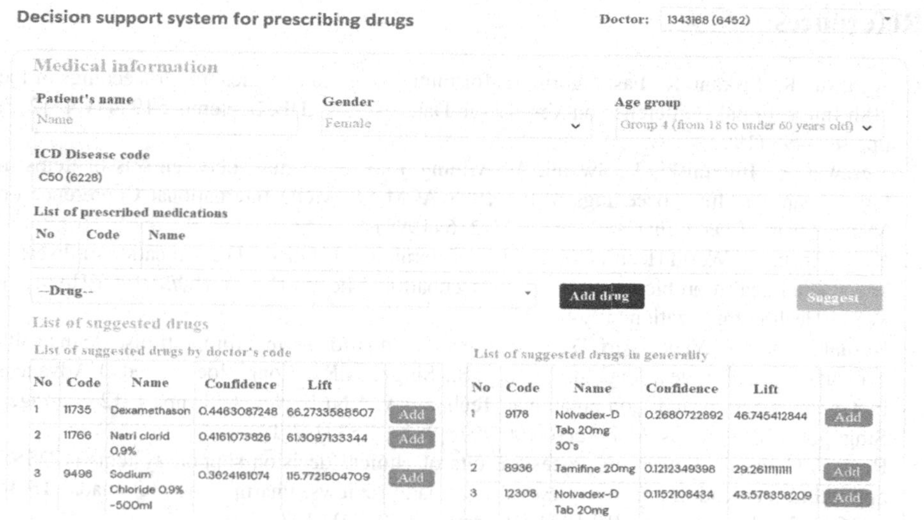

Decision support system for prescribing drugs   Doctor: 1343188 (6452)

Medical information

Patient's name   Gender   Age group
Name   Female   ⌄   Group 4 (from 18 to under 60 years old) ⌄

ICD Disease code
C50 (6228)

List of prescribed medications
No   Code   Name

--Drug--   ▾   [Add drug]   [Suggest]

List of suggested drugs

List of suggested drugs by doctor's code   List of suggested drugs in generality

No	Code	Name	Confidence	Lift		No	Code	Name	Confidence	Lift	
1	11735	Dexamethason	0.4463087248	66.2733588507	[Add]	1	9178	Nolvadex-D Tab 20mg 30's	0.2680722892	46.745412844	[Add]
2	11766	Natri clorid 0.9%	0.4161073826	61.3097133344	[Add]	2	8936	Tamifine 20mg	0.1212349398	29.261111111	[Add]
3	9496	Sodium Chloride 0.9% -500ml	0.3624161074	115.7721504709	[Add]	3	12308	Nolvadex-D Tab 20mg	0.1152108434	43.578358209	[Add]

Fig. 11. Web-based application interface to mining association rules

The web-based interface of our clinical decision support systems is presented in Fig. 11. The doctor will enter information such as doctor code, patient name, gender, age group, disease code, and drug list (if any) and click Suggest. First, the System will suggest prescriptions according to the doctor's own habits in the left drug recommendation, and the recommendations according to the popularity of all doctors in the right recommendation. The system can support both experienced and new doctors. Finally, when the doctor prescribes a drug, the system will put that drug on the left side and continue to suggest the next prescription.

4   Conclusion

This study has developed a decision support system for a prescription recommendation based on prescription data from a provincial hospital in Ca Mau. The system has proposed prescription recommendations for different patient profiles, including age, gender, primary disease codes, and recently prescribed medications. The system can suggest personalized prescriptions based on individual doctors' prescribing history and the aggregate prescribing history of all doctors, thereby supporting prescription recommendations for both new and experienced doctors.

By removing the noisy attributes, choosing appropriate thresholds of support value, and using the LIFT measure for evaluating obtained rules, the proposed system has the ability to help doctors make drug prescription decisions with high accuracy and low risks for patients.

References

1. Agrawal, R., Srikant, R.: Fast algorithms for mining association rules. In: Proceedings of the 20th International Conference on Very Large Databases, VLDB, September 1994, vol. 1215, pp. 487–499 (1994)
2. Agrawal, R., Imieliński, T., Swami, A.: Mining association rules between sets of items in large databases. In: Proceedings of the 1993 ACM SIGMOD International Conference on Management of Data, June 1993, pp. 207–216 (1993)
3. Karjalainen, A., World Health Organization: International statistical classification of diseases and related health problems (ICD-10) in occupational health (No. WHO/SDE/OEH/99.11). World Health Organization (1999)
4. Komal Kumar, N., Vigneswari, D.: A drug recommendation system for multi-disease in health care using machine learning. In: Hura, G.S., Singh, A.K., Siong Hoe, L. (eds.) Advances in Communication and Computational Technology. LNEE, vol. 668, pp. 1–12. Springer, Singapore (2021). https://doi.org/10.1007/978-981-15-5341-7_1
5. Reis, W.C., et al.: Impact on process results of clinical decision support systems (CDSSs) applied to medication use: overview of systematic reviews. Pharm. Pract. (Granada) 15(4), 1036 (2009). https://doi.org/10.18549/PharmPract.2017.04.1036
6. Schedlbauer A., et al.: What evidence supports the use of computerized alerts and prompts to improve clinicians' prescribing behavior? J. Am. Med. Inform. Assoc. 16(4), 531–538 (2009)
7. Tong, L.R., Zhang, J., Ma, L., Xin, L., Hu, S., He, J.F.: An improved Apriori algorithm based on Linkedlist for the prevention of clinic pharmaceutical conflict. Appl. Mech. Mater. 513, 651–656 (2014)
8. World Health Organization: International Statistical Classification of Diseases and Related Health Problems: Alphabetical Index, vol. 3. World Health Organization (2004)

Author Index

N. Thai-Nghe et al. (Eds.): ISDS 2023, CCIS 1950, pp. 313–315, 2024.
https://doi.org/10.1007/978-981-99-7666-9

Printed in the United States
by Baker & Taylor Publisher Services